STATE WILDLIFE MANAGEMENT AND CONSERVATION

Wildlife Management and Conservation

Paul R. Krausman, Series Editor

State Wildlife Management and Conservation

EDITED BY

THOMAS J. RYDER

Published in Association with *THE WILDLIFE SOCIETY*

JOHNS HOPKINS UNIVERSITY PRESS | BALTIMORE

Johns Hopkins University Press
2715 North Charles Street
Baltimore, Maryland 21218-4363
www.press.jhu.edu

Library of Congress Cataloging-in-Publication Data

Names: Ryder, Thomas J., editor.
Title: State wildlife management and conservation / edited
    by Thomas J. Ryder.
Description: Baltimore : Johns Hopkins University Press,
    2018. | Series: Wildlife management and conservation |
    Includes bibliographical references and index.
Identifiers: LCCN 2017017327| ISBN 9781421424460
    (hardcover : alk. paper) | ISBN 1421424460 (hardcover :
    alk. paper) | ISBN 9781421424477 (electronic) | ISBN
    1421424479 (electronic)
Subjects: LCSH: Wildlife management—United States
    | Wildlife management—United States—Finance. |
    Wildlife conservation—United States.
Classification: LCC SK361 .S7264 2018 | DDC 333.95/4—
    dc23
LC record available at https://lccn.loc.gov/2017017327

A catalog record for this book is available from the British
Library.

*Special discounts are available for bulk purchases of this
book. For more information, please contact Special Sales at
410-516-6936 or specialsales@press.jhu.edu.*

Johns Hopkins University Press uses environmentally
friendly book materials, including recycled text paper that
is composed of at least 30 percent post-consumer waste,
whenever possible.

Contents

Preface

I've had the privilege of working for more than 30 years for a state wildlife agency, serving the public as a habitat biologist, wildlife biologist, supervisor of wildlife biologists, deputy wildlife chief, and the Wyoming governor's wildlife policy advisor. Throughout my career, I was struck that no one had ever chronicled the many and varied contributions of state agencies to the conservation and management of wildlife. State agencies were included as an important component of the total conservation landscape in seminal publications like Leopold's *Game Management* (1933), Trefethen's *An American Crusade for Wildlife* (1975), Krausman and Cain's *Wildlife Management and Conservation* (2013), and others. In addition, many individual states documented their own histories and contributions to wildlife management. However, a detailed and all-encompassing book describing state agencies' total contributions to wildlife conservation does not exist. The topic was a good fit for The Wildlife Society's scientific management book series with Johns Hopkins University Press, and after discussion and hundreds of hours of work, the authors and I are proud to present the following treatise to tell our story.

State wildlife agencies are as diverse as the species they're responsible for conserving. In Pennsylvania, Utah, and Nevada, terrestrial wildlife is the sole responsibility of the agency, but 23 other agencies oversee wildlife and fisheries management, and seven also administer state parks or forestry. Wildlife agencies in 15 additional states have broad responsibility for all natural resources, including environmental protection. Owing to the broad and complex array of responsibilities among agencies, this volume documents only the states' role in management of terrestrial wildlife. We fully recognize the critical importance of state management of fisheries, parks, forests, and environmental protection. Because of the nature of this book series, however, we believe that other state responsibilities should be covered in future, program-specific books.

This tome consists of 15 chapters. The first four chapters present the history and current structure of state wildlife agencies, the legal basis for state management, the role of states in wildlife policy decisions, and the evolution of funding for state wildlife management. Chapter 5 details the crucial importance of state wildlife law enforcement. The next five chapters relate how states manage big game, small and upland game, furbearers, migratory game birds, and nongame wildlife. Chapters 11 and 12 outline the role of states in managing human–wildlife conflict and wildlife diseases. The use of human dimensions and field wildlife research by state agencies are discussed in chapters 13 and 14. The book concludes with a summary of the most important major challenges facing state wildlife agencies in the future. Intended for a variety of audiences, we believe that this book would be valuable as a standalone text for upper-level undergraduate or graduate courses designed to introduce students to state wildlife agencies.

For individuals who have a passion for wild places and wild things, don't mind long days in the field, and revel in the constant challenge of managing a publicly held trust resource, a career in state wildlife management might just be for you. This book can provide salient information to help you decide.

Thomas J. Ryder
Deputy Wildlife Chief (retired)
Wyoming Game and Fish Department

Acknowledgments

This book was only possible through the collaboration and contributions of many people. I am especially indebted to the chapter authors and reviewers listed below. Their hard work and subject-matter expertise have humbled me and made this book a powerful treatise chronicling state wildlife conservation and management in our great country. I'd also like to thank three anonymous reviewers for their thoughtful and helpful critique of the entire draft text.

I appreciate the guidance and support offered by Dr. Vincent Burke, Ms. Meagan Szekely, and Ms. Tiffany Gasbarrini of Johns Hopkins University Press. Dr. Paul Krausman, my good friend, mentor, and editor of The Wildlife Society's book series, provided many timely suggestions that helped me navigate the editorial process. Dr. Ken Williams, executive director of The Wildlife Society, similarly provided strong moral support and encouragement for the project.

I owe a huge thank-you to my former bosses, Director Scott Talbott and Chief Game Warden Brian Nesvik of the Wyoming Game and Fish Department, for the agency's generous contribution to help defray costs associated with publishing this book. Without the department's financial support, the book wouldn't have been able to hit the streets.

Lastly, I extend my most heartfelt thanks to my wife for her incredible support, patience, and understanding during my many job-related trips to the backcountry and meeting rooms of Wyoming during my career with the Wyoming Game and Fish Department. I love you, Lisa!

Chapter Authors

M. Carol Bambery
Counsel
Association of Fish and Wildlife Agencies

Gordon R. Batcheller
Director
Northeast Association of Fish and Wildlife Agencies

Chad J. Bishop
Director, Wildlife Biology Program
University of Montana

Vernon C. Bleich
Wildlife Biologist (retired)
California Department of Fish and Wildlife

Dale Caveny
Colonel (retired)
North Carolina Wildlife Bureau

David K. Dahlgren
Extension Associate Professor
Department of Wildlife Resources and Jack H.
Berryman Institute
Utah State University

Daniel J. Decker
Unit Associate Director
Human Dimensions Research Unit
Cornell University

Karie L. Decker
Assistant Wildlife Division Administrator
Nebraska Game and Parks Commission

Thomas A. Decker
Wildlife Biologist
US Fish and Wildlife Service

Billy Dukes
Chief of Wildlife
South Carolina Department of Natural Resources

John D. Erb
Furbearer Biologist
Minnesota Department of Natural Resources

John R. Fischer
Director
Southeast Cooperative Wildlife Disease Study

Ann B. Forstchen
Human Dimensions Coordinator
Florida Fish and Wildlife Conservation Commission

Jonathan W. Gassett
Southeast Field Representative
Wildlife Management Institute

Parks Gilbert
Legal Strategy Attorney
Association of Fish and Wildlife Agencies

Colin M. Gillin
State Wildlife Veterinarian
Oregon Department of Fish and Wildlife

Tim L. Hiller
Furbearer Ecologist
Wildlife Ecology Institute

Daniel Hirchert
Wildlife Biologist
Animal and Plant Health Inspection Service
Wildlife Services
US Department of Agriculture

Michael W. Hubbard
Research Scientist
Missouri Department of Conservation

Mark Humpert
Wildlife Diversity Director
Association of Fish and Wildlife Agencies

Scott Hygnstrom
Professor
University of Wisconsin–Stevens Point

Robert P. Lanka
Supervisor, Statewide Habitat and Wildlife Technical
Services
Wyoming Game and Fish Department

Richard E. McCabe
Director Emeritus
Wildlife Management Institute

Jennifer Mock-Schaeffer
Government Affairs Director
Association of Fish and Wildlife Agencies

Brian Nesvik
Wildlife Division Chief and Chief Game Warden
Wyoming Game and Fish Department

Shaun L. Oldenburger
Webless Migratory Game Bird Program Leader
Texas Parks and Wildlife Department

John F. Organ
Director
Cooperative Fish and Wildlife Research Units
US Geological Survey

Ronald J. Regan
Executive Director
Association of Fish and Wildlife Agencies

Michael A. Schroeder
Upland Bird Research Scientist
Washington Department of Fish and Wildlife

William F. Siemer
Research Associate
Human Dimensions Research Unit
Cornell University

Christian Smith
Western Field Representative
Wildlife Management Institute

Randy Stark
Chief Game Warden (retired)
Wisconsin Department of Conservation

Gary J. Taylor
Legislative Director Emeritus
Association of Fish and Wildlife Agencies

J. Scott Taylor
Wildlife Division Administrator
Nebraska Game and Parks Commission

Daniel J. Thompson
Large Carnivore Section Supervisor
Wyoming Game and Fish Department

Kurt VerCauteren
Wildlife Biologist
Animal and Plant Health Inspection Service
Wildlife Services
US Department of Agriculture

Mark P. Vrtiska
Migratory Game Biologist
Nebraska Game and Parks Commission

H. Bryant White
Furbearer Research Coordinator
Association of Fish and Wildlife Agencies

Steven A. Williams
President
Wildlife Management Institute

Chapter Reviewers

Charles Anderson
Colorado Division of Wildlife and Parks

Gordon R. Batcheller
Northeast Association of Fish and Wildlife Agencies

Nichole Bjornlie
Wyoming Game and Fish Department

Jason Boulanger
University of North Dakota

Mark Bruscino
Wyoming Game and Fish Department (retired)

Carol Chambers
University of Northern Arizona

James Cordoza
Massachusetts Division of Fisheries and Wildlife

Paul Curtis
Cornell University

Daniel J. Decker
Cornell University

Thomas A. Decker
US Fish and Wildlife Service

Thomas Flowers
Montana Department of Fish, Wildlife, and Parks

Becky Humphries
National Wild Turkey Federation

Cynthia Jacobson
US Fish and Wildlife Service

Paul R. Krausman
University of Montana (retired)

David Krementz
Arkansas Cooperative Fish and Wildlife Research Unit

Stephen Mealey
Idaho Fish and Game Department (retired)

John F. Organ
US Geological Survey

Robert Rolley
Wisconsin Department of Natural Resources

STATE WILDLIFE MANAGEMENT
AND CONSERVATION

1

John F. Organ and
Richard E. McCabe

History of State Wildlife Management in the United States

Each state wildlife agency has its own history, a history of time, place, conventions, personalities, and issues—biological, political, and administrative, not necessarily in that order. Each state wildlife agency has a unique complexity of organization, authorities, and programs. None is static in those regards, for neither are the social, political, economic, and environmental conditions and circumstances within its jurisdiction. One commonality is their statutory charge, variously worded, to maintain and steward natural resources of the state in the best interest of the citizens of the state, now and in the future. Some state wildlife agencies are responsible for more kinds of natural resources, such as marine resources, than others.

A second commonality is that all of these agencies were established in the interest of conservation—preserving, managing, and protecting desirable wildlife populations and their inherent values. That commonality, that history, is part and parcel of the larger history of conservation in America; that context, in brief, is the focus of this chapter.

New Frontiers

The history of America may be characterized as one of overcoming or merely overwhelming its frontiers—geographical, social, and political—a theory first advanced by Frederick Jackson Turner in 1893 (Turner 1894; Billington 1978; Faragher 1999). Each of these frontiers contributed to the ever-changing character

of the country, including contemporary regard for the sustainability of towering forests, transparent waters, enormous schools of fish, flocks of birds, and herds of game. The first frontier began with continental probes by European and Russian explorers, Spanish and French gold and fur seekers, and Christian missionaries into the uncharted homeland of perhaps 3 million Native Americans (continental inhabitants for at least 14,000 years [Stanford and Bradley 2013] and possibly as much as 23,000 years [Raghavan et al. 2015]), in search of imperial fortunes and souls. Next was clumsy settlement of the New World's perimeters by unprepared European pilgrims.

By most exploration accounts, America teemed with wildlife and fishes of extraordinary variety and abundance. Europeans (English, Dutch, and Swedes) who colonized the New World's eastern coast found just that, but not always and not everywhere. The earliest arrivals were hemmed along the coasts by vast marshes, swamps, mature pine and northern hardwood forests, and trepidation about the beyond. "The European colonist, poised on the shores of a wild continent, recognized no value in wilderness" (Trefethen 1975:32). Fish and wildlife provided primary subsistence, but local supplies tended to be seasonal and some were too readily depleted. As settlements found purchase and growth, new locations were established along rivers and other watercourses, yet within reach of the oceans and Old World commerce.

The new Americans generally were imbued with

religious fervor and desperate desire for personal freedoms. Many tenaciously clung to Old World customs and cultural traditions, but far from their European origins they found prospect in self-determination and relief from social hierarchy and its elitist restraints on taking game. So bountiful were the forests, fields, and waters that fur trade became the primary business and economy of the colonies. It exceeded the importance of timber and fish (Matthiessen 1964; Dolan 2010).

In most of the toehold colonies, "game and fish were free and common to any Person who can shoot or take them, with any lett, hindrance of Opposition whatsoever" (Gabriel 1912:19). However, "any Person" tended to apply only to America-born Caucasian males. Again, in most colonies, those entitled to hunt were restricted only from property that was purposely fenced for their exclusion (Lund 1980). The freedom from trespass was based on *ferae naturae*, a legal doctrine that made wild animals owned by those who captured them, not by the person on whose property the animals were captured. These freedoms were a far cry from the European system, which essentially held a hunting and fishing monopoly for the blue bloods.

Game On

"Colonists killed game for food, for market, for bounties and for an escape from the drudgery of farming" (Herman 2001:23; see also Beverley 1705). In addition, deer were taken at times as protection against crop depredations. In 1624, John Smith (1907[1]:78) observed that Jamestown planters "do so traine up their servants and youth in shooting deere, and fowle." Little thought was given to checking the take of wild game for subsistence or otherwise, at least until desired species, e.g., white-tailed deer and wild turkey, were extirpated or nearly so.

As early as 1630, the General Court of Massachusetts Colony declared, "euy Englishe man that killeth a wolfe in any pte within the limits of this patent shall have allowed him 1d [penny] for any beast & horse & ob. For every weaned swine & goat in euy plantcon to be levied by the constables of sd plantacons" (Shurtleff 1853:81; Young and Goldman 1944; deCalesta 1976). Other colonies followed suit. Virginia did so two years later, offering goods and privileges in lieu of money.

Early observers tended to perceive wildlife declines as matters of dislocation. But not all. William Elliott (1846:166–167) of Beaufort, South Carolina, observed, "There are causes in operation which have destroyed . . . the game to the extent that in another generation, this manly pastime [of hunting] will no longer be with our reach . . . Undoubtedly the most obvious cause of the disappearance of the deer and other game is the destruction of the forests. . . . [As a result of] the uncalled-for destruction of forests . . . the trampling and cropping of the shrubs and undergrowth [by livestock] . . . and the practice of burning the woods in spring to give these cattle more luxuriant pasturage," wildlife habitat was gone and game had disappeared. Elliott further remonstrated commercial deer hunts for killing the animals left. He also saw fault with Americans generally for failing to give any protection to wildlife, "seeing any such move as aristocratic and a threat to the 'rights' of the people" (Reiger 1975:228).

John J. Brown (1857:231–232), author of *The American Angler's Guide*, rhetorically asked his readers whether they thought sport angling would survive:

> You who have trod the mossy bank in pursuit of trout, and warred against the swift current when the striped basse [sic] was the object of your sport, will answer emphatically *no*. You are painfully assured that the well-known haunts where in happy boyhood you took many a "silver side," are deserted, and the overarching banks of your favorite streams conceal your spotted friends no longer. You know . . . you take few and still fewer, and that some of your former places are now never visited by the sought for game. It is the commonest complaint of the old anglers that fishing nowadays is uncertain, that . . . "times are not as they used to be," and so also says the gunner of his favorite sort of game.

Even though hunting had great utility for colonists, not all participated or approved of those who were hunters. For one thing, only a minority of colonists owned firearms. Puritans, in particular, rejected blood sports (Herman 2001). British aristocracy in early America were wont to decry those who hunted (and especially those who hunted rather than farmed), among other pejoratives, as "indolent," "barbaric," "idle and disorderly"—English savages who "range the Country, with their Horse and Gun, without Home or Habitation. . . . For, they having no Sort of Education, naturally follow Hunting—Shooting—Racing—Drinking—Gaming, and ev'ry Species of Wickedness. Their Lives are only one continual Scene of Depravity of Manners, and Reproach to the Country; being more abandoned to Sensuality, and more Rude in Manners, than the Poor Savages round us" (Woodmason 1953:226).

The earliest hunting restrictions were promulgated by local jurisdictions. In 1646, the town of Portsmouth, Rhode Island, closed deer hunting from May 1 to November 1 and called for a violation penalty of 5 pounds. This ordinance established a pattern of laws adopted by most colonies before 1720 (Trefethen 1975). Following a 1741 law that prohibited killing whitetails between December 31 and August 1, each New Hampshire town was required to appoint two persons to search houses for possession of venison and fresh deer hides. Massachusetts hired wardens, or "reeves," around the same time. Fifty years later, New York passed a law prohibiting the killing of heath hens between April 1 and February 5. This was a year after New Jersey imported an exotic game species, the Hungarian partridge (Phillips 1928).

By 1800, all of the original 13 colonies had enacted some type of restriction on the taking or use of whitetailed deer, although enforcement of some sort was concomitant only in North Carolina (1738) and elsewhere lagging by decades or even centuries (Trefethen 1975). "Blue law" closures to Sunday hunting went into effect in colonies in the 1700s (11 eastern states still are affected). The first multiyear closure of hunting was for deer in Massachusetts, beginning in 1718.

A Vast and Empty Chaos

The colonial frontier was followed by immigrant dispersal from coastal areas by those seeking to wrest more and better from the "vast and empty chaos" of the interior and its variously indignant inhabitants (Cushman 1963:91; Cronan 1983). It happened concurrently with the frontier of national independence and the experiment of a democratic republic.

Although many of the new Americans favored a strong central government, Jeffersonian democracy prevailed—advocating for states' rights, free education, opposition to slavery—during the early years of the

1800s, staving off Federalists. A political landscape was evolving, giving adult males equal right to express their wishes. The parties that emerged tended to be ideologically entrenched over such issues as immigration and naturalization, social reform, slavery, a central bank, national expansion, foreign trade, and more. Despite opposing political ideologies (often becoming seriously contentious), the American people were learning to flex their will (Kennedy 1988).

Growth Spurts

The infant republic's next formidable frontiers were the Louisiana Purchase (1803) and the War of 1812. The former—an acquisition of approximately 828,000 square miles of wilderness—came at a cost of only $0.03 per acre. The latter, lasting until early 1815 and considered the "second American war of independence" (Langguth 2006), was a virtual stalemate with powerful Great Britain. However, it gained the upstart country international stature militarily and politically. Notably, the conflict all but terminated Indian resistance in the East, and it elevated the profile of General Andrew Jackson, whose Tennessean soldiers won significant battles in the South. These events, along with Jackson's election to the presidency (1829–1837), availed a surge of westward expansion and helped forge a distinctive American ethos—one of ambition, independence, intrepidness, and self-reliance, all of which would be reflected in the exploitation of and eventual notion of husbanding natural resources (McWilliams 2015).

Jacksonian democracy supplanted Jeffersonian democracy and exacerbated political division. Jackson favored agriculture over industry and states' rights over central government, as had Jefferson several decades before. But Jackson favored rapid national expansion, supported slavery, and vetoed a national bank. His Indian policy, following the Removal Act of 1830, forcibly displaced the "Five Civilized Tribes" westward across the Mississippi River—the "Trail of Tears" tragedy in 1831 (Choctaw), 1832 (Seminoles), 1834 (Creek), 1837 (Chickasaw), and 1838 (Cherokee) (Foreman 1989; Wallace 1993). He also sent the military to chase, subdue, and relocate Sac, Fox, and Kickapoo Indians (the "Blackhawk War" in 1832; Trask

2006; Jung 2007), as well as the Seminoles in Florida (the "Second Seminole War" from 1832 to 1835; Mahon 1967; Knetsch 2003; Missal and Missal 2004). In response to Jackson's "tyranny" and political stances, a new political party—the Whigs—emerged. It represented business and particularly favored strength in the national government, and it opposed rapid national expansion. More so, it undermined the two-party system. After several decades of success, it lost power over the slavery issue. Northern Whigs merged with the new Republican Party (which also had absorbed the vestiges of the Federalist Party), while southern Whigs joined the Democratic Party.

In 1842, a US Supreme Court case ratified the conservation keystone doctrine of public trust in *Martin v. Waddell*, which provided that wildlife is owned by no one and is held in trust and maintained by government of, by, and for the people. This landmark case is covered in great detail in chapter 2 of this book.

The significant link of that period, or frontier, to conservation and to state wildlife agencies, both yet decades away, was the beginning of a *tradition* of partisan political contentiousness. Party lines were drawn in the proverbial sand over matters unrelated to natural resources and subordinated to party allegiance the priority, if any, of the welfare of those resources.

Westering and War

The next "frontier" was an optimistic presumption called "manifest destiny" (O'Sullivan 1839, 1845; Pratt 1927). It embraced a spirit of self-determination, a provocation of patriotism and a justification for apocryphal dominion over the landscape and its resources, and further systematic dispossession of indigenous inhabitants' homelands and cultures (DeVoto 1947; Brown 1970).

Concern about reduction of game in New York prompted sportsmen to form the New York Sporting Association in 1844 to do what it could to protect wildlife, mainly setting certain harvest limits and hunting periods. The organization grew and later became the New York Association for the Protection of Game. Between 1844 and 1900, at least 374 other self-styled game protective organizations were established.

In 1849, the California Territorial Legislature adopted the common law of England as the rule in all state courts. Before that time, Spanish and then Mexican laws applied (California Department of Fish and Game 1999). The most significant legal incident was the Mexican government decree in 1830 that California "mountain men" were illegally hunting and fishing. Captain John Sutter, among others, had been responsible for enforcing Mexican fish and game laws.

The next frontier was a cruelly uncivil Civil War splitting ideologies and loyalties between North and South, and the toll on American humanity was beyond pitiable. At no time before or after was *e pluribus unum* put more to the test.

Even before the era of Reconstruction was under way to fix a sundered national identity and a marred countryside, George Perkins Marsh (1864:35) advanced the concept of landscape ecology and warned of its increasing fragility: "man . . . has too long forgotten that the earth was given to him for usufruct alone, not for consumption, still less for profligate waste."

Chief among resource concerns for Marsh (1874) and most other early conservationists was the loss of forests. Approximately 90 percent of the eastern United States was forested at the time of colonization. As the principal source of fuel and building material, trees were cut, burned, or girdled to clear land for plow and pasture, for building ships and part or all of other structures, and for paper and trade. Greed and shortsightedness were the mortal enemies of a healthy productive land, wrote Marsh. He noted that drastic changes of landscape at the hands of humans would adversely change biological communities (Udall 1963).

The Gilded Age

On the heels of national wounding was the Industrial Revolution—a half-century technological frontier that overlapped the predatory nationalism of manifest destiny and exceeded in capability all that of humankind theretofore. Industrial America was a newly urbanizing frontier, altering the socioeconomic milieu of the nation and imposing by a thousandfold new demands on its virtually unprotected natural resources. It was an era that saw a transcontinental railroad (1863–1869) facilitated by the collusion of Congress and robber barons, by Bessemer process steel, and by multinational immigrant labor fed, among other things, bison procured by the likes of Buffalo Bill Cody (Lavender 1970; Brown 1977). The iron horse accommodated the end of western Indian economies, lifestyles, and resistance; the demise of bison and nearly so for elk, wild sheep, grizzly bears, and pronghorn; the attrition of open range; and the introduction of cattle ranching and barbed wire. The railroad and compressed refrigeration aided unhindered market hunting. Repeating rifles and the railroad gave a boost to unregulated sport hunting in the West. The Homestead Act of 1862, the Morrill Land Grant Act of 1862, the Desert Land Act of 1877, and the Timber and Stone Act of 1878 privatized more than 500,000 acres of the public domain, and an additional 180 million acres were doled out to railroads for rights-of-way. The selling off of public lands by states added to the toll (Porterfield 2005).

Agricultural implement innovations increased farmland productivity and efficiency. The Empire grain drill, for example, planted seed in rows and covered it. The John Deere steel plow, with correctly shaped moldboard, cut cleanly through caking prairie soil. The Empire threshing machine separated wheat from chaff and bagged the grain. And the Wilder sulky enabled farmers to ride as they plowed. Although droughts periodically and seriously undermined farming attempts, farms increased from 1.5 million in 1860 to 6.4 million half a century later (Kohlmeyer 1962; Rasmussen 1975).

From the First

The first state conservation agencies were in Massachusetts and New Hampshire in 1866 (Trefethen 1975; Cardoza 2015). They began as fisheries commissions. California followed with a fisheries commission in 1870. "Game" wasn't added to the New Hampshire and California charges or names until 1878. Massachusetts received such authority in 1886.

In 1871, Congress established the US Fishery Commission, and under the direction of Dr. Spencer Baird it began to study and propagate fishes. Within a few years most of the states had created similar agencies. The responsibilities of these early fish commissions were expanded gradually to cover birds and mammals. A similar trend took place in the federal agencies when, at the request of the American Ornithologists' Union

(AOU), Congress, on March 3, 1885, authorized the Department of Agriculture to study the economic importance of birds. The office, originally the Division of Economic Ornithology and Mammalogy, became the Bureau of Biological Survey in 1896 "to inform the public of wildlife problems, but was powerless to do much about them" (Trefethen 1964:9–10). The federal renewable resource agencies, however effective, provided an organizational template for corresponding state agencies.

By 1880, all extant states had game laws, in force by statute. Massachusetts passed laws in 1818 protecting "useful" birds; New Jersey followed in 1820. In 1838, New York ruled that multiple guns (batteries) could not be used against waterfowl, a law later repealed, to the relief of commercial hunters. Additional protection for nongame birds came in the form of restrictions against damaging nests or eggs in Connecticut and New Jersey in 1850. New York was the first state to require a hunting license, in 1864; the first to impose a nonresident license was New Jersey, also in 1864; the first daily bag limit (25 prairie-chickens) was in Iowa in 1878; the first rest day was required in Maryland in 1872; and market hunting became taboo in Arkansas in 1875.

Swanson (1940:199) reported that state hunting and fishing laws in the period 1860–1890 were "not taken very seriously." She also wrote than many newspapers in Minnesota "implied" that hunters and anglers entirely disregarded the laws.

Power of the Pen and Clubs

After the Civil War, outdoor adventure writing became a popular genre. *Forest and Stream* (1873–1939) served during George Bird Grinnell's 35 years of editorship and ownership to convey conservation messages and concerns, and even attempted to provide data on game populations, as well as outdoor adventures. Other national and regional outdoor sporting serials and periodicals bridging the centuries—e.g., *American Turf Register and Sporting Magazine*; *Recreation*; *Turf, Field, and Farm*; *American Sportsman*; *Outing*; *Shield's Magazine*; *Sports Afield*; *American Field*; *Century*; *Field and Stream*; *Outdoor Life*—gained substantial, vicarious audiences by reporting what, where, when, and how to take the most, and made those recreations noble, romantic, and/or dangerous, but always thrilling, manly, and heroic. They prompted a Leatherstocking revival, as well as impetus for opposition to commercial exploitation of fish and wildlife resources (Reiger 1975). Their readerships, mostly young men, were introduced to the writings of such men as John J. Brown, Elisha Lewis, Emerson Hough, George Armstrong Custer, Theodore Roosevelt, George Shields, Frank Buck, Nash Buckingham, Randolf Marcy, George Sears, Erle Stanley Gardner, Stewart Edward White, "Frank Forester" (William Henry Herbert), Zane Grey, Edgar Rice Burroughs, Archibald Rutledge, and Ned Buntline (E. Z. C. Judson). Adventure books by Hough, Roosevelt, Buck, Grey, Burroughs, Jack London, Gardner, Herman Melville, Colonel Henry Patterson, and others filled the minds of lads of all ages with outdoor thrills and expectations of things and places wild. Also popular were hunting novels by Edward S. Ellis (Deerfoot and Boy Pioneer series), "Harry Castlemon" (Charles Austin Fosdick; Sportsman's Club, Rod and Gun, and Boy Hunter series), and Thomas W. Knox (Young Nimrods in North America; Knox 1881), which memorialized the courage and manliness of hunters. Anthropomorphic books by Ernest Thompson Seton and Thorton W. Burgess also were widely read.

In the latter decades of the 1800s, rod and gun clubs proliferated. From a mere handful in 1875, 986 were listed in The Sportsman's Directory in the early 1890s. Less than 20 years later (Pond 1891), more than half (563) were in the Midwest. Many were social fraternities as much as recreational outlets; nearly all provided hunting and/or fishing opportunities and a voice against legislative restrictions on the pursuits. Some were of the simple, seasonal log cabin camp variety; others were privileged and operated in catered opulence and monopolized the best woods and waters. The latter were inclined to enlist waterfowl market gunners to protect and maintain club lands and waters, which the gunners found much more lucrative and safe. As Reiger (1975:159) pointed out, "Hunting was regulated by club rules which preceded state laws."

"The destruction of birds of all sizes and shapes had reached proportions of a national pastime in the last quarter of the nineteenth century, and supported a number of minor industries as well" (Matthiessen 1964:165). Grinnell, substantially through his weekly

Forest and Stream conservation advocacy editorials and stories, founded the Audubon Society in 1886, to instigate a public response to bird-part fashion (e.g., feather and bird nest hats). He called for reform "inaugurated by women . . . [whose] tender hearts will be quick to respond" (Grinnell 1886:41). J. A. Allen (1886a, 1886b), an incorporator of the New York Audubon Society and first president of the AOU (an outgrowth of the Nuttall Ornithological Club), also mainly faulted the country's millinery business and importation of bird wings and skins in bales to England and France for the same purpose. He deplored the taking of insectivorous birds and their eggs for sport and food and estimated that the annual loss of birds was on the order of 5 million.

Two years earlier, in 1884, the AOU drafted a "model law" for the protection of nongame birds and of the nests and eggs of nongame and game birds alike. It was a prototype intended for state bird-protection legislation throughout the country. Pennsylvania and New York adopted the model law immediately, and five other states followed suit by 1900. Geer v. Connecticut in 1896 (see chap. 4) settled the matter of authority for wildlife within state borders (until Hughes v. Oklahoma in 1979), and the Lacey Act of 1900 and Federal Tariff Act of 1913 further crippled the millinery trade.

By the end of the nineteenth century, the AOU had withdrawn to insular scientific matters and essentially turned over conservation concerns to local and state Audubon chapters, first emerging in Massachusetts. State chapters (excluding the Massachusetts Audubon Society) founded the National Audubon Society in 1905.

In Forest and Stream, Grinnell (1884:301) editorialized for a New York "association of men bound together by their interest in game and fish, to take active charge of all matters pertaining to the enactment and carrying out of the laws on the subject. There is abundant material for such a body. Why can it not be organized?" Fortuitously, Grinnell's criticism of Theodore Roosevelt's (1885) book Hunting Trips of a Ranchman prompted the two men to meet. In short order, they developed a deep mutual friendship, and the national conservation movement had its progenitors, foremost activists, and patron saints.

Roosevelt and Grinnell initiated the Boone and Crockett Club in December 1887. Its five purposes (penned by Grinnell, Roosevelt, and Archibald Rogers) were to "promote manly sport with rifle," "promote travel and exploration in the wild and unknown, or but partially known, portions of the country," "work for the preservation of the large game of this country, and, so far as possible, to further legislation for that purpose, and to assist in enforcing the existing laws," "promote inquiry into, and to record observations on the habits and natural history of, the various wild animals," and "bring about among the members the interchange of opinions and ideas on hunting, travel, and exploration; of the various kinds of hunting rifles; and the haunts of game animals, etc." (*Forest and Stream* 1888:124; Grinnell 1910). The Boone and Crockett Club began with a cadre of wealthy, influential, and politically connected men who confronted the conservation challenges of the day, with vigor and success (Ward and McCabe 1988). Its early focus was on big game issues. What the Nuttall Club, AOU, and Audubon societies were to wild birds, the Boone and Crockett Club was to large game.

The Progressive Era

Overlapping the Industrial Revolution was the Progressive Era (~1890–1920)—a frontier of social and political reform, and a time when

> Americans were ready to be concerned about their environment . . . and a general tendency to look favorably on conservation. There were several other factors that figured in the growth and character of conservation before World War I. One was the existence in the U.S. of a technological capacity capable of at least entertaining the large-scale ideas for environmental engineering. In addition, the national indignation, growing since the 1870s, at concentrated wealth, conditioned some Americans to accept the ideology of conservation. And the movement would have taken a different, and probably a less potent, form had it not coincided with the widespread acceptance of the philosophy that the central government should be strong and willing to use its strength in the public interest. Along with the passing of the era of easy resources, such developments conditioned Americans to accept Progress conservation. (Nash 1968:37, 38)

Game wardens from Wisconsin, Minnesota, North and South Dakota, Illinois, and Iowa convened in St. Paul, Minnesota, in December 1892, to consider a uniform game law for those states. Legislation was drafted, but it failed to be adopted in some of the states (Sweeney 1908). Similar meetings occurred elsewhere in the country thereafter, but nothing major came of them for another decade.

The Frontier Thesis

At the 1893 World's Columbian Exposition in Chicago, historian Frederick Jackson Turner presented a paper titled "The Significance of the Frontier in American History," the subject of which became controversially known as his "frontier thesis," referenced at the beginning of this chapter. Turner indicated that expansion was the most important factor in American history. He explained that an area of free land, its continuous recession, and the westward advance of American settlement accounted for the country's development and character. However, since all land of the country was claimed by then, there was no more frontier. Turner questioned whether America would continue to develop as a culture and whether its citizens would retain "that coarseness and strength combined with acuteness and acquisitiveness . . . that dominant individualism now that the frontier was gone" (Turner 1894:226–227).

Turner's thesis was both compelling and controversial at the time and remains so. One is hard-pressed to deny that the national character was framed on the basis of geographic sprawl. However, the thesis failed to acknowledge that frontiers aren't necessarily geographic. Ironically, it was presented at a time when the conservation frontier was fomenting in response to declines in wildlife, pristine forests, and water quality, in favor of stump farms, heavily grazed rangeland, and industrial pollution. What the thesis also failed to identify were the tenacity and resilience of the American character it described. Despite "dominant individualism," Americans, as groups, communities, or a national public, readily responded with unity in times of crisis. By the Progressive Era, natural resource use, abuse, and monopolistic misuse were at crisis points, along with a change in the national temper. The hope of the antebellum frontier had eroded. Such issues as rampant immigration, urban life, economic turmoil, and confusing governments, along with physical changes to the environment, under-mined the perception of American tradition and the promise of individualism. The public was discontented and ready to put its faith in change (Nash 1966).

Prime Movers

Fortunately, conservation reform had a well-positioned and influential catalyst and architect in George Bird Grinnell. And by consequence of a national tragedy in the assassination of President William McKinley, it gained its foremost champion in Theodore Roosevelt. Along with Gifford Pinchot, hired in 1898 as chief of the flagging Division of Forestry in the Department of Agriculture, and with the support of a legion of like-minded persons of foresight, they not only awakened public conscience to the deteriorating situations but also rallied support for state and federal infrastruc-tures to deal with the problems. To be sure, very few of the reforms occurred quickly and without strenuous opposition. For one thing, Roosevelt, Grinnell, Madi-son Grant, John Burnham, William Hornaday, William Wadsworth, and others (mugwumps and middle- and upper-class, Protestant, Anglo-Saxon, American-born men) sought to protect the country by means of pro-gressive reforms, including segregation of true Ameri-cans (excluding Native Americans)

> from insolent hordes of racial and ethnic others, while guarding the old America of individualism and self-help from the excesses of Gilded Age capitalism. To save hunting as a rite of Americanness, however, hunters had to save game, and to save game, they had to rely on gov-ernment. . . . In accepting the challenge of saving game, hunters made themselves stewards of the American environment. Stewardship, indeed, had been implicit in sport hunting from its Jacksonian inception. In identi-fying Americanness with wild animals and wilderness, hunters had made themselves American Natives, men with a special appreciation for the continent's fauna, geography, and sublime scenery. Demanding govern-ment aid to protect wilderness and wildlife was a way to reaffirm this identity and to save hunting as a rite (and right) of the democratic many. (Herman 2001:278)

In 1898, George Shields, editor/publisher of *Recrea-tion* magazine, founded the League of American Sports-

men. The league's intent was to advance conservation through the support of its members. The organization faltered when Shield's fortunes did a decade later, but it was another supportive voice and sounding board for natural resource protections while it lasted.

Forest and Stream was a decidedly powerful commu-nication vehicle for Grinnell, who used it assiduously to promote such achievements as establishment of the forest reserve system, beginning with the Yellowstone Park Timberland Reserve (now Shoshone National For-est) in 1891; passage of the 1891 Act to Repeal Timber Culture Laws, which revised public land laws, includ-ing removing timberlands from the public domain, thus creating opportunity for a federal forestry bureau; the Yellowstone Game Protection Act of 1894; the Lacey Game and Wild Bird Preservation and Disposition Act of 1900; the Reclamation Act of 1902; establishment of Glacier National Park in 1910; and virtually every other major piece of federal conservation legislation prior to World War I. Reiger (1975:149) wrote, "During the formative 1885–97 period, Roosevelt absorbed not only Grinnell's ideas, but also his point of view." Reiger also identified three themes in Grinnell's thought evolution about natural resources. First was a scientific view of changes in the West. Grinnell was an experienced zo-ologist, geologist, naturalist, and ethnographer, as well as a writer, editor, and publisher. Second was an ethical code for outdoorsmen, particularly hunters. The third theme was that of managing natural resources as a busi-ness. These themes were conveyed to Roosevelt even before he became the nation's chief executive.

Grinnell was disappointed when Roosevelt was in-vited to accept the nomination for US vice president in 1900. He felt that the position would relegate Roosevelt to political obscurity when his term or terms ended; he preferred that Roosevelt prepare for an eventual run at the presidency (Reiger 1975). Nevertheless, Roosevelt reluctantly joined the ticket with William McKinley in 1900. Grinnell's sentiments aside, most historians have agreed that Roosevelt was unlikely ever to gain the presidency by election. The bombastic cowboy / Rough Rider / statesman was considered too egocentric and an uncompromising political loose cannon (e.g., Morris 1979, 2001; Brands 1997; Goodwin 2013).

At the Pan-American Exposition in Buffalo, New York, on September 6, 1901, President William McKinley was fatally wounded by anarchist Leon Czolgosz. He died on the morning of September 14. At 3:00 p.m. that day, Theodore Roosevelt was sworn in as the 26th US President, and "a new era in the history of the American land began" (Reiger 1975:147).

Walking Softly

Roosevelt was president until March 4, 1909. It was America's halcyon conservation period in many respects. "The lack of direction in American development appalled Roosevelt and his advisors. They rebelled against a belief in the automatic beneficence of unrestricted economic competition, which, they believed, created only waste, exploitation, and unproductive economic rivalry" (Hays 1959:266). During Roosevelt's term, he used his executive authority and the momentum of progressiveness to create 5 national parks, 4 national game preserves, 18 national monuments, 150 national forests, and 51 federal bird reservations. Twenty-four reclamation projects were undertaken during his tenure. In sum, 230 million acres were set aside mainly for conservation.

Herman (2001) indicated that the frontier of conservation in the United States was less than democratic, by virtue of leading to the abolishment of market hunting, subsistence hunting in most instances, and hunting in certain national or state "commons," such as designated parks, forests, and preserves. Conservation laws governing the take of wildlife made poachers, game hogs, and pothunters of those who theretofore had legally pursued game for subsistence or markets that fulfilled public need and want. "Just as the price for enclosure in England had been paid by commoners, so in America the price for saving game was paid by those too poor, or too 'other,' to gain the ear of the U.S. Congress and state legislatures" (Herman 2001:279).

Establishment of national or state commons in the interest of conservation was neither wrong nor elitist, although conflicts that resulted from their creation were callous and, in many cases, unnecessary, according to Warren (1997). Yet, if America was to retain wildlife and the tradition of hunting, federal and state commons were essential. Furthermore, those commons enabled millions of people of all stripes to become recreational hunters. Hunting and angling were egalitarian activities, although all participants were required to abide by restrictions that intentionally favored none.

Wildlife officials from Colorado, Minnesota, Montana, Wyoming, Oregon, and Utah met in July 1902 at Yellowstone National Park's Mammoth Hot Springs. There, they founded the National Association of Game and Fish Wardens and Commissions. Its primary objective would be advancing interstate communication and cooperation. In addition, it would serve for exchanging information, identifying improvements for state and national conservation programs, and, through resolutions, supporting favorable or objecting to unfavorable federal legislation and actions. "The discussions at this [1902] meeting," Sweeney (1908:469) wrote, "had a marked effect upon subsequent legislation and have resulted in more stringent enforcement of game laws." As Trefethen (1975:137–138) observed of the organization that would eventually morph into today's Association of Fish and Wildlife Agencies (AFWA), "Its membership represented the leading authorities on fish and wildlife administration in the nation, with ready access to the governors and legislatures to whom they were responsible."

Conservative Use

Use of the word "conservation" in the context of natural resources came about in 1907, when either Pinchot or Overton Price, one of his assistants, coined the term's new meaning. Until then, "conservation" had meant something vaguely concerned with the canning of foods. As applied to natural resources, however, it meant "wise use, without waste" of timber, fish, wildlife, grasslands, and other self-replenishing natural products that moderate use would not deplete. Roosevelt adopted "conservation" as a catchword of his administration (Pinchot 1947:326; Trefethen 1964:12–13). Despite the fact that conservation entered the American lexicon as something noble and urgent, misuse and overuse of the country's natural resources remained evident. "Progressives . . . agreed passionately on the need for honesty and a social conscience in the administration of resources. . . . Conservationists were convinced that hostility toward materialism and toward money men and special interests usually was warranted. . . . If nothing else united the conservationists, there was this hatred of the boodler, the rank materialist, the exploiter" (Bates 1957:30).

A Comprehensive Character

Trefethen (1964:13) described early state efforts at managing their natural resources: "The states' scattered efforts toward conservation programs of their own received direction in 1908, when Roosevelt [at US Chief Forester Gifford Pinchot's instigation] called a Conference of Governors for that purpose. This stimulated the formation of state park, forest, and wildlife conservation programs across America. By then, however, the nation's wildlife resources were at an all-time low. In fact, the larger wild mammals seemed doomed. . . . The beaver appeared about to follow the passenger pigeon into limbo. The Conference of Governors was the turning point in this trend."

At this extraordinary gathering of federal and state officials and political, judicial, and industry leaders, little discussion specifically referenced conservation of fish and wildlife. Rather, the presentations dealt mainly with forest and water resources and soils. A basic theme was exploitation without extermination. In his opening address, Roosevelt advised,

> This Conference on the conservation of natural resources is in effect a meeting of the representative of all the people of the U.S. called to consider the weightiest problem now before the Nation . . . that the natural resources of our country are in danger of exhaustion if we permit the old wasteful methods of exploiting them longer to continue. . . . It is safe to say that the prosperity of our people depends directly on the energy and intelligence with which our natural resources are used. It is equally clear that these resources are the final basis for national power and perpetuity. (McGee 1909:3, 7)

During that conference, which achieved conservation consensus among the many participants, Nebraska congressman William Jennings Bryan rendered his view on the matter:

> I am jealous of any encroachment upon the rights of the State, believing that the States are as indestructible as the Union is indissoluble. . . . I do believe, that

it is just as imperative that the general Government shall discharge the duties delegated to it, as it is that the States shall exercise the powers reserved to them. There is no twilight zone between the Nation and the State, in which exploiting interests can take refuge from both, and my observation is that most—not all, but most—of the contentions over the line between Nation and State are traceable to predatory corporations which try to shield themselves from deserved punishment, or endeavoring to prevent needed restraining legislation. (McGee 1909:202)

The applause that greeted these comments showed that federalist and anti-federalist sentiments were alive and well, albeit under different names. Furthermore, by no means was everyone in attendance and throughout Congress and the nation progressively minded. Reformers had altruism on their side, plus a proactive president with executive clout, but there were many industrialists and western capitalists who railed and rallied against change and bully government.

Riesch (1952:340) wrote, "The principle of conservation was no invention of the [Franklin Delano] Roosevelt Administration. With respect to many resources, conservation activities had been introduced by earlier governments both federal and state. Notable examples are the withdrawal of land for National Forests and the National Conservation Conference of Governors called by Theodore Roosevelt. But never before did conservation acquire such a comprehensive character."

Growing Pains and Gains

Trefethen (1964:13–14) described the continuing evolution of state management: "Between 1908 and 1920, state wildlife agencies grew and developed rapidly. In 1895, North Dakota passed a law requiring all hunters to buy state licenses. Between 1910 and 1920, most of the states adopted similar laws, earmarking funds for their fish and game agencies. States developed law enforcement programs and further tightened laws in an effort to increase the supply of wildlife. But there was little scientific knowledge of the needs of the various species. If a particular practice seemed to work in one state, others promptly adopted it." Early wildlife management by states emphasized stocking, law enforcement, predator eradication, and occasional winter feeding of revered game species such as white-tailed deer and elk. Artificial propagation and release of game and fish dominated their trial-and-error efforts. "Spurred by the successful introduction of Chinese ring-necked pheasants into Oregon in 1881, every state experimented, usually futilely, in importing other foreign birds" (Trefethen 1964:14).

Typical of the situation in more than a few states, early game laws in Minnesota were described as "a heterogeneous mass of special enactments, passed at the *suggestion* of various members of the legislature without coherence or design. If carried out . . . [it] was questionable whether they would benefit or harm wildlife" (Day 1875:22).

In the absence of sound scientific understanding of wildlife ecology and habitat needs, attempts to propagate populations of game birds such as wild turkeys, waterfowl, and bobwhites by releasing pen-raised stock failed almost from the beginning. Translocations of deer, elk, and other big game usually met with failure by virtue of faulty capture techniques and equipment, transportation mortality, and clumsy release procedures (e.g., Kennamer et al. 1992).

In Pennsylvania, New York, Wisconsin, and elsewhere, irrupting white-tailed deer herds exceeded carrying capacity on winter ranges and, consequently, experienced substantial die-offs and habitat destruction. Closures to hunting, buck-only hunting, and deer-yard supplemental feeding efforts were popular notions and knee-jerk practices. In the absence of widespread understanding of the species' biology and given the reluctance of most sportsmen to accept limitations to tradition, democracy and political sway trumped what little science state game agencies could muster (Leopold et al. 1947).

Despite many failures, recoveries of wildlife, such as ruffed grouse, cottontail rabbits, raccoons, quail, and other game and nongame edge-habitat species, were occurring in parts of the East and Upper Midwest, following clear-cutting of mature virgin forests. "This was generally attributed to tightened game laws; few recognized that early stage regrowth of forests and a plethora of abandoned farmsteads with destroyed woodlots were largely responsible" (Trefethen 1964:14).

State initiatives were the next frontier of conserva-

tion history. The first state game farm was established in Illinois in 1905 (Palmer 1912). The first state refuge for upland game was established in Pennsylvania, also in 1905. Limited licensing—attempting to limit take to the annual increase—was initially done in Wyoming for moose in 1915 (Blair 1987). Creation of wildlife food plots first occurred on Pennsylvania refuges in 1917. These were all positive steps, but the agencies were attempting to work in a vacuum of science and wherewithal.

Trefethen (1975) noted that many of the early fish and game administrators served in official capacity without compensation (although not all were independently wealthy); some even helped defray agency expenses from their own pockets. Some administrators gained their position by nepotism or cronyism. However, as far as the authors have determined, the majority of first- and second-generation agency heads were sportsmen and dedicated conservationists.

In 1911, representatives of the sporting firearms manufacturers decided to intervene to control commercial hunting. The industry first offered the New York Zoological Society $25,000 per year for five years to initiate a program to protect game. As a vocal opponent of recreational hunting, the society's president, William Hornaday, turned down the considerable sum, not wanting to associate with firearms manufacturers (Dehler 2013). Next, industry made the same offer to the National Association of Audubon Societies, essentially doubling the societies' annual income. On behalf of the organization, T. Gilbert Pearson agreed to the proposal to "check the relentless slaughter of game-birds and mammals" (Wildlife Management Institute 1982:2; see also Haskell 1937). However, the societies' board of directors reacted to outside criticism (principally Hornaday; Dehler 2013) that it would be selling out "to the gun people who wanted to kill all the birds of the country." The check was returned. Industry responded by creating its own organization, the American Game Protective and Propagation Association (AGPPA). It was led by John Burnham, former chief game protector of New York, Klondike gold rusher, and business manager of *Forest and Stream*.

The charge of the newly formed AGPPA was to promote wildlife restoration on national and international scales. It took on what were perceived as the most urgent needs at the time, namely, enforcement of game laws, creation of state conservation agencies, game propagation, and stocking. Among its activities its first year, the AGPPA assisted with codifying game laws for New York, Vermont, and Kentucky. It helped establish the Virginia Commission of Game and Inland Fisheries. It acquired a 5,000-acre property in Plymouth County, Massachusetts, for the East Head Game Farm and sanctuary. And it launched a program to get federal legislation to protect migratory birds (Belanger 1988).

Winging It

Very early in the twentieth century, legislative efforts got under way, led by Congressman George Shiras III, to give the federal government authority over migratory birds, particularly hunting bag limits, elimination of spring hunting, and effective enforcement of the 1900 Lacey Act. It was not until 1913 that the Weeks–McLean Act achieved that end. However, opposed by states—particularly those with spring migration concentrations—the act was declared unconstitutional two years later (Trefethen 1975).

At both state and federal levels, the conservation movement lost its momentum after Roosevelt's departure from the White House; President Taft abandoned a number of Roosevelt's progressive programs and plans, including some related to conservation advances. Public enthusiasm for progressive change also flagged.

William Tecumseh Hornaday was almost certainly the most vitriolic of the conservation frontier's crisis mongers, even though many other scientists of the day echoed his concerns that mass species extinctions were imminent. Hornaday (1913, 1931) focused much of his verbosity on the newfangled pump shotgun, the arms and ammunition companies, automobiles, immigrants ("aliens") who killed songbirds for food, duck hunting clubs, and "game hogs" (a term he may have coined), which included, by his estimate, 85 percent of American hunters and, apparently, anyone or any institution that had the temerity to disagree with him. However, the bristly, enigmatic Victorian zookeeper (and hunter) wasn't always off the mark with his doom-and-gloom augury and finger pointing. His writings and his speaking engagements generally were well received, and many of his associates and audiences saw

him as a charismatic martyr, as did William T. himself. His rants raised and merited a great deal of awareness about wildlife issues (Dehler 2013).

In 1916, a year after the Weeks–McLean Act was declared unconstitutional, the United States entered into a treaty with Great Britain to protect migratory birds in the United States and Canada. Congress passed, and President Woodrow Wilson signed, an enabling act in 1918 that gave force to the Migratory Bird Treaty Act. The treaty gave federal protection in the two countries for all migratory birds, including regulatory authority for the hunting of waterfowl, shorebirds, doves, and other migratory game birds. States retained responsibility for protecting and regulating hunting of nonmigratory species. Migratory bird harvest regulations could be set by individual states, but those regulations could not exceed federal limits. Once again, some states asserted federal preemption of state rights (Trefethen 1975; Bean 1983).

Frank W. McAllister, the attorney general of Missouri, advised hunters to ignore the federal daily bag limit (25) and spring hunting restriction, and instituted a test case when he and four companion hunters (including the national committeeman for the Missouri Democratic Party) were arrested on February 25, 1919, by federal district inspector (warden) Ray P. Holland for spring hunting and overbagging. Suit was brought in federal district court by the state of Missouri, charging that Holland's enforcement of the Migratory Bird Treaty Act in Missouri was unconstitutional by virtue of violation of the Tenth Amendment of the US Constitution, to wit, that the federal government possesses only those powers delegated to it by the Constitution. All remaining powers, Missouri argued, are reserved for the states or the people. The district court upheld the act. Missouri appealed to the US Supreme Court—the landmark *Missouri v. Holland* case (https://supreme.justia.com/cases/federal/us/252/416/). In a 7–2 decision, the Supreme Court held that the national interest in protecting migratory birds could be guaranteed only by federal action, superseding state authority. Accordingly, it validated the exercise of treaty power as supreme law of the land and found no violation of the Tenth Amendment (Trefethen 1975; Bean 1983).

In rendering the decision, Justice Oliver Wendall Holmes observed, "To put the claim of the State upon title is to lean upon a slender reed. Wild birds are not in the possession of anyone; and passion is the beginning of ownership. The whole foundation of the states' rights is the presence within their jurisdictions of birds that yesterday had not arrived, tomorrow may be in another state and in a week a thousand miles away. . . . But for the treaty and the statute, there soon might be no birds for any powers to deal with" (252 U.S. 416, 40 S. Ct. 382, 64 L. Ed. 641 [1920]; Trueblood 1970; Bean 1977).

Regression and Recession

The next frontier for conservation was rather dismal. The movement turned juggernaut by Grinnell, Roosevelt, and others was halted by the ignominy of a horrific world war and the subsequent election of two presidents for whom conservation was not a priority.

Warren G. Harding became president in 1921 and served until his death in 1923. "In his view, the conservation issue was unimportant. He stood for rapid resource development within an unfettered private enterprise system" (Swain 1963:160).

State fish and wildlife agencies were struggling to make headway in the absence of science, trained personnel, and funding. The agencies that existed then were funded almost exclusively by hunting and fishing license revenue, which didn't amount to much and was often used by governors for nonconservation purposes. Americans were tired of government-imposed wartime restrictions, and many were disinclined to purchase licenses to exercise their recreational freedoms (Trefethen 1975). Furthermore, government in general and natural resource matters in particular provoked public disfavor following Teapot Dome—reportedly the most corrupt and sensational government scandal until Watergate (Cherny 2010).

Calvin Coolidge succeeded to the presidency in 1923 and "had almost no aptitude for the subtleties of conservation policy. . . . He wanted the states to discharge their public functions 'so faithfully that instead of an extension on the part of the Federal Government there can be a contraction'" (Swain 1963:162). This translated to an expectation that states would assume management of their own resources, despite an obvious lack of funds to do so effectively and apathy toward federal conservation activity.

Herbert Hoover became president in 1929, and, with respect to conservation, the era of executive laxity ended. Although a states' rights advocate and federal decentralist like his predecessor, Hoover believed that natural resources of the commons should not be plundered in the name of individualism. He crusaded for fish hatcheries, flood control, waterways development, oil and soil conservation, and volunteerism. Swain (1963) asserted that President Hoover's primary conservation contribution was enabling a revival of interest in orderly use and development of natural resources. However, it may be argued that his administration's most significant contribution was its support of scientific and technological research, which paved the way for wildlife management to emerge as a science-based discipline. Also, the recovery of confidence in a conservation-sensitive administration heralded public adoption of the frontier of Franklin D. Roosevelt's New Deal programs.

American Game Policy and the Advent of Professional Wildlife Management

In 1930, state agency wildlife conservation programs were focused on laws, seasons, bag limits, and enforcement. The American Fisheries Society had been formed in 1870, but a formal wildlife management profession did not exist. Leading conservationists observed a continuing decline in game and other wildlife despite the existing conservation programs. The American Wildlife Institute appointed a committee, headed by Aldo Leopold and including the likes of future US congressman and senator A. Willis Robertson, the Virginia game commissioner, to outline a course of action for conservation in the nation. The American Game Policy presented at the 17th American Game Conference described the problem of declining wildlife and outlined steps that were needed to reverse the trend (Leopold 1930). The policy called for the formal establishment of game management as a profession, the staffing of agencies with trained professionals, stable and equitable funding for the agencies, and an active program (including research) of restoration of populations and habitat.

State fish and wildlife agencies in the 1930s faced daunting tasks of enhancing wildlife except by often unsuccessful trial-and-error stocking, translocations, and bounty systems, without much in the way of funds, trained personnel, research, experience, and public understanding. Circumstantial and erratic recoveries of some game following the Dust Bowl era, most notably small game and waterfowl, heightened sportsman enthusiasm (Trefethen 1975). Great were the expectations for more fish, game, and harvest opportunity; understanding was minimal of the variables of wildlife population ecology and the constraints on management to accommodate those expectations. Wrote Leopold (1933:410), "As long as game administration consisted merely of limiting the citizen's shooting privileges, there was little room for experimentation . . . in better cropping methods."

Universities began working hand in hand with state conservation agencies, both in research and in preparing students for professional careers. Early in the century, courses in game management or wildlife conservation were available at several universities. Most of these courses were embedded in forestry, zoology, or entomology programs. Cornell University, the University of Michigan, the State University of New York at Syracuse, the University of Minnesota, and the University of California–Berkeley were among the first universities to offer courses in wildlife management (Organ 2013). The work of prominent professors in the conservation movement, especially Aldo Leopold at the University of Wisconsin–Madison, ensured that universities and fully developed curricula would be part of the emerging profession (Meine 1988). The linkage between the wildlife management profession and universities was solidified when J. N. "Ding" Darling established the first Cooperative Wildlife Research Unit (Coop. Units) in Iowa in 1932 (Goforth 1994).

Darling also served on the three-person "Beck Committee," with Leopold and committee chairman Thomas Beck, editor of *Collier's* magazine and chair of the Connecticut State Board of Fisheries and Game. They were appointed by President Franklin Roosevelt "to devise a wildlife program that would dovetail with [the president's] submarginal land-elimination program" (Lendt 1979). The committee's report met with mixed success. On one hand, it was the first enunciation of needs for a program to counter drought and drainage on waterfowl nesting areas. On the other hand, it failed

to generate previously promised financial support. Yet, there were two other noteworthy outcomes. First was a firm friendship with Leopold, with whom Darling sided on the committee's report recommendations. Second, Darling's tenacious advocacy for conservation measures, his political connections, and his experience with the Iowa Fish and Game Commission enabled him to be appointed director of the faltering US Bureau of Biological Survey in 1934.

Also in 1934, states began adopting a model game law prepared by the International Association of Game, Fish, and Conservation Commissioners, which called for a fish and wildlife agency administrator to be selected on the basis of merit by a nonpartisan board of commissioners (Shoemaker 1935). The boards or commissions also would set fishing and hunting regulations based on recommendations by biologists working for the administrator. Appointees would serve staggered terms, so governors could not flood the commission with personal or party favorites. And biologists, managers, conservation officers, and other salaried employees would be selected on merit and protected by civil service.

As director of the Bureau of Biological Survey, Darling established the Cooperative Wildlife Research Units on a national scale in 1935, beginning with a partnership among nine state land-grant universities, the school's corresponding state fish and wildlife agency, the Department of the Interior, and the American Wildlife Institute. Their mission was (and is) to meet actionable science needs of the agencies, develop the future agency workforce through graduate education and mentoring, and train agency practitioners in new approaches to science (Whalen and Thompson 2015).

Prior experiences with hunting and fishing were important influences for those entering the profession (see Angus 1995). Early leaders recognized that some "pre-existing skill in woodmanship, hunting, and fishing" was important for professional preparation (Leopold 1939:158). But they also warned, "There is danger though of confusing the aptitude for hunting and fishing, which is desirable, with the aptitude for laborious studies of animals and plants, which is indispensable" (Leopold 1939:158). Not surprisingly, state agencies became dominated by a hunting subculture, even though nonhunted wildlife species were featured in some programs. A client-based user-pay/user-benefit system of wildlife conservation operated throughout much of the profession (Decker et al. 1996; chap. 4 of this book).

Ding Darling made innumerable improvements to the Bureau of Biological Survey during his 20-month tenure as director. One in particular was raising morale of the agency's staff. Another was successfully pressing for Ira N. Gabrielson, a brilliant, charismatic, wildlife biologist, to be his successor and the first director of the renamed US Fish and Wildlife Service. "Dr. Gabe" would prove to be an influential proponent of federal/state cooperation.

Dollars and Sense for Conservation

Professional positions within government agencies and curricula within universities developed concurrently and, together with formation of The Wildlife Society in 1937 (Swanson 1987), led to the emergence of a recognizable wildlife management profession. Funding for key elements of the agencies was linked to earmarked fees paid by hunters. Until 1937, as noted earlier, "state fish and game departments had subsisted entirely on hunting and fishing license revenues. Of their limited funds, little was left over from routine expenses and law enforcement costs to conduct wildlife restoration. Several of the funds also were being diverted to work unrelated to conservation" (Trefethen 1964:18). Most significantly, the 1937 Federal Aid in Wildlife Restoration Act (better known as the Pittman–Robertson Act, after its congressional sponsors) and other conservation funding laws were passed, as discussed in chapter 4 of this book.

Ding Darling did not "fade away" after his successful stint with the Bureau of Biological Survey. He was the prime mover and first president of the General Wildlife Federation, founded in 1936 at the First North American Wildlife Conference, called into session by President Roosevelt (it was a continuation of the American Game Conference started in 1915). The organization intended to unite and mobilize conservation organizations under a central state organization, supported and overseen by a flagship national organization. It was modeled after the coordination of Indiana sportsmen's clubs by C. R. "Pink" Gutermuth, director

of education for the state's Department of Conservation. The federation (its name changed a few years later to National Wildlife Federation) was to provide a coordinated, unified, grassroots voice on conservation issues at the state and federal levels (Trefethen 1975; Lendt 1979).

What work the state agencies were able to afford from 1933 until 1942 was supplemented by the Civilian Conservation Corps (CCC). For unemployed, unmarried men aged 18–23 (later, 17–28), the CCC was a popular New Deal public relief national program during the Great Depression (Ermentrout 1982). The enlistees—a total of approximately 3 million over the nine years—planted more than 3 billion trees, constructed more than 800 parks nationwide, upgraded existing state parks, fought wildfires, built service buildings, and extended public roadways into remote areas (www.ccclegacy.org). Many of the latter two are still in use today, having been turned over to state fish and wildlife agencies.

World War II was calamitous. Eventual victories in Europe and the Pacific were costly, given the death toll on all sides, the temporary devolution of humanity, and the unnecessary cost of natural resources plundered for the effort (Kennedy 1988). Not least was the loss of time or lives of men who left schooling in wildlife science to join the military.

Thousands of veterans who returned from World War II and the Korean War took advantage of the GI Bill and went to college to study fisheries and wildlife conservation. Millions more returning veterans bought hunting and fishing licenses and went afield (Trefethen 1975). Pittman–Robertson Act funding increased accordingly, and restoration programs broadened to address a wider array of game and furbearer species.

North American Model of Wildlife Conservation

At this time in the evolution of state wildlife conservation, a set of principles that formed the bedrock for policies and programs was in place. This has been described as the North American Model of Wildlife Conservation (Organ et al. 2012). Seven principles were outlined in this construct, listed below with their pertinence to state wildlife management:

1. *Wildlife resources are a public trust.*—Each state exerted ownership of wildlife by way of their constitution, legislation, or common law except for circumstances whereby federal ownership as dictated by the Constitution prevailed (Batcheller et al. 2010). Threats to public ownership exist and include efforts to privatize and commercialize wildlife resources or to confine them for special use (Organ and Batcheller 2010).

2. *Markets for game are eliminated.*—State and federal laws and policies prevent or restrict commercial sale of game meat. Markets for legal and illegal wildlife (e.g., reptiles and exotic birds) do exist, and efforts to curb many of these are under way.

3. *Allocation of wildlife is by law.*—Every state has a system of laws that provides for legal and equitable allocation and protection of wildlife. Inconsistencies exist in many cases across taxa (Organ et al. 2012).

4. *Wildlife can be killed only for a legitimate purpose.*—State laws define seasons and limits for those wildlife species and populations that can be legally harvested. Many states have wanton waste laws that require maximum utilization without waste of game meat from harvested animals. Certain species (e.g., rattlesnakes, prairie dogs, and coyotes) are unprotected in some states, and the application of this principle is deficient (Organ et al. 2012).

5. *Wildlife is considered an international resource.*—This principle is exemplified through the system of waterfowl management and flyway councils, the Marine Mammal Protection Act of 1972, and the Convention on Trade in Endangered Species of Wild Fauna and Flora of 1973. International conservation efforts and partnerships involving states are increasing.

6. *Science is the proper tool to discharge wildlife policy.*—State wildlife agencies have strong science foundations. The integration of science into policy and the political process varies by state, and challenges abound (Pielke 2007; Organ et al. 2012).

7. *Democracy of hunting is standard.*—Every state has a licensing system in place whereby any citizen in good (legal) standing can participate. Access to certain species and areas may be restricted by lottery or auction, or through fee-for-access hunts.

Hunter Education

The increase in hunters after World War II also brought an increase in hunting accidents and fatalities (Jones et al. 1987). As a result, the first formal hunter education and safety training program was initiated in 1949 in New York. Today, all state wildlife agencies have hunter education programs that foster safety, responsibility, and ethics. Certification of hunters is mandatory in all states except in remote portions of Alaska. The International Hunter Education Association develops curriculum standards, and certification in one state is honored by all other states. Amendments to the Pittman–Robertson Act in 1970 and 1972 established excise taxes on handguns and archery equipment, and half of these funds were made eligible to support state hunter education programs.

Postwar Growing Pains

Despite the substantial gains and positive portent for conservation in the postwar frontier, not all was right with progress. In 1946, E. Sydney Stephens, chairman of the Missouri Conservation Commission, lambasted footdraggers in the profession during a plenary address at the 11th North American Wildlife Conference. On standards of adequate legal authority, employment of trained personnel, development of wildlife environment, education, practical research, cooperation with landowners, and support of citizen organizations, here is how Stephens (1946:25) rated state wildlife administrations "on the basis of their own statements":

> Twenty-five are lacking in adequate legal authority to administer wildlife resources or to regulate their use. Regulation is essential to conservation . . . it is an essential administrative function. Sixteen states employ no trained technicians whatever, or are not better than 20 per cent equipped or manned. Fourteen give no attention to the improvement or development of environment. Twenty-one carry on no cooperation with any group or individual. Fourteen make no effort whatever in the field of education, and twenty others do not claim to be more than 50 per cent efficient in that vital field; none is more than 70 per cent efficient. Twenty-three, or practically one half of the states, do not carry

on research of any kind. Nineteen do not cooperate with any landowner or land-use agency. Five states maintain no forestry departments or agencies and six states have no cooperative forest fire prevention and control programs—all this despite the fact that forests are inextricably related to wildlife, that trees prevent soil erosion and thus contribute to flood control and the further fact that the value of standing timber in this Nation is about 10 billion dollars. Twenty-three complain of the absence of adequate support of organized groups. The turnover in directors is faster than a jet-propelled plane. Their average tenure in office is 5 years and 25 days.

By these standards the departments of 12 states are less than 25 per cent efficient, and 30 rank below 50 per cent; and only 5 have a "passing" grade of 60 or better. The 12 states which rank less than 25 per cent efficient collect from sportsmen and expend $2,345,100 annually. Since they are so pitifully deficient in the application of so many sound practices; since they are expending money for outmoded and even detrimental practices; since they are dominated by politics, the money which they expend is wasted—all to the detriment of wildlife. They should be painlessly but promptly put to death. The next 18 might be given a stay of execution on their promise to reform.

The Environmental Movement

The United States in the aftermath of World War II became markedly different in a number of respects from the America of the Dust Bowl and Depression eras, when modern state fish and wildlife agencies were spawned (Kennedy 1988). An increasingly urban and suburban population base was a product of a population boom and a soaring economy that was shifting from resource extraction to service orientation. People had less direct connection with nature in their day-to-day lives, but more leisure time to vacation outdoors. Technological advancements in media allowed nature and wildlife to be presented to the public in idealized and often inaccurate ways. Public interest in wildlife and environmental issues grew: the number of nongovernmental wildlife organizations expanded from 56 in 1945 to more than 300 by the mid-1970s and to more than 400 by the 1980s (Dunlap 1988).

The ecology movement, which Schoenfeld (1968) characterized as "The Third American Revolution," began to swell with the publication of Rachel Carson's *Silent Spring*, and groups formed to oppose seal hunting in Canada and exploitation of bobcats and exotic spotted cats. The public was confronted with evangelical messages advising that nature was fragile, exploitation of seals and exotic cats was cruel and unnecessary, furs were the mark of the fashionably elite, and government programs were irresponsible (e.g., nuclear waste and predator control). At the same time, people were witnessing the effects of urban and industrial pollution on air quality, water quality, and scenic vistas. The breakdown of traditional social structures and institutions during the 1960s and subsequent rise of individualism as a governing moral code contributed to what sociologists call the "differentiation" of the value structure of American society (Muth 1991). The Malthusian, "tragedy of the commons" (Hardin 1968), and "sky is falling" messages that confronted the public were manifestations of conflicting values in society at large.

Societal changes that accelerated during the 1960s and 1970s led to an expansion of wildlife agency mandates, broadened interest in state agency programs from wildlife stakeholders other than traditional clients, and brought into the profession people who were not influenced primarily by hunting and fishing (Muth 1991; Decker et al. 1996; Organ et al. 1998). Animal rights activism became a social movement that challenged the philosophical underpinnings of the wildlife profession and called into question whether wildlife management was being unduly influenced by consumptive users (Herscovici 1985; Francione 1996). The North American Wildlife Policy of 1973 (Wildlife Management Institute 1973) reflected changes that had occurred in society and the environment since the 1930 American Game Policy. The change in nomenclature from "game" to "wildlife" was indicative of shifts within the profession (Organ and Fritzell 2000). The policy stated that "the first big job is to prevent irreversible losses of species, populations and life communities" (Wildlife Management Institute 1973:10). It further stated that an equal challenge is to prevent a cultural loss and made reference to pioneer skills, which was a pillar of Theodore Roosevelt's and George Bird Grinnell's advocacy for sport hunting (Cutright 1985; Brands 1997). The profession was evolving. Most professionals who were grounded in the consumptive-use tradition had no problem embracing broader mandates, but some did. Many newcomers with protectionist ideologies had difficulty understanding and appreciating the role of consumptive use in wildlife conservation (Pfaffko 2014). One thing remained constant—funding for state wildlife conservation programs was derived primarily from consumptive-user fees. Conservation biology emerged as a discipline from the academic community and private organizations in the 1980s. The new discipline was cited, in part, as a need for an interdisciplinary approach to prevent the serious loss of biological diversity (Iltis 1970; Soulé 1985). It differed from wildlife management (or other resource-oriented fields) by focusing on the preservation of biological systems across landscapes, rather than the sustainable use of natural resources. Conservation biology became an alternative for students and professionals who were interested in wildlife conservation but whose primary influence was not hunting related (Organ and Fritzell 2000).

This frontier of societal changes impacted state fish and wildlife agencies. Trends of the late 1960s and mid-1970s to the late 1990s revealed a decrease (from 30 in 1976 to 25 in 1997) in the number of organizationally independent state fish and wildlife agencies and an increase (from 20 in 1968 to 34 in 1997) in the number receiving general revenues (Wildlife Management Institute 1987, 1997). As a result, a rapid loss of autonomy in establishing funding priorities occurred. The number of states in which the governor or cabinet officer establishes priorities increased from 6 in 1987 to 27 in 1997, and the number of agencies able to use dedicated funds without legislative approval decreased from 11 in 1987 to 7 in 1997 (Wildlife Management Institute 1997).

Program Expansion

Changes that occurred during the 1960s and 1970s, with traditional fish and wildlife agencies being subsumed into broader natural resources agencies, along with an increased breadth of stakeholders, placed greater demands on state agencies. Most states relied on traditional funding—license fees and Pittman–Robertson Act funds (Essig et al. 2012). Most agency programs reflected the funding base and were focused on game species. As states endeavored to expand their programs to address rare and declining species outside of the traditional realm and offer services to the non-hunting public, funding became a limiting factor.

Missouri achieved a milestone in conservation funding history in 1976 when voters adopted a state constitutional amendment raising the general sales tax by one-eighth of a cent and dedicating that revenue to the Department of Conservation (www.moga.mo.gov /statutes/144.htm). This achievement did not come easily; it required a public coalition of conservationists to exert political pressure. Essential to success was the proactive efforts of the Department of Conservation in redirecting resources to expand programs and reach out to nontraditional public sectors. This built trust and demonstrated the department's commitment to program expansion (Jacobson 2008).

Other states have attempted to adopt the Missouri approach, but with little success. A notable exception is Arkansas. In 1996, 10 years after a failed initial attempt, voters approved Amendment 75 to the state's constitution, which resulted in the Game and Fish Commission receiving 45 percent of one-eighth of 1 percent of the state sales tax (www.sos.arkansas.gov/elections/Documents/Constitution%202011%20Amendments.pdf). Virginia, Minnesota, and Iowa have also successfully achieved expanded funding, albeit much less than Missouri and Arkansas. Income tax refund check-off contributions and designer license plate sales succeeded initially in generating nongame funds in a number of states, but none has provided a substantial sustainable source of revenue (Jacobson et al. 2010).

Governance Structure

Leopold (1930) advocated a state wildlife governance structure whereby a board or commission of trustees appointed by the governor would preside over decision-making. This body would be appointed to staggered overlapping terms so that no individual governor could exert undue political influence. Most states eventually adopted this structure, where the agency director reports to the commission, while others have a director who reports up the chain of command to the governor. In some states (e.g., Kansas), the agency director is a member of the governor's cabinet, while in others the director reports to a superagency head, who reports to the governor. Both models—board/commission oversight and governor oversight—have been criticized. Jacobson and Decker (2008) and Nie (2004) advocate for a more participatory governance structure than the traditional representative approach of commissions, and Organ and Batcheller (2010) advocate for true trusteeship under a commission structure. Decker et al. (2016) developed a set of wildlife governance principles for potential adoption that integrate public trust and good governance principles.

Within state wildlife agencies, organizational structures vary. For example, law enforcement responsibilities vary across agencies from a separate administrative division with a chief reporting to the agency director or commissioner (e.g., Maine) to being embedded as part of a biologist's or land manager's responsibilities. In some states wildlife law enforcement is contained in an entirely separate agency (e.g., Massachusetts).

Most state agencies have a dedicated wildlife management or conservation section. The scope of responsibilities within wildlife sections across states varies, though. For example, research may be an integral part of an agency's responsibilities (e.g., Arizona Game and Fish Department), while others, such as the Wyoming Game and Fish Department, do not have research staff and rely instead on their Cooperative Wildlife Research Unit.

Wildlife diversity programs developed in states during the 1970s, and today, every state wildlife agency has staff dedicated to the conservation of rare species and those not hunted for sport. In some states (e.g., Massachusetts and New Jersey), the wildlife diversity program has a dedicated assistant director and is a parallel program to wildlife management. Other states have various structures, ranging from a separate diversity branch within a wildlife conservation section to a totally integrated approach where, with the exception of species listed as endangered or threatened, organizational structure is focused on taxa (e.g., birds, mammals, herpetofauna) rather than whether they are harvested or not (e.g., Maine).

Despite the diversity in organizational structures, there is much continuity among state wildlife agencies. Through the Association of Fish and Wildlife Agencies,

common ground in addressing transboundary common conservation challenges is facilitated and effective. The variations in structure cited above represent the diverse needs, cultures, and institutions inherent in our diverse nation.

Succeeding chapters in this book succinctly describe state wildlife funding sources and management practices since the mid-1970s. As you'll discover, the states' management role in conserving wildlife of the United States has never been more daunting, or more important.

LITERATURE CITED

Allen, J. A. 1886a. Destruction of birds for millinery purposes. Science 7(160):196–197.

———. 1886b. The present wholesale destruction of bird-life in the U.S. Science 7(160):191–195.

Angus, S. 1995. Women in natural resources: Stimulating thinking about motivations and needs. Wildlife Society Bulletin 23:579–582.

Batcheller, G. R., M. C. Bambery, L. Bies, T. Decker, S. Dyke, D. Guynn, M. McEnroe, M. O'Brien, J. F. Organ, S. J. Riley, and G. Roehm. 2010. The public trust doctrine: Implications for wildlife management in the United States and Canada. Technical Review 10-01. The Wildlife Society, Bethesda, Maryland, USA.

Bates, J. L. 1957. Fulfilling American democracy: The conservation movement, 1907–1921. Mississippi Valley Historical Review 44:29–30, 38, 42, 47–48, 53–54, 57.

Bean, M. 1977. The evolution of national wildlife law. US Government Printing Office, Washington, DC, USA.

———. 1983. The evolution of national wildlife law, 2nd edition. US Government Printing Office, Washington, DC, USA.

Belanger, D. 1988. Managing American wildlife: A history of the International Association of Fish and Wildlife Agencies. Washington, DC, USA.

Beverley, R. 1705. The history and present state of Virginia. R. Parker, London, UK.

Billington, R. A. 1978. America's frontier heritage. University of New Mexico Press, Albuquerque, New Mexico, USA.

Blair, N. 1987. The history of wildlife management in Wyoming. Wyoming Game and Fish Department, Cheyenne, Wyoming, USA.

Brands, H. W. 1997. TR: The last romantic. Basic Books, New York, New York, USA.

Brown, D. 1970. Bury my heart at Wounded Knee. Holt, Rinehart & Winston, New York, New York, USA.

———. 1977. Hear that lonesome whistle blow. Henry Holt, New York, New York, USA.

Brown, J. J. 1857. The American angler's guide; or, complete fisher's manual, for the United States; containing the

opinions and practices of experienced anglers of both hemi-
spheres. D. Appleton, New York, New York, USA.

California Department of Fish and Game. 1999. Department of
Fish and Game celebrates serving California for 130 years.
Outdoor California 60(6).

Cardoza, J. E. 2015. The Massachusetts division of fisheries and
wildlife: 1866–2012. Massachusetts Division of Fisheries
and Wildlife, Westborough, Massachusetts, USA.

Cherny, R. W. 2010. Graft and oil: How Teapot Dome became
the greatest political scandal of its time. Gilder Lehrman
Institute of American History. www.gilderlehrman.org/
history-by-era.

Cronan, W. 1983. Changes in the land: Indians, colonists, and
the ecology of New England. Hill & Wang, New York, New
York, USA.

Cushman, R. 1963. Reasons and considerations touching
the lawfulness of removing out of England into parts of
America. In D. B. Heath, editor. Mourt's relation: A journal
of the English plantation settled at Plymouth in New England, by certain English adventurers both merchants and
others. Applewood Books, Bedford, Massachusetts, USA.

Cutright, P. R. 1985. Theodore Roosevelt, the making of a
conservationist. University of Illinois Press, Urbana, Illinois, USA.

Day, D. 1875. First annual report of the State Fish Commissions
of Minnesota. Printed for the State Legislature, St. Paul,
Minnesota, USA.

deCalesta, D. S. 1976. Predator control: History and policies.
Extension Service Circular 710. Oregon State University,
Corvallis, Oregon, USA.

Decker, D. J., C. C. Krueger, R. A. Baer, Jr., B. A. Knuth, and
M. E. Richmond. 1996. From clients to stakeholders:
A philosophical shift for fish and wildlife management.
Human Dimensions of Wildlife 1:70–82.

Decker, D. J., C. Smith, A. Forstchen, D. Hare, E. Pomeranz,
C. Doyle-Capitman, K. Schuler, and J. Organ. 2016. Governance principles for wildlife conservation in the 21st
century. Conservation Letters 9(4):290–295. doi:10.1111
/con1.12211.

Dehler, G. J. 2013. The most defiant devil: William Temple
Hornaday and his controversial crusade to save American
wildlife. University of Virginia Press, Charlottesville, Virginia, USA.

DeVoto, B. 1947. Across the wide Missouri. Houghton Mifflin,
Boston, Massachusetts, USA.

Dolan, E. J. 2010. Fur, fortune and empire. W. W. Norton, New
York, New York, USA.

Dunlap, T. R. 1988. Saving America's wildlife. Princeton University Press, Princeton, New Jersey, USA.

Elliott, W. 1846. Carolina sports by land and water; including
incidents of devil-fishing, wild-cat, deer, and bear hunting,
etc. Beaufort, South Carolina, USA.

Ermentrout, R. A. 1982. Forgotten men: The Civilian Conservation Corps. Exposition Press, Smithtown, New York,
USA.

Essig, R. J., J. F. Organ, and S. S. Stevens. 2012. Sources and
trends of fish and wildlife conservation funding. Transactions of the North American Wildlife and Natural Resources Conference 77.

Faragher, M. J. 1999. Commentary in rereading Frederick Jackson Turner: The significance of the frontier in American
history and other essays. Yale University Press, New Haven,
Connecticut, USA.

Foreman, G. 1989. Indian removal: The emigration of the Five
Civilized Tribes. University of Oklahoma Press, Norman,
Oklahoma, USA.

Forest and Stream. 1888. The Boone and Crockett Club. Forest
and Stream 30(3):124.

Francione, G. L. 1996. Rain without thunder: The ideology
of the animal rights movement. Temple University Press,
Philadelphia, Pennsylvania, USA.

Gabriel, T. 1912. An historical and geographical account of
Pennsylvania and of West-New Jersey. In A. C. Myers,
editor. Narratives of early Pennsylvania and of West New
Jersey and Delaware, 1630–1707. Charles Scribner's Sons,
New York, New York, USA.

Goforth, W. R. 1994. The cooperative fish and wildlife research
unit program. Special Publication, US National Biological
Survey, Washington, DC, USA.

Goodwin, D. K. 2013. The bully pulpit. Simon & Schuster, New
York, New York, USA.

Grinnell, G. B. 1884. The game protection fund. Forest and
Stream 22(16):301.

———. 1886. The Audubon society. Forest and Stream 26(3):41.

———, editor. 1910. A brief history of the Boone and Crockett
Club with officers, constitution and list of members for the
year 1910. Forest and Stream Publishing, New York, New
York, USA.

Hardin, G. 1968. The tragedy of the commons. Science
162(3859):1243–1248.

Haskell, W. S. 1937. The American Game Protective and Propagation Association: A history. Camp Fire Club of America,
New York, New York, USA.

Hays, S. F. 1959. Conservation and the gospel of efficiency: The
progressive conservation movement, 1890–1920. Harvard
University Press, Cambridge, Massachusetts, USA.

Herman, D. J. 2001. Hunting and the American imagination.
Smithsonian Institution Press, Washington, DC, USA.

Herscovici, A. 1985. Second nature: The animal rights controversy. CBC Enterprises, Montreal, Quebec, Canada.

Hornaday, W. T. 1913. Our vanishing wild life. Charles Scribner's Sons, New York, New York, USA.

———. 1931. Thirty year war for wild life. Charles Scribner's
Sons, New York, New York, USA.

Iltis, H. H. 1970. Man's forgotten necessity eco-variety. Field
and Stream 74(2):44–47.

Jacobson, C. A. 2008. Wildlife conservation and management in the 21st century: Understanding challenges for institutional transformation. PhD diss., Cornell University, Ithaca, New York, USA.

Jacobson, C. A., and D. J. Decker. 2008. Governance of state wildlife management: Reform and revive or resist and retrench? Society and Natural Resources 21:441–448.

Jacobson, C. A., J. F. Organ, D. J. Decker, G. R. Batcheller, and L. Carpenter. 2010. A conservation institution for the 21st century: Implications for state wildlife agencies. Journal of Wildlife Management 74:203–209.

Jones, J., F. Eyman, F. Disbrow, and H. Moe. 1987. Hunter education: Safety and responsibility. Pages 209–217 in H. Kallman, editor. Restoring America's wildlife. US Department of the Interior, Washington, DC, USA.

Jung, P. J. 2007. The Blackhawk war of 1832. University of Oklahoma Press, Norman, Oklahoma, USA.

Kennamer, J. E., M. Kennamer, and R. Brenneman. 1992. History. Pages 6–17 in J. G. Dickson, editor. The wild turkey: Biology and management. Stackpole Books, Harrisburg, Pennsylvania, USA.

Kennedy, P. M. 1988. The rise and fall of the great powers: Economic change and military conflict from 1500 to 2000. Random House, New York, New York, USA.

Knetsch, J. 2003. Florida's Seminole wars: 1817–1858. Arcadia, Charleston, South Carolina, USA.

Knox, T. J. 1881. Young nimrods in North America: A book for boys. Harper & Brothers, New York, New York, USA.

Kohlmeyer, F. W. 1962. Homestead centennial symposium, Lincoln, Nebraska, June 11–14, 1962: A report of the proceedings. Agricultural History 36:222–224.

Langguth, A. J. 2006. Union 1812: The Americans who fought the second war of independence. Simon & Schuster, New York, New York, USA.

Lavender, D. 1970. The great persuader: The biography of Collis P. Huntington. Doubleday, Garden City, New York, USA.

Lendt, D. L. 1979. Ding: The life of Jay Norwood Darling. Iowa State University Press, Ames, Iowa, USA.

Leopold, A. 1930. Report to the American game conference on an American game policy. Transactions of the American Game Conference 17:281–283.

———. 1933. Game management. Charles Scribner's Sons, New York, New York, USA.

———. 1939. Academic and professional training in wildlife work. Journal of Wildlife Management 3:156–161.

Leopold, A., L. K. Sowls, and D. L. Spencer. 1947. A survey of over-populated deer ranges in the United States. Journal of Wildlife Management 11:162–177.

Lund, T. 1980. American wildlife law. University of California Press, Berkeley, California, USA.

Mahon, J. K. 1967. History of the second Seminole War. University Press of Florida, Gainesville, Florida, USA.

Marsh, G. P. 1864. Man and nature; or, physical geography as modified by human action. Charles Scribner, New York, New York, USA.

———. 1874. The Earth as modified by human action. Scribner, Armstrong, New York, New York, USA.

Matthiessen, P. 1964. Wildlife in America. Viking Press, New York, New York, USA.

McGee, W. J., editor. 1909. Proceedings of a conference of governors in the White House, Washington, DC, May 13–15, 1908. Arno Press, New York, New York, USA.

McWilliams, J. 2015. How the battle of New Orleans birthed the American character. www.newyorker.com/news/news-desk/battle-new-orleans-birthed-american-democracy.

Meine, C. 1988. Aldo Leopold: His life and work. University of Wisconsin Press, Madison, Wisconsin, USA.

Missal, J., and M. Missal. 2004. The Seminole wars: America's longest Indian conflict. University Press of Florida, Gainesville, Florida, USA.

Morris, E. 1979. The rise of Theodore Roosevelt. Coward, McCann & Geoghegan, New York, New York, USA.

———. 2001. Theodore Rex. Random House, New York, New York, USA.

Muth, R. M. 1991. Wildlife and fisheries policy at the crossroads: Contemporary sociocultural values and natural resources management. Transactions of the Northeast Section of The Wildlife Society 48:170–174.

Nash, R. 1966. The American cult of the primitive. American Quarterly 18:717, 524–25, 534–37.

———. 1968. The progressive conservation crusade. Pages 37–38 in R. Nash, editor. The American environment: Readings in the history of conservation. Addison-Wesley, Reading, Pennsylvania, USA.

Nie, M. 2004. State wildlife policy and management: The scope and bias of political conflict. Public Administration Review 64:211–233.

Organ, J. F. 2013. The wildlife professional. Chapter 3 in P. R. Krausman and J. W. Cain III, editors. Wildlife conservation and management. Johns Hopkins University Press, Baltimore, Maryland, USA.

Organ, J. F., and G. R. Batcheller. 2010. Toward the state wildlife management institution of the future: Key elements. Transactions of the North American Wildlife and Natural Resources Conference 75:139–142.

Organ, J. F., and E. K. Fritzell. 2000. Trends in consumptive use and the wildlife profession. Wildlife Society Bulletin 38:780–787.

Organ, J. F., V. Geist, S. P. Mahoney, S. Williams, P. R. Krausman, G. R. Batcheller, T. A. Decker, R. Carmichael, P. Nanjappa, R. Regan, R. A. Medellin, R. Cantu, R. E. McCabe, S. Craven, G. M. Vecellio, and D. J. Decker. 2012. The North American Model of Wildlife Conservation. Technical Review 12-04. The Wildlife Society, Bethesda, Maryland, USA.

Organ, J. F., R. M. Muth, J. E. Dizard, S. J. Williamson, and T. A. Decker. 1998. Fair chase and humane treatment:

Balancing the ethics of hunting and trapping. Transactions of the North American Wildlife and Natural Resources Conference 63:528–543.

O'Sullivan, J. 1839. The great nation of futurity. United States Democratic Review 6(23).

———. 1845. Annexation. United States Democratic Review 17(85).

Palmer, T. S. 1912. Chronology and index of American game protection, 1776–1911. Bureau of Biological Survey Bulletin 41. US Department of Agriculture, Washington, DC, USA.

Pfaffko, M. 2014. Lessons in hunting and conservation: Exploring the Conservation Leaders for Tomorrow program. Wildlife Professional 8(4):54–57.

Phillips, J. C. 1928. Wild birds introduced or transplanted in North America. Technical Bulletin Number 61. US Department of Agriculture, Washington, DC, USA.

Pielke, R. A., Jr. 2007. The honest broker: Making sense of science in policy and politics. Cambridge University Press, Cambridge, UK.

Pinchot, G. 1947. Breaking new ground. Harcourt, Brace, New York, New York, USA.

Pond, F. E. [pseud. Will Wildwood], comp. 1891. The Sportsman's directory: Containing a carefully classified descriptive record of the principal American manufacturers of and dealers in guns, ammunition, fishing tackle, and sporting goods; dog breeders, kennel clubs, state sportsmen's associations, fish commissioners, game wardens, racing and trotting associations, athletic and aquatic clubs, sporting journals, books and publishers, etc., etc. Pond & Coldey, Milwaukee, Wisconsin, USA.

Porterfield, J. 2005. The Homestead Act of 1862: A primary source history of the settlement of the American heartland in the late 19th century. Rosen, New York, New York, USA.

Pratt, J. 1927. The origin of "manifest destiny." American Historical Review 32(4):795–798.

Raghavan, M., M. Steinriicken, K. Harris, S. Schiffels, S. Rasmussen, M. DeGiorgio, A. Albrechtsen, C. Valdiosera, M. C. Avila-Arcos, A. Malaspinas, A. Eriksson, I. Moltke, M. Metspalu, J. R. Homburger, J. Wall, O. E. Cornejo, J. V. Moreno-Mayar, T. S. Korneliussen, T. Pierre, M. Rasmussen, P. F. Campos, P. de Barros Damgaard, M. E. Allentoft, J. Lindo, E. Metspalu, R. Rodriguez-Varela, J. Mansilla, C. Henrickson, A. Seguin-Orlando, H. Malstrom, T. Stafford, Jr., S. S. Shringarpure, A. Moreno-Estrada, M. Karmin, K. Tambets, A. Bergstrom, Y. Xue, V. Warmuth, A. D. Friend, J. Singarayer, P. Valdes, F. Balloux, H. Leboreiro, J. L. Vera, H. Rangel-Villalobos, D. Pettener, D. Luiselli, L. G. Davis, E. Heyer, C. P. E. Zollikofer, M. S. Ponce de Leon, C. I. Smith, V. Grimes, K. Pike, M. Deal, B. T. Fuller, B. Arriaza, V. Standen, M. F. Luz, F. Ricaut, N. Guidon, L. Osipova, M. I. Voevoda, O. L. Posukh, O. Balanovsky, M. Lavryashina, Y. Bogunov, E. Khusnutdinova, M. Gubina, E. Balanovska, S. Federova, S. Litvinov, B. Malyarchuk, M. Derenko, M. J. Mosher, D. Archer, J. Cybulski, B. Pet-

zelt, J. Mitchell, R. Worl, P. J. Norman, P. Parham, B. M. Kemp, T. Kivisild, C. Tyler-Smith, M. Sandhu, M. Crawford, R. Villems, D. G. Smith, M. R. Waters, T. Goebel, J. R. Johnson, R. S. Malhi, M. Jakobsson, D. J. Meltzer, A. Manica, R. Durbin, C. D. Bustamante, Y. S. Song, R. Nielsen, and E. Willersley. 2015. Genomic evidence for the Pleistocene and recent population history of Native Americans. Sciencexpress 23 July 2015:1–10.

Rasmussen, W. D. 1975. Agriculture in the United States: A documentary history. Random House, New York, New York, USA.

Reiger, J. 1975. American sportsmen and the origins of conservation. Winchester Press, New York, New York, USA.

Riesch, A. L. 1952. Conservation under Franklin D. Roosevelt. PhD diss., University of Wisconsin–Madison, Madison, Wisconsin, USA.

Roosevelt, T. R. 1885. Hunting trips of a ranchman: Sketches of sport on the northern cattle plains. G. P. Putnam's Sons, New York, New York, USA.

Schoenfeld, C. 1968. The third American revolution. University of Wisconsin Center for Environmental Communications and Education Studies, Madison, Wisconsin, USA.

Shoemaker, C. D. 1935. The model state game and fish administrative law. Transactions of the American Game Conference 21:177–183.

Shurtleff, N. B., editor. 1853. Records of the governor and company of the Massachusetts Bay in New England. Volume 1, 1628–1641. W. White, Boston, Massachusetts, USA.

Smith, J. 1907. The generall historie of Virginia, New England and the Summer Isles together with the true travels, adventures and observation, and a sea grammar. Volume 1. Maclehose, Glasgow, UK.

Soulé, M. E. 1985. Conservation biology and the "real world." Pages 1–12 in M. E. Soulé, editor. Conservation biology: The science of scarcity and diversity. Sinauer Associates, Sunderland, Massachusetts, USA.

Stanford, D. J., and B. A. Bradley. 2013. Across Atlantic ice: The origin of America's Clovis culture. University of California Press, Berkeley, California, USA.

Stephens, E. S. 1946. Where are we and what time is it? Transactions of the North American Wildlife Conference 11:21–27.

Swain, D. C. 1963. Federal conservation policy, 1921–1933. University of California (Berkeley) Publications in History 76:160–170.

Swanson, E. B. 1940. The use and conservation of Minnesota game 1850–1900. Thesis, University of Minnesota, St. Paul, Minnesota, USA.

Swanson, G. A. 1987. Creation and early history. Wildlife Society Bulletin 15:14–22.

Sweeney, Z. T. 1908. Biennial report of the commissioner of fisheries and game for Indiana. William B. Burford, Indianapolis, Indiana, USA.

Trask, K. A. 2006. Black Hawk: The battle for the heart of America. Henry Holt, New York, New York, USA.

Trefethen, J. T. 1964. Wildlife management and conservation. D. C. Heath, Boston, Massachusetts, USA.

———. 1975. An American crusade for wildlife. Winchester Press, New York, New York, USA.

Trueblood, T. 1970. Ray P. Holland. Field and Stream 75(2):82, 195–196, 198–199.

Turner, F. J. 1894. The significance of the frontier in American history. Pages 197–227 in The Annual Report of the American Historical Association for 1893. Government Printing Office, Washington, DC, USA.

Udall, S. L. 1963. The quiet crisis. Holt, Rinehart & Winston, New York, New York, USA.

Wallace, A. 1993. The long bitter trail: Andrew Jackson and the Indians. Hill & Wang, New York, New York, USA.

Ward, G. B., and R. E. McCabe. 1988. Trail blazers in conservation: The Boone and Crockett Club's first century. Pages 47–122 in W. H. Nesbitt and J. Reneau, editors. Records of North American big game. The Boone and Crockett Club, Dumfries, Virginia, USA.

Warren, L. E. 1997. The hunter's game: Poachers and conservationists in twentieth century America. Yale University Press, New Haven, Connecticut, USA.

Whalen, K. G., and J. D. Thompson. 2015. The cooperative research units' model: Enabling past and future-based conservation. Transactions of the North American Wildlife and Natural Resources Conference 80.

Wildlife Management Institute. 1973. The North American wildlife policy, 1973. Wildlife Management Institute, Washington, DC, USA.

———. 1982. Wildlife Management Institute. Wildlife Management Institute, Washington, DC, USA.

———. 1987. Organization, authority and programs of state fish and wildlife agencies. Wildlife Management Institute, Washington, DC, USA.

———. 1997. Organization, authority and programs of state fish and wildlife agencies. Wildlife Management Institute, Washington, DC, USA.

Woodmason, C. 1953. The Carolina backcountry on the eve of the revolution. R. J. Hooker, editor. University of North Carolina Press, Chapel Hill, North Carolina, USA.

Young, S. P., and E. A. Goldman. 1944. The wolves of North America. American Wildlife Institute, Washington, DC, USA.

2

Public Trust Doctrine and the Legal Basis for State Wildlife Management

Gordon R. Batcheller,
Thomas A. Decker, and
Robert P. Lanka

Early Beginnings

Ironically, the legal status of eastern oysters (*Crassostrea virginica*) had enormous bearing on the conservation and management of wildlife species, the focus of this book. By all accounts, the nineteenth-century oyster industry was lucrative. Eastern oysters thrived in coastal bays from New Brunswick to the Gulf of Mexico. William Strachey, an early English settler, wrote in 1612 of the Chesapeake Bay, "Oysters there be in whole bancks and bedds, and those of the best: I have seene some thirteen inches long [*sic*]" (Hakluyt Society 1849:127). Francis Louis Michael of Switzerland journeyed to Virginia in 1701–1702 and wrote, "The abundance of oysters in [*sic*] incredible. There are whole banks of them so that the ships must avoid them" (Hinke 1916:35).

It was not long before English settlers and early Americans began harvesting oysters for the growing markets of Boston, New York, Philadelphia, and Baltimore. The lower Hudson River estuary had more than 300 square miles of oyster beds, and New York Harbor itself may have contained half the world's oysters. So on January 1, 1835, William C. H. Waddell, a resident of New York, rented 500 acres of property in the township of Perth Amboy, New Jersey, including

The findings and conclusions in this chapter are those of the authors and do not necessarily represent the views of the US Fish and Wildlife Service.

100 acres of lands covered by water containing prime oyster beds, to a "John Den." It turns out that Den was simply a legal construct, not a real oyster farmer. As a wealthy New York businessman and investor, Waddell sought a definitive legal judgment on disputed lands and disputed laws, and he set up a test case that would be heard in the federal courts. Merit Martin and his associates also sought to test the law, and one can visualize the merriment that may have ensued as he and six others on January 2, 1835, "with force and arms, &c., entered in the tenements with the appurtenances, in which said John Den [*sic*] was so interested in the manner . . . and ejected him . . . then and there, to the great damage of John Den" (*Martin et al. v. Waddell's Lessee* 1842).

The Decision

The case was first heard in the Federal Circuit Court for New Jersey in 1835, and Waddell prevailed. Merit Martin believed that their excursion to harvest oysters on the open waters and underlying mudflats owned by Waddell wasn't trespass at all, and their attorneys took the case, on appeal, to the US Supreme Court in 1842. They won (i.e., the US Supreme Court overruled the lower court's decision that upheld Waddell's claims).

Based largely on an earlier case heard by the New Jersey State Supreme Court (*Arnold v. Mundy* 1821), Chief Justice Roger B. Taney wrote the seminal opinion in *Martin et al. v. Waddell* firmly establishing the public

Legal access to oyster beds was the central issue that led to the first major public trust ruling by the US Supreme Court, *Martin et al. v. Waddell's Lessee* (1842). *Photo courtesy of the Library of Congress.*

trust doctrine (PTD) as a key legal principle in natural resources management. The question before the court was whether an individual, William C. H. Waddell, could control access to oyster-bearing mudflats via the open water of Raritan Bay, New Jersey. The answer was no, he could not. Citing the Magna Carta and Letters of Patent from the King of England, Charles II, to his brother the Duke of York, the court found that principles of public stewardship that applied in England for centuries must also be applied to the trust responsibilities of several states. As noted by Justice Taney in his opinion, "In the case before us, the rivers, bays, and arms of the sea, and all prerogative rights within the limits of the charter, undoubtedly passed to the Duke of York, and were intended to pass, except those saved in the letters patent." He went on to note, "And when the people of New Jersey took possession of the reins of government [following the American Revolution], and took into their own hands the powers of sovereignty, the prerogatives and regalities which before belonged either to the crown or the parliament, became immediately and rightfully vested in the state." Judge Taney went on to write, "The question here depends . . . upon the charters granted by the British crown; under which certain rights are claimed by the state, on the one hand and by private individuals, on the other." Finally, the court made its pivotal finding: "Of this decision of the Supreme Court of New Jersey, we are of the opinion

Chief Justice Roger B. Taney authored the US Supreme Court decision in *Martin et al. v. Waddell's Lessee* (1842), firmly establishing the public trust doctrine as a legal principle germane to the governance of public use of natural resources. *Photo courtesy of Gordon R. Batcheller, Northeast Association of Fish and Wildlife Agencies.*

The Magna Carta sealed at Runnymede in 1215 by King John of England established enduring principles of freedom and liberty, protecting the rights and property of the people against the tyranny of kings. In *Martin et al. v. Waddell's Lessee* (1842), Chief Justice Taney cited the Magna Carta: "The policy of England since Magna Charta [*sic*] . . . has been carefully preserved to secure the common right of piscary for the benefit of the public." *Image courtesy of the US National Archives and Records Administration.*

that the proprietors [William C. H. Waddell] are not entitled to the rights in question; and the judgment of the Circuit Court must, therefore, be reversed." So, Merit Martin and his colleagues were not guilty of trespass on the open waters of Raritan Bay, and they could indeed harvest oysters consistent with laws of the state

of New Jersey, irrespective of Waddell's ownership and control of adjacent lands. This decision squarely established, for the first time, the principle that state governments, not private individuals, have a public trust responsibility vis-à-vis natural resources.

Dissent

It could have been otherwise. *Martin v. Waddell* was not a unanimous opinion. The dissenters, Justices Thompson and Baldwin, wrote a strong dissent that, if upheld, would have upended a key structural underpinning of the PTD. In dissent, Thompson struck a distinction between the oyster fishery and "floating fish," noting the natural occurrence of oysters on beds of "land" versus fish of the sea. "The latter is entirely local and connected with the soil. . . . They are planted and cultivated by the hand of man like other productions of the earth." Referencing a case heard in England concerning the Burnham River, the dissenters found that the actions of Martin and his men were nothing less than "trespass for breaking and entering the several oyster fishery of the plaintiffs." Thompson opined that the chief justice erred in ruling for the defendant, *Martin et al.*, and the state of New Jersey had no prerogative for trustee status of the oyster beds of Raritan Bay.

Evolution of the Public Trust Doctrine

While the decision in *Martin v. Waddell* essentially held that state governments have a fundamental trust responsibility for public resources based on underpinnings of the Magna Carta and English common law, the Supreme Court in *Geer v. State of Connecticut* (1896) went even further to address responsibilities of several states toward wildlife resources. In *Geer*, the court found that "wild fowling" was within the state's purview as a trustee of wildlife, firmly establishing the principle that state governments should be responsible for conservation and management of wildlife species. However, the role of the federal government in regulating certain wildlife species became clear with ratification of the 1916 Convention for the Protection of Migratory Birds and the subsequent 1918 passage of the Migratory Bird Treaty Act. When the state of Missouri challenged the authority of the federal government to enforce certain provisions of the act, the US Supreme Court sided strongly with the federal government (*Missouri v. Holland* 1920). As Michael Bean put it, this decision "dealt a stunning blow to those who had felt the state ownership doctrine was a bar to federal wildlife regulation" (Bean 1977:25).

Although *Hughes v. Oklahoma* (1979) partially reversed *Geer*, the upshot of the former was primarily to ensure a role for the federal government in conserving wildlife on federal lands and waters. It did not undo the premise of the obligation of state governments to conserve wildlife within their boundaries. Indeed, an active role for the federal government is embedded in the seminal decision of *Martin v. Waddell* as highlighted by Organ et al. (2012:12): "The trustee status of states in regard to wildlife is transferred to the federal government . . . when wildlife falls within the parameters of the U.S. Constitution's Supremacy Clause . . . , Commerce Clause, and Property Clause." As noted by Chief Justice Taney in *Martin v. Waddell*, the powers assumed under the PTD by states were "subject only to the rights since surrendered by the Constitution."

Joseph J. Sax (1970) outlined the evolution of key principles and legal precedents leading to the current form of the PTD in the United States. He reached back to ancient Roman law, English common law, and several US Supreme Court cases, including *Martin v. Waddell* and *Illinois Central Railroad Company v. State of Illinois* (1892), to suggest that the PTD should be applied to environmental issues. According to his obituary published in the *New York Times* on March 10, 2014, Sax "fuel(ed) the environmental movement by establishing the doctrine that natural resources are a public trust requiring protection."

Bonnie J. McCay (1998) described the PTD in the context of the "oyster wars" that prevailed in northern New Jersey during the 1800s. She further explained how ecological, economic, political, and social variables intersected in complex ways and led to pivotal state and federal court decisions that collectively constitute the underpinning legal history of the PTD in the United States. Together, these scholars provide a solid grounding for anyone seeking a detailed history on the evolution of the PTD in the United States.

Public Trust and the North American Model of Wildlife Conservation

Through tradition, practice, court rulings, and state and federal laws, the PTD firmly establishes the responsibility of government, both state and federal, to conserve and manage wildlife for public benefit. This central principle is key to the North American Model of Wildlife Conservation and indeed is a fundamental tenet for the wildlife management profession.

The importance of the PTD to the North American Model of Wildlife Conservation is well documented (Organ et al. 2012) and discussed more fully in chapter 1. Wildlife professionals consider the PTD as foundational to state wildlife conservation programs throughout the United States. Batcheller et al. (2010) provide a technical review highlighting historical antecedents of the PTD as analyzed and described by Sax (1970). As Sax (1970) explains, the PTD has its roots in sixth-century Roman law (the "Justinian Institutes"): "By the law of nature these things are common to all mankind: the air, running water, the sea, and consequently the shores of the sea." The English Magna Carta in 1215 affirmed primacy of Roman law. Acceptance of these principles within English common law occurred in approximately 1641. Justice Taney later affirmed these principles in his now famous ruling in *Martin v. Waddell*.

State Implementation of the Public Trust Doctrine

While both federal and state court cases addressed various aspects of the PTD, state legislative bodies have not been idle. Batcheller et al. (2010) summarized how states have affirmed jurisdiction over wildlife stewardship. As of 2010, 38 states had implemented a clear statement of state ownership of wildlife within statute. New Hampshire law states, "It shall be the policy of the state to maintain and manage [wildlife] resources for future generations" (N.H. Rev. Stat. § 212-B:2). North Carolina statutory language is direct concerning purposes of wildlife: "The enjoyment of the wildlife resources of the state belongs to all of the people of the state" (N.C. Gen. Stat. § 113-133.1). Georgia has very strong language: "Wildlife is held in trust by the state for the benefit of its citizens and shall not be reduced to private ownership except as specifically provided for in this title" (Ga. Code § 27-1-2).

As of 2014, 44 states were members of the Interstate Wildlife Violator Compact, which includes the following in Section 11: "The participating states find that wildlife resources are managed in trust by the respective states for the benefit of all their residents and visitors." In joining the compact, states have agreed to foundational principles central to the PTD. In New York, for example, State Environmental Conservation Law (ECL) Title 25 was added to join the compact and includes this clear statement: "Wildlife resources are managed in trust by the respective states for the benefit of all residents and visitors." ECL § 11-0105's statement, "The State of New York owns all fish, game, wildlife, shellfish, crustacean and protected insects in the state, except those legally acquired and held in private ownership," firmly established key principles of the PTD as state policy.

Two states, Louisiana and Alaska, have enshrined principles of the PTD in their state constitutions. Alaska's constitution is unambiguous: "Wherever occurring in their natural state, fish, wildlife, and waters are reserved to the people for common use" (Article VIII, § 3). Louisiana's constitution is equally clear: "to protect, conserve, and replenish all natural resources, including the wildlife and fish of the state, for the benefit of its people" (Article IX, § 1).

However, only 15 states have addressed the PTD in case law. Batcheller et al. (2010) concluded, "The benefits of strengthening the PTD for the benefit of wildlife resources are clear. Codifying the Doctrine in statute, or amending state constitutions to include it may secure its future, clarify its purpose, and ensure that its principles are more consistently applied and less subject to interpretation." To that end, "model statutory language" is provided by Batcheller et al. (2010) to seek enduring protection and recognition that the PTD is a foundational pillar of the North American Model of Wildlife Conservation. Two key components of such language are as follows: (1) the state declares that wildlife is held in trust by the state for benefit of its citizens, and (2) individual citizens have an equitable right of access to wildlife as a public trust resource. Batcheller et al. (2010) further recommended that state statutes clearly prohibit government from transferring wildlife resources to private ownership, except as provided by law.

Models of State Governance

Within state wildlife agencies, there are several governance models for establishing policy, ranging from appointed commissions or boards to appointed agency directors. In some cases, there are appointed directors that report to appointed commissions or boards. In this sense, a director functions as "chief executive officer," while appointed commissioners or board members function as policy makers or trustees of the wildlife resource. These, in turn, are typically subject to policies established by elected officials in state legislatures, with concurrence of the respective governor.

As highlighted by Smith (2011:1540), "A trustee must either possess or have effective ownership control of the corpus of the trust to make decisions regarding management of the trust and distribution of proceeds from the trust in the interest of the beneficiaries." In terms of wildlife management, this means that the body or person(s) overseeing the PTD on behalf of wildlife resources must stress the status of wildlife as a preeminent responsibility when allocating the use of these resources (e.g., legal harvest).

A trust manager, conversely, must have technical expertise to inform policy makers accordingly, and a primary responsibility of a state wildlife agency is to ensure that technical personnel understand their role vis-à-vis the PTD. As Smith (2011:1540) argues, "State wildlife agency professionals are trust managers, not trustees." This should not be viewed as insignificant even though state agency professionals typically do not have policy-making authority. Smith (2011:1543) states, "State wildlife agency professionals can best advance application of the PTD by informing elected and appointed officials about their roles and responsibilities as trustees, informing the public about their rights and responsibilities as beneficiaries under the PTD, and working to embed the PTD in codified law. If the public fully understands the PTD, the citizens of a state will be more effective in holding elected and appointed officials accountable than SWA [state wildlife agency] professionals can be." Thus, state wildlife agency biolo-

Biologists conduct field research to understand the status of wildlife populations, thereby fulfilling their obligations as "trust managers" (Smith 2011). By doing so, biologists are equipped to advise the trustees (e.g., Game Commissions) of necessary management actions to ensure the well-being of wildlife populations. *Photo courtesy of Mark Gocke, Wyoming Game and Fish Department.*

gists are subservient to policy-making bodies and individuals, but they remain critical to policy formulation.

Public Trust Doctrine Initiatives

There have been several initiatives to ensure that the PTD is understood by state wildlife agency professionals. The subject has been discussed at the North American Wildlife and Natural Resources Conference (e.g., Boggess and Jacobson 2013; Decker et al. 2013) and Western Association of Fish and Wildlife Agencies (2006). Training on the PTD is now included in standard curricula for the Conservation Leaders for Tomorrow (CLfT.org) program (D. Windsor and Z. Lowe, personal communication).

In some cases, advocacy groups have emerged to specifically promote and protect the PTD. For example, "Enhancing Montana's Wildlife & Habitat," an organization based in Bozeman, Montana, includes "advocating for the PTD which establishes the management of land, wildlife, fish and waterways for the benefit of the public and future generations" as central to their purpose (emwh.org).

Within state wildlife agencies, policy makers have instituted measures to inculcate an understanding of the PTD among agency professionals through internal training and outreach and to specifically cite the PTD in their strategic plans (Vermont Fish and Wildlife Department 2011; Pennsylvania Game Commission

2015). Collectively, these and other efforts are raising awareness among state wildlife agency professionals of their roles and responsibilities vis-à-vis the PTD.

Jacobson et al. (2010) describe a fundamental challenge for state wildlife agencies. Who are the beneficiaries of the wildlife trust? For many state agencies, key funding is derived from the sale of fishing, hunting, and trapping licenses, along with federal funding via the Pittman–Robertson Wildlife Restoration program, derived from the sale of firearms and ammunition. Based solely on a user-pay/user-benefit model, the "trust beneficiaries" of the trust could be perceived as rather narrow, namely, only the persons who fund state wildlife agencies. Jacobson et al. (2010) argue that this is fundamentally flawed, and that broad-based funding benefiting all citizens is needed for long-term sustainability of the "conservation institution" for the twenty-first century. Consequently, all citizens must be considered not only "stakeholders" in the conventional sense but also true trust beneficiaries who must accept the responsibility to hold the trustees and, by implication, the trust managers accountable for their actions. Not only do Jacobson et al. (2010) make the case for broad-based funding, but they also argue that "trustee-based governance" would strengthen conservation institutions. However, achieving this ideal model for state wildlife agencies is illusive given the highly political nature of today's state agency policy makers.

According to the Association of Fish and Wildlife Agencies, most state agency directors hold these positions for less than three years, often relieved of duty for blatantly political purposes. That fact begs the question, in such a politically charged and risky employment atmosphere, how can an agency policy maker truly embrace the responsibility of serving as a trustee for the benefit of all citizens? Jacobson et al. (2010) posit that this is only possible when and if broad-based funding is secured and all citizens hold the trustees accountable for their actions on behalf of wildlife. They warn, "The sustainability of fish and wildlife populations in the long term would be questionable without stability in programs to protect trust resources. Accomplishment of such reform in governance likely can only be achieved through advocacy of a strong coalition of partners willing to speak with one voice and exert the requisite political pressure" (Jacobson et al. 2010:206).

Decker et al. (2014) provide guidance for state wildlife agencies and their trust managers (i.e., agency biologists) in the context of public outreach and citizen engagement efforts. They argue that effective stakeholder engagement is essential for ensuring that trustees and trust managers understand the needs of stakeholders (beneficiaries) in managing the trust (wildlife resources). A key consideration in designing responsible stakeholder engagement practices is to ensure that all potential beneficiaries are enfranchised. Decker et al. (2014:175) make it clear: "a trustee is expected to make policy decisions from a broader perspective." However, since most state agencies are primarily funded via a narrow subset of stakeholders, it is difficult to maintain comprehensive objectivity and fulfill the responsibilities of a trustee. Decker et al. (2014) suggest that the failure to maintain objectivity may infer "unfair privilege" given to some beneficiaries (e.g., hunters and trappers). They recommend the use of effective stakeholder engagement practices that identify broadly supported policies through a consensus-building approach. A corollary to this recommendation is to ensure that state agencies "vigorously self-enforce inappropriate pursuit of personal interests of trustees and managers in their execution of trust administration" (Decker et al. 2014:177). For example, it would be a violation of the public trust for a trustee (e.g., game commissioner) or trust manager (e.g., wildlife professional) to advocate for a game management regulation that clearly benefits them personally and singularly.

Trustees and trust managers need to understand the principles of conflict of interest and ethical behavior to ensure that they always act in the best interests of the trust (i.e., wildlife resources) and the broad spectrum of beneficiaries. These are often legal requirements in each respective state, adopted via various forms of ethical standards or "public officer law" where conflict of interest is defined and enforced.

In support of these professional standards, The Wildlife Society has adopted a "Code of Ethics" and "Standards for Professional Conduct" for wildlife biologists that are integral to its Certification Program (http://wildlife.org/learn/professional-development -certification/certification-programs).* Those reinforce key principles for public employees engaged in wildlife biology and management. A key component of strengthening and ensuring the integrity of the PTD should be training to ensure that state wildlife agency trustees and managers understand these core principles. As summarized by Decker et al. (2014), the PTD can be upheld by (1) practicing effective stakeholder engagement and public participation in decision-making and (2) practicing good governance principles, including understanding and avoiding conflict of interest in the development and administration of wildlife management policies.

Threats to the Public Trust Doctrine

There is consensus within the wildlife profession that the PTD is foundational to the North American Model of Wildlife Conservation. However, Batcheller et al. (2010) also documented several clear threats to the PTD specific to the work of state wildlife agencies.

Animal Rights Philosophy

Fundamental to the PTD is the concept that wildlife resources are public property, based on legal analysis

---

* The American Fisheries Society has adopted "Standards of Professional Conduct" similar to those of The Wildlife Society. See https://fisheries.org/about/governance/standards-of-professional -conduct/.

The Public Trust Doctrine

Here is your country. Cherish these natural wonders, cherish the natural resources, cherish the history and romance as a sacred heritage, for your children and your children's children. Do not let selfish men or greedy interests skin your country of its beauty, its riches or its romance.
Theodore Roosevelt

WWW.EMWH.ORG

Theodore Roosevelt, one of the most important men in early US conservation history. *Reprinted by permission from Kathryn QannaYahu, Enhancing Montana's Wildlife & Habitat, www.emwh.org/emwh.htm.*

provided by Sax (1970). This contrasts sharply with the "animal rights" philosophy that holds that animals may not be considered property in the conventional sense of being "owned," but instead have inherent personal rights that are sacrosanct and infallible. Francione (1996:4) stated it bluntly: "Animal rights theory . . . contains a nascent blueprint for the incremental eradication of the property status of animals." Since this philosophy is not compatible with the PTD or the North American Model of Wildlife Conservation, legislative or other policy changes that recognize animal rights as legitimate are a clear threat to state wildlife agencies and wildlife conservation.

Antipathy toward Wildlife

Sustaining a trust (i.e., wildlife resources) requires the presence of beneficiaries who care about the long-term health and sustainability of those resources. If people harbor antipathy or hostility toward wildlife, they will not be concerned with maintaining integrity of the PTD. When people are passionate about wildlife and have equitable access to enjoy wildlife, they pay attention to the work of state wildlife agencies and insist on policies and action to protect that trust.

One of the most influential conservation policy makers was President Theodore Roosevelt. His accomplishments are thoroughly described in *The Wilderness*

Warrior (Brinkley 2009), including the permanent protection of about 300 million acres of public lands. As multiple biographies of Roosevelt depict, he was a passionate outdoorsman and nature enthusiast, beginning at an early age. His passion for nature was translated into a passion for protecting the public trust once he assumed elected office.

Richard Louv (2008) described a "staggering divide" between children and nature that prevails in modern society, and he coined the term "nature-deficit disorder" to explain that a child separated from nature is a child deprived of health and happiness. The question for those who care about wildlife resources, manifested as a public trust, is whether future policy makers will care enough to ensure that the integrity of the trust is sustained. Where are our future Roosevelts? Will future leaders have sufficient passion for wildlife to ensure that the PTD remains the foundation for the North American Model of Wildlife Conservation? Indeed, creation of a new generation of wildlife enthusiasts must become the preeminent concern of state wildlife agencies.

Overabundant Wildlife

In an unexpected twist to restoration of wildlife, Batcheller et al. (2010) describe the conundrum of overabundant wildlife and the resulting impact on public attitudes toward wildlife conservation: "Public antipathy towards wildlife may increase when people experience property damage (e.g., deer–vehicle collisions) or the competition it creates for their livelihood (e.g., wildlife depredations on crops or forage)." If the public's primary exposure to wildlife is through property damage or as a perceived vector of human disease, it is difficult to foresee sustainable support for robust wildlife populations or conservation of their habitat. This loss of support could easily translate to a long-term decline in integrity of the PTD.

Commercialization of Wildlife

An insidious and growing threat to the PTD is the interest in commercialization of wildlife. In many states native or exotic wildlife may be owned, and in the case of cervids and bovids, they may be kept within fenced enclosures. Under a commercial business model, these animals may be sold for shooting purposes or slaughtered for commercial meat production. According to Organ et al. (2012:13), "The legal status of animals held in captivity under these conditions is equivocal." The key question is whether animals that are genetically "wild" and held under captive conditions should be considered as wildlife, and therefore subject to the responsibilities and prerogatives associated with the PTD. What happens if a bona fide wild animal enters into a fenced enclosure? Does it remain wildlife in the legal context, and therefore under the jurisdiction of the relevant state wildlife agency? Or, does it become de facto captive and subject to the legal regime of a state agriculture agency?

Demarais et al. (2002) reviewed the social issues related to confinement of wild ungulates, including an assessment of implications with regard to the PTD. For private landowners, the stakes are high. Through genetic breeding and selection, ungulates have been produced with enormous antlers typically not encountered or possible within true wild populations. Yet there is a market for those animals, and private landowners and businesses have successfully marketed the shooting of these animals as "trophies." Wildlife, especially deer, may be "claimed" by private landowners, confined and bred for certain genetic characteristics,

and released again to a "game ranch" typically at great profits. Clearly, economic incentives are powerful in these instances and give rise to national debates and legal challenges that may threaten the PTD.

Public Access to Wildlife

Organ et al. (2012) underscored the importance of public access to private and public lands for the purposes of hunting, trapping, and wildlife observation. Evidence suggests that gaining access to private property to enjoy wildlife is becoming more difficult (Responsive Management / National Shooting Sports Foundation 2008). In the case of leased lands, high fees may be charged by private landowners to provide exclusive use of those lands for hunting. In essence, leased lands are removed from use for all but the most willing or able to pay, since a large number of hunters cannot afford to pay high user fees (Duda et al. 1998).

Privatization of Wildlife

Today, the PTD continues to be threatened by the desire of some to make wildlife the property of private individuals. While the rights of landowners are well known and respected with regard to trespass, no one should be able to confine wildlife to private property under the auspices of the PTD. As illustrated in the case studies below, the greatest challenge to the PTD today is not unlike the scenario that emerged in the early 1800s. Who has the authority to protect, manage, and regulate the take of wildlife on private lands—private individuals or the public via appointed or elected trustees?

The future integrity of the PTD requires strict adherence to principles associated with public ownership of wildlife. While citizens and governments are both practitioners of the PTD, it remains incumbent on the public (beneficiaries) to hold the trustee (government) accountable for their greater interest in wildlife resources. Citizens must not allow governments to fetter away these prized trust resources to special interests in granting de facto ownership of wildlife through either inaction or legislative intervention.

Selected Case Studies

Today, similar elements and pressures from early colonial America are still at work in relation to public ownership of wildlife. The following case studies involve not oysters but members of the deer family and illustrate the constant threat that exists to the PTD.

Under most state laws, possession of live wildlife is allowed only under the authority of a permit or license. Such permits are for explicit purposes (e.g., animal products, education, "hunting," research, rehabilitation) and are seldom granted to enable conversion of public wildlife resources to "pet" status.

Beginning in the mid-1990s, there has been increased interest in diversifying agricultural practices to allow farming of deer species for antlers, hides, meat, and urine. Justified as "alternative agriculture practices," proponents claim that diversifying agriculture allows farmers to maintain economic viability by deriving income from sources other than traditional livestock or row crops. In several states, captive rearing of deer has expanded beyond agricultural purposes and now includes "hunting" of these animals within "high fences," an approximately $1 billion annual industry.

Consequently, some private land owners have challenged state laws and regulations in an effort to possess and own wildlife on their land. They wish to claim ownership of deer bought from other private businesses or of wild deer once they are enclosed within "high fences." Subsequently, they seek to claim private ownership of these animals to "hunt," to offer "hunts" for a fee, or for other purposes absent oversight from state wildlife agencies. The following two case studies illustrate successful efforts by sportsmen and concerned citizens to overturn efforts to privatize wildlife.

Case Study 1: Captive Cervid Facilities in Vermont

In this case, the Vermont State Legislature altered the authority over hunting wildlife by transferring jurisdiction of hundreds of wild animals held within an enclosure by a private landowner from the state wildlife agency to the agricultural agency (i.e., Agency of Agriculture, Food and Markets). This was the most significant affront to the PTD in Vermont in the past century. Subsequently, beneficiaries of the trust (the public), through their actions and efforts, held the government accountable and orchestrated a reversal to this legal change. The reversal as it stands today further clarified, codified, and strengthened the PTD for Vermont citizens.

Vermont was an independent republic prior to entering the United States as the fourteenth state. In 1777, Vermont's constitution was the first in the United States to grant the right to hunt and fish. Chapter 2, § 67 states the following: "The inhabitants of this State shall have liberty in seasonable times, to hunt and fowl on the lands they hold, and on other lands not inclosed [sic], and in like manner to fish in all boatable and other waters (not private property) under proper regulations, to be made and provided by the General Assembly."

This section of law followed along the lines and common practice of other European colonies that granted access to areas and allowed the public to harvest fish, fowl, and other wildlife resources. It clearly rejected the traditional premise of English law that animals were owned by landowners, frequently the Crown itself. This state constitutional right clearly established that Vermont landowners do not own wildlife; rather, wildlife is held in trust by the state and available to the public "to hunt and fowl."

In a manner reminiscent of cases posed in *Arnold v. Mundy* (1821) and *Martin et al. v. Waddell* (1842), market forces associated with captive deer brought a key public trust issue to the fore of Vermont public policy. The question facing policy makers and concerned citizens was whether an individual can own wild animals and treat them as their own property when cultivated in an artificial/agricultural setting.

In the mid-1990s, the Vermont Department of Agriculture (later renamed the Agency of Agriculture, Food and Markets) had legislative authority over captive deer operations, a place where domesticated deer were privately or publicly maintained or held for economic or other purposes within a perimeter fence or confined space. Deer under this section of law were typically exotic imported deer, fallow deer (*Dama dama*). Native white-tailed deer (*Odocoileus virginiana*) and moose (*Al-*

Captive shooting operations are a threat to the public trust doctrine because they call into question ownership of wildlife behind fences. Similar to the "Oyster Wars" of the early 1800s, today's game farmers confine captive and sometimes wild cervids within enclosures seemingly beyond the public domain, thereby threatening key principles associated with the public trust doctrine. *Photo courtesy of Chris D. Grondahl, North Dakota Game and Fish Department.*

ces americana) could not be possessed as farm animals. All animals held within these facilities were for agricultural purposes and could not be hunted. Hunting within these facilities was under the jurisdiction of the Vermont Department of Fish and Wildlife and not authorized. As private enclosures expanded, both white-tailed deer and moose were confined. Local observers have speculated that some animals were unlawfully introduced within private enclosures.

Wapiti (or North American elk, *Cervus elaphus*) were also allowed to be imported for farming purposes. Under state law, wapiti were not classified as wild animals and could be imported for farming practices. Thus, prior to documentation of chronic wasting disease (CWD) in the western United States, wapiti were lawfully brought into Vermont ostensibly for agricultural purposes. White-tailed deer and moose were never allowed to be legally kept for agricultural purposes. After outbreaks of CWD in the United States and Canada, the importation of live elk into Vermont was prohibited, and markets for elk products (e.g., antlers and meat) disappeared.

The desire to "hunt" captive native wildlife (white-tailed deer and moose) and captive exotic wildlife such as elk, wild boar (*Sus scrofa*), and Spanish goats (*Capra pyrenaica*) increased at two cervid facilities that housed these animals. In the late 1990s, the Vermont legislature passed a law directing the Vermont Fish and Wildlife Department to regulate these facilities if they were to engage in "hunting." In accordance with state

law, on three separate occasions beginning in 2000, the Vermont Fish and Wildlife Board (a regulatory board composed of 14 citizens appointed by the governor) attempted to develop regulations regarding captive cervid "hunt" facility operations. Proposals addressed fencing requirements, marking of animals, preventing and responding to escapes into the wild, disease prevention and testing, and other animal welfare measures. At the same time, CWD came to national attention with outbreak of this disease in captive cervid facilities in the western United States and subsequent occurrence in the wild. In Vermont, as in many other states, different agencies shared responsibilities for regulation of captive cervid facilities (depending on purposes for which a species was held).

The Fish and Wildlife Department and the Fish and Wildlife Board took steps beginning in 2000 to promulgate rules related to "hunting" in these facilities. Development of the regulation involved extensive engagement with the agency of agriculture, the deer farming industry, captive cervid facility owners, the hunting community, and others. Disease concerns were very high, as CWD had not been found in Vermont and white-tailed deer are the most important big game in the state.

Testimony and public comment concerning these proposed regulations were very contentious. The Fish and Wildlife Department proposed grandfathering two facilities to allow hunting of cervids, but it would not allow further expansion of these facilities in Vermont. The proposed regulations did not advance owing to threat of legal initiatives from captive deer facility owners, attempts to standardize captive hunt facility requirements under agriculture's rules, threats of lawsuits, changes in directives from the legislature, and a lengthy public input process.

In the last days of the 2010 legislative session, the 2010 Appropriations Bill (H.789) transferred full jurisdiction and authority for regulatory oversight of captive hunt facilities from the Fish and Wildlife Department to the Vermont Agency of Agriculture, Food and Markets (formerly the Department of Agriculture). The bill also permitted an individual citizen landowner to own wild animals outright and defined them as a "special purpose herd." At the time in which the bill was enacted into law, there were fewer than 10 captive cervid facilities in Vermont. Two made it clear that they intended to provide "hunts" of native elk, moose, and white-tailed deer.

One facility had enclosed a number of native moose and white-tailed deer when they constructed a large fence for rearing imported non-native deer species. The fence encompassed a 700-acre forested area. Over time this enclosed herd grew, and by 2011, there were 150–175 white-tailed deer and 6 moose. The ownership of these animals was granted to the private landowner as a result of this legislation, a direct assault on key tenets of the PTD.

The initial reaction to the new legislation generated a considerable outcry from hunting and other organizations and citizens. Numerous legislators stated that they were not aware of the "transfer" provision in a general budget bill. Hunting organizations stated that there was no public comment or testimony to this important change. And professional organizations stated objections to the reversal of PTD elements. For example, The Wildlife Society sent a strongly worded letter of concern to all members of the Vermont legislature.

In the ensuing months, several key legislators and representatives from The Wildlife Society, Orion—The Hunter's Institute, and state hunting organizations formed a working coalition to coordinate legislative action to reverse changes to the 2011 budget bill. They also sought to strengthen the PTD within Vermont law. Working collaboratively with appropriate policy makers in the Fish and Wildlife Board and the Fish and Wildlife Department, a new law (Act 54) was passed by the legislature and signed by the governor in April 2011. This revision to Vermont law (10 V.S.A. § 4081) ultimately clarified and strengthened the PTD in Vermont and awarded full authority for management of wildlife to the department: "Policy (a)(1) As provided by Chapter 11, 67 of the Constitution of the State of Vermont, the fish and wildlife of Vermont are held in trust by the state for the benefit of the citizens of Vermont and shall not be reduced to private ownership. The State of Vermont, in its sovereign capacity as a trustee for the citizens of the State, shall have ownership, jurisdiction and control of all of the fish and wildlife of Vermont."

After decades of legislative directives, regulatory hearings, panel reviews, three separate drafts of regulations by the Fish and Wildlife Board, reversals of captive cervid regulations, and last-minute passage of regulatory language awarding ownership of deer to a private landowner, public policy with regard to the ownership of wildlife had come full circle and was back on a secure legal basis. Private ownership of wildlife species was revoked, and management authority was restored to the Fish and Wildlife Department. The board subsequently adopted regulations to address the issue of captive cervid "hunting" in Vermont, grandfathering the two facilities, but forbidding further captive cervid "hunt parks" from opening.

For the two grandfathered captive cervid hunting facilities, all moose and white-tailed deer within enclosures were destroyed and tested for CWD. The disease was not confirmed in any specimen. Further, Vermont prohibited importation and possession of wild boar, and both hunting facilities subsequently closed.

If the legal provision granting ownership of a "special purpose herd" had withstood challenge, the conservation paradigm for many species of wildlife in Vermont would have been significantly altered. This reversal of law in 2011 and the further refinement and clarification provided within 10 V.S.A. § 4081 were supported by several key legislators, The Wildlife Society, Orion—The Hunter's Institute, and state hunting associations. The expanded policy section of state law is unambiguous, securing the principle that all "fish and wildlife of Vermont are held in trust by the state for the benefit of the citizens of Vermont and shall not be reduced to private ownership."

Case Study 2: Captive Cervids in Wyoming, Three Legal Cases Addressing Private Ownership

Between August 1989 and February 1990 a citizen of Wyoming applied to the Wyoming Game and Fish Commission to import and possess 14 species of exotic wildlife and 5 species of native wildlife. Exotic wildlife included aoudad (or Barbary sheep, *Ammotragus lervia*), blackbuck (*Antilope cervicapra*), ibex (*Capra ibex*), markhor (*Capra falconeri*), European mouflon sheep (*Ovis vignei*), alpine chamois (*Rupicapra rupicapra*), Marco Polo's sheep (or argali, *Ovis ammon*), axis deer (*Axis axis*), western red deer (*Cervus elaphus*), sika (*Cervis nippon*), fallow deer, muntjac (*Muntiacus* spp.), European roe (*Capreolus capreolus*), and wild boar. Native wildlife included gray wolf (*Canis lupus*), pronghorn (*Antilocapra americana*), bighorn sheep (*Ovis canadensis*), moose, and wapiti. These species were to be used for breeding and rearing, exhibition, weed control research, meat production, and controlled hunting.

Wyoming statute clearly states the legislature's intent that wildlife is publically owned and that private ownership of native ungulates and trophy game (i.e., grizzly bear [*Ursus arctos*], black bear [*Ursus americanus*], mountain lion [*Puma concolor*], and gray wolf) is prohibited: "For the purpose of this act, all wildlife in Wyoming is the property of the state. It is the purpose of this act and the policy of the state to provide an adequate and flexible system for control, propagation, management, protection and regulation of all Wyoming wildlife. There shall be no private ownership of live animals classified in this act as big or trophy game animals or any wolf or wolf hybrid" (WS 23-1-103).

The legislature delegated authority to the commission to "regulate or prohibit" the importation of other native wildlife and all exotic species and further clarified their position for ungulates and trophy game by stating that they would "permit the importation of big or trophy game animals into Wyoming only for exhibition purposes or for zoos" (WS 23-23-1-302(a)(xxvi)).

Since the Wyoming Game and Fish Department had, at best, a passing knowledge of the exotic species applied for, a committee was appointed in February 1990 to gather information from literature and from scientists and regulators working with or knowledgeable about these species. Based on existing literature and the preponderance of expert opinion, the committee found that (1) escapes are inevitable; (2) even with quarantine measures, disease transmission was a real concern that could result in significant expenditures of public funds to control; (3) competition for forage and space was a real concern; (4) hybridization and genetic pollution were real concerns; (5) high wire fences, particularly in strategic locations, could inhibit the ability of native free-ranging wildlife to migrate between seasonal habitats; (6) privatization of wildlife provides a mechanism for illegal take of public wildlife; (7) take of wildlife in confined situations could result in an erosion in public support for hunting; and (8) privatization of wildlife was inconsistent with the views of the people of the state. Based on its findings, the committee recommended that the commission deny all applications.

NATIVE WILDLIFE

Given statutory prohibitions on private ownership of native ungulates and wolves, in March 1990 the commission denied all requests for native wildlife. The applicant then filed suit with the United States District Court for the State of Wyoming alleging that Wyoming state law denied his constitutional right to engage in interstate commerce and his right to due process. This court granted the state's request for summary judgment, finding that protection of the priceless resources of the state outweighed the individual's economic interest. That decision was appealed by the applicant to the Tenth Circuit Court of Appeals, which reversed the District Court decision and remanded the case back to it. In 1992, the applicant withdrew his complaint, which can be refiled at any time.

EXOTIC WILDLIFE

Using its statutory authority, information provided in the committee report, and public input overwhelmingly opposed to the applications, the commission also denied all applications for exotic wildlife in March 1990. The commission decided that these applications, if granted, were not in the best interest of Wyoming's native wildlife. Since the commission, under Wyoming law, had the right to regulate or prohibit applications for exotic wildlife, the commission's denial was appealed by the applicant to the Wyoming District Court in Crook County. The judge overturned the commission's decision and ordered a contested case hearing in front of a hearing officer appointed by the court. The hearing officer had no authority other than to hear evidence, consider it, and make a recommendation to the commission on how to proceed. In October 1992, the hearing officer recommended that the commission permit the application for all species except western red deer and wild boar. In February 1993, the commission considered the recommended decision in open session; heard arguments, including several department objections to the hearing officer's recommendations; and again decided to deny all applications. The applicant did not exercise his right to appeal the commission's decision, thereby concluding the case.

HYBRID WILDLIFE

In August 1990, the applicant, without permit from the commission, imported 12 wildlife hybrids. These animals were the result of crosses between wapiti, western red deer, and sika. The applicant argued that these animals were domestic since there was no more than 50 percent of any one species. This argument found its genesis in an existing commission regulation that made a cross between a dog and a wolf domestic so long as there was less than 50 percent wolf blood. Taxonomic classification of the deer family accepted by the department at that time considered wapiti and western red deer one species. Consequently, progeny resulting from breeding between wapiti and western red deer were considered one species (*Cervus elaphus*), and a cross between this progeny and sika were 50 percent native wildlife, thereby under jurisdiction of the commission. More importantly, the department, in their September 1990 suit, argued that hybrids between wildlife are wildlife and that the simple act of hybridization, no matter how unlikely in the natural world, did not in and of itself result

in wildlife becoming domestic animals. This case was heard by the same Crook County judge who heard the exotic wildlife case. In July of 1991, he found that these animals were wildlife and that they were exotic wildlife since this hybrid is not found in the wild in Wyoming. Since they were exotic wildlife, the judge further found that they fell under jurisdiction of the commission. Since they were imported without a permit, he ordered their removal, which was done.

These three Wyoming cases were important in affirming the PTD. They took several years to work through the commission and state and federal courts. Significant defense of the PTD occurred at two levels. The first was governmental. The commission, while under intense public scrutiny, considered the information before them and decided that upholding the PTD was in the best interest of the people of Wyoming and the state's wildlife. The commission also provided the Game and Fish Department with authority and substantial financial resources necessary to contest, in court, actions and suits brought by the applicant. Currently, Wyoming has only one game farm, permitted for only one species of native wildlife, which the legislature grandfathered in 1973 while codifying state ownership of wildlife. Without the commission's foresight and commitment of resources defending the PTD in these cases, it is likely that game farming in Wyoming would look very different today.

Secondly, Wyoming citizens (beneficiaries of the PTD) played an extremely important role in this process. This issue was in the news for over three years, resulting in intense public interest. The support of a large majority of citizens for keeping wildlife in public ownership was critical. Citizen involvement was and remains a crucial element of ensuring the benefits of the PTD to all beneficiaries. The beneficiaries must be willing to step forward and work with state fish and wildlife agencies, holding government trustees accountable and ensuring that benefits of the PTD are not eroded by privatization and private interests. Without active public support for public wildlife, it is easy for legislatures and agencies to allow public wildlife to become one more private commodity in commerce.

LITERATURE CITED

Arnold v. Mundy 1821, 6 N.J.L. 1.

Batcheller, G. R., M. C. Bambery, L. Bies, T. Decker, S. Dyke, D. Guynn, M. McEnroe, M. O'Brien, J. F. Organ, S. J. Riley, and G. Roehm. 2010. The public trust doctrine: Implications for wildlife management and conservation in the United States and Canada. Technical Review 10-01. The Wildlife Society, Bethesda, Maryland, USA.

Bean, M. J. 1977. The evolution of national wildlife law. Environmental Law Institute.

Boggess, E., and C. A. Jacobson. 2013. Getting back to basics: How employing public trust doctrine principles can help garner support for state fish and wildlife agencies. Transactions of the North American Wildlife and Natural Resources Conference 78:18–25.

Brinkley, D. 2009. The wilderness warrior: Theodore Roosevelt and the crusade for America. HarperCollins, New York, New York, USA.

Decker, D. J., A. B. Forstchen, C. A. Jacobson, C. A. Smith, J. F. Organ, and D. Hare. 2013. What does it mean to manage wildlife as if public trust really matters? Transactions of the North American Wildlife and Natural Resources Conference 78:46–53.

Decker, D. J., A. B. Forstchen, E. F. Pomeranz, C. A. Smith, S. J. Riley, C. A. Jacobson, J. F. Organ, and G. R. Batcheller. 2014. Stakeholder engagement in wildlife management: Does the public trust doctrine imply limits? Journal of Wildlife Management 79:174–179.

Demarais, S., R. W. DeYoung, L. J. Lyon, E. S. Williams, S. J. Williamson, and G. J. Wolfe. 2002. Biological and social issues related to confinement of wild ungulates. Technical Review 02-3. The Wildlife Society, Bethesda, Maryland, USA.

Duda, M., S. Bissell, and K. Young. 1998. Wildlife and the American mind: Public opinion on and attitudes toward fish and wildlife management. Responsive Management National Office, Harrisburg, Virginia, USA.

Francione, G. L. 1996. Rain without thunder: The ideology of the animal rights movement. Temple University Press, Philadelphia, Pennsylvania, USA.

Geer v. State of Connecticut 1896, 161 U.S. 519.

Hakluyt Society. 1849. The historie of travaile into Virginia Britannia; expressing the cosmographie and comodities of the country, togither with the manners and customs of the people. Gathered and observed as well by those who went first thither as collected by William Strachey, Gent, First Secretary of the Colony. R. H. Major, editor. London, UK.

Hinke, W. J. 1916. Report of the journey of Francis Louis Michel from Berne, Switzerland, to Virginia, October 2, 1701–December 1, 1702. Virginia Magazine of History and Biography 24:1–43.

Hughes v. Oklahoma 1979, 441 U.S. 322.

Illinois Central Railroad v. State of Illinois 1892, 146 U.S. 387.

Jacobson, C. A., J. F. Organ, D. J. Decker, G. R. Batcheller, and L. Carpenter. 2010. A conservation institution for the 21st century: Implications for state wildlife agencies. Journal of Wildlife Management 74:203–209.

Louv, R. 2008. Last child in the woods: Saving our children from nature-deficit disorder. Algonquin Books, Chapel Hill, North Carolina, USA.

Martin et al. v. Waddell's Lessee 1842, 41 U.S. 367.

McCay, B. J. 1998. Oyster wars and the public trust: Property, laws, and ecology in New Jersey. University of Arizona, Tucson, Arizona, USA.

Missouri v. Holland 1920, 252 U.S. 416.

Organ, J. F., V. Geist, S. P. Mahoney, S. Williams, P. R. Krausman, G. R. Batcheller, T. A. Decker, R. Carmichael, P. Nanjappa, R. Regan, R. A. Medellin, R. Cantu, R. E. McCabe, S. Craven, G. M. Vecellio, and D. J. Decker. 2012. The North American Model of Wildlife Conservation. Technical Review 12-04. The Wildlife Society, Bethesda, Maryland, USA.

Pennsylvania Game Commission. 2015. Pennsylvania Game Commission Strategic Plan 2015–2020. Pennsylvania Game Commission, Harrisburg, Pennsylvania, USA.

Responsive Management / National Shooting Sports Foundation. 2008. The future of hunting and the shooting sports: Research-based recruitment and retention strategies. Harrisonburg, Virginia, USA.

Sax, J. L. 1970. The public trust doctrine in natural resource law: Effective judicial intervention. Michigan Law Review 68:471–566.

Smith, C. A. 2011. The role of state wildlife professionals under the public trust doctrine. Journal of Wildlife Management 75:1539–1543.

Vermont Fish and Wildlife Department. 2011. Vermont Fish and Wildlife Department Strategic Plan 2011–2016. Vermont Fish and Wildlife Department, Montpelier, Vermont, USA.

Western Association of Fish and Wildlife Agencies. 2006. Proceedings of the Western Association of Fish and Wildlife Agencies. Bismarck, North Dakota, USA.

3

JENNIFER MOCK-SCHAEFFER,
M. CAROL BAMBERY,
PARKS GILBERT, AND
GARY J. TAYLOR

State Fish and Wildlife Agencies and Conservation

A Special Relationship

The United States is unique in the world with respect to how fish and wildlife conservation is accomplished, which is principally by the state fish and wildlife agencies in cooperation with the federal fish and wildlife and land management agencies. A second unique characteristic is that it is primarily funded by hunters and anglers. This is manifested through the sale of state hunting licenses, stamps, and permits matched by federal excise taxes on sporting arms, ammunition, bows, and arrows, for wildlife conservation and shooting sports. Likewise, the sale of fishing licenses, stamps, and permits is matched by federal excise taxes on fishing equipment, import duties on boats, and a portion of the federal gasoline excise tax attributable to outboard motors and small engines, for sport fish conservation, boating access, and boating safety. Federal law prohibits the diversion of state hunting and fishing license, stamp, and permit revenues and federal excise taxes to any program other than sport fish and wildlife conservation (which also includes hunter education, shooting ranges, boating access, and boating safety) at the risk of losing the state's share of its federal excise taxes. While funding through excise taxes is specifically prescribed for sport fish, states' authority to manage fisheries resources is for all fish.

Most state fish and wildlife agencies receive only modest or no appropriations from the state treasury. In contrast, federal land managing agencies, the US Fish and Wildlife Service (USFWS), and the National Marine Fisheries Service (NMFS) receive congressionally appropriated dollars to manage certain fish and wildlife species concurrently with the states and to manage habitat on land such as the US Forest Service (USFS), Bureau of Land Management (BLM), and USFWS wildlife refuges. Appropriated funds are also used to administer private land conservation programs in the case of the Natural Resources Conservation Service (NRCS), the US Farm Service Agency (FSA), and the USFS.

While Canada embraces the North American Model of Wildlife Conservation, provincial hunting and fishing license fees are deposited in the general treasury for appropriation back to the provincial government. There are no federal excise taxes on sporting arms, ammunition, and fishing tackle in Canada.

State fish and wildlife agencies are the principal, front-line managers of fish and wildlife within their borders, including on most federal lands. They strive to proactively maintain healthy fish and wildlife populations, initiating conservation strategies to preclude the need to list species under the federal Endangered Species Act and to maintain state management authority for as many species as possible. They have broad trustee powers grounded in state statutes, constitutions, or both. Despite state fish and wildlife agencies having primary legal authority to manage most fish and wildlife in the United States, in some instances the states have concurrent jurisdictional authority with federal agencies and share fish and wildlife management responsibilities with federal agencies. And, since the federal land management agencies own the land, they manage the habitat, so a cooperative state–federal agency relationship is imperative. However, even with strong trustee powers grounded in state laws and regulations, state agencies' legal authority to manage fish and wildlife is often under challenge from Congress and the federal executive branch agencies. Some of this is due to a lack of understanding of state authority and some may be by design, but if successful, the end result would be a diminution of state authority to manage fish and wildlife in favor of federal authority and oversight.

Reacting and responding to these challenges is the foundational and fundamental mission of the Association of Fish and Wildlife Agencies (AFWA), founded in 1902. This vigilance is the purpose of a small but dedicated professional staff at the AFWA office in Washington, DC, who monitor congressional and federal executive branch activity to ensure that there is no diminution or derogation of state authority to conserve fish and wildlife. The AFWA also seeks and facilitates the enhancement of state–federal relationships in order to ensure that more conservation is delivered on the ground.

State Agencies' Structure and General Duties

Laws establish state agencies, determine their duties, and set and inform their organizational structure. This section will provide an overview of the role of law in state fish and wildlife agencies' structure and function.

State fish and wildlife agencies exist as entities because state constitutional provisions, statutes, or both created them and gave them the legal authority to manage fish and wildlife resources. Such constitutional provisions and statutes recognize, either implicitly or explicitly, that states have authority over fish and wildlife resources and that management of these resources is the states' responsibility. Fish and wildlife are held in trust by the states for present and future generations of our citizens.

For example, the state of New York has a statute setting out the policy of the Department of Environmental Conservation that includes management and protection of fish and wildlife:

1. It shall be the responsibility of the department . . . by and through the commissioner to carry out the environmental policy of the state. . . . In so doing, the commissioner shall have power to . . .
c. Provide for the propagation, protection, and management of fish and other aquatic life and wildlife and the preservation of endangered species. (New York Environmental Conservation Law § 3-0301)

Georgia has a similar statute:

The ownership of, jurisdiction over, and control of all wildlife, as defined in this title, are declared to be in the State of Georgia, in its sovereign capacity, to be controlled, regulated, and disposed of in accordance with this title. Wildlife is held in trust by the state for the benefit of its citizens and shall not be reduced to private ownership except as specifically provided for in this title. All wildlife of the State of Georgia is declared to be within the custody of the department for purposes of management and regulation in accordance with this title. (Official Code of Georgia Annotated § 27-1-3(b))

And Alaska's constitution has a provision containing much the same principle, though it does not in this provision specifically identify a state agency as the manager of fish and wildlife:

Fish, forests, wildlife, grasslands, and all other replenishable resources belonging to the State shall be utilized, developed, and maintained on the sustained yield principle, subject to preferences among beneficial uses. (Alaska Constitution § 4. Sustained Yield)

Aside from demonstrating how state agencies derive their authority, these examples establish an important point about the varying types of governmental structures that states employ to exercise authority over fish and wildlife. Take New York, for instance. The Department of Environmental Conservation is headed by a commissioner, who is appointed by the governor and approved by the state senate. The department's Division of Fish, Wildlife, and Marine Resources, which is headed by a director, has the responsibility to manage fish and wildlife. The department writes regulations that set a framework for its management work and other related duties. In other states, such as Alaska, the state fish and wildlife agency is a stand-alone cabinet-level entity (not nested under a larger environmental or natural resources department, such as in New York), and the commissioner or director reports directly to the governor. Also in Alaska, the Board of Game sets the wildlife harvest allocations, while the department makes management decisions to achieve those allocations. Other states, such as Georgia, also vest regulatory authority in a board or commission. While the structure and distribution of authority may differ from state to state, the end result is the same: state agencies have legal authority to manage fish and wildlife.

The above constitutional and statutory provisions offer a solid grounding in how state agencies are created and their general duty, namely, to manage fish and wildlife for the benefit of the citizens of the state. These provisions more or less codify the public trust principle treated earlier in this book, as will be traced in common law (law created by judicial decisions rather than passed by a legislature and codified in statute or written as a rule by an executive branch agency) in a subsequent section of this chapter.

From this duty to manage fish and wildlife grows a number of specific responsibilities and functions for the state agencies. Generally, these include

- managing and acquiring habitat;
- educating the public about conservation and hunter safety, boating safety, etc.;
- carrying out effective fish and wildlife biological research and management work;
- maintaining recreational facilities;
- selling hunting and fishing licenses, permits, stamps, and tags;
- writing and enforcing conservation regulations;
- conducting fish and wildlife surveys and inventories to establish population trends;
- assessing the quantity and quality of habitat for individual species or suites of species;
- setting population and habitat objectives to ensure sustainable fish and wildlife resources;
- establishing seasons, bag limits, and means and methods of take for game and sport fish;
- recommending voluntary habitat management prescriptions for private landowners interested in conservation, including providing state and federal incentive programs;

- monitoring and managing fish and wildlife diseases; and
- assessing the effects of development on species and habitats and commenting on permits and environmental impact statements in order to avoid, minimize, or mitigate those effects.

As noted above, state agencies (either directly or through their boards or commissions) promulgate regulations to set the structure for their work underneath the general statutes and constitutional provisions that give them authority to manage fish and wildlife. Here is an example of a hunting regulation from New York:

> Any person who hunts or takes deer during a bowhunting, muzzleloading or regular season must possess, on his or her person, a license and carcass tag valid to hunt deer during that particular season. (New York Codes, Rules, and Regulations, 6 CRR-NY 1.11(c))

A Common-Law Basis for State Authority to Manage Fish and Wildlife

The previous section explained the constitutional and statutory law that codifies state agencies' authority. However, these laws were not written in a vacuum. They codify a concept that dates back to the Roman Empire and that came to North America via English common law. The concept is the public trust in fish and wildlife (which has been covered elsewhere in this book), which assigns ownership of fish and wildlife to states.

The public trust is, in many ways, a creature of property law. It recognizes that fish and wildlife are property, as opposed to having their own legal status under the law, as some animal rights groups posit. Further, fish and wildlife are not simply considered property owned by no one, or *res nullius*, nor does ownership of them go to the owner of the land on which they live. Instead, state governments hold title to fish and wildlife, as trustees of these resources, for their citizens. Like a trustee of real property or financial assets, states are to manage the fish and wildlife so that the populations, which can be compared to trust principal, are sustained for the current and future citizens of their states, who are the beneficiaries of the trust.

The public trust in fish and wildlife can be traced to Roman law recognizing a public right to fish, although Roman law did not recognize wildlife as belonging to anyone:

> By the law of nature these things are common to all mankind—the air, running water, the sea, and consequently the shores of the sea. No one, therefore, is forbidden to approach the seashore, provided he respects habitation, monuments, and the buildings, which are not, like the sea, subject only to the law of nations. (Institutes of Justinian, J. INST. 2.1.1)

Likewise, English common law recognized a right to fish (and therefore that the public collectively owned the resource). However, the Crown and landowners had special prerogatives to take game, such as deer, and the common people did not share this right. Poaching was punished, sometimes severely. Some call this the European model of wildlife ownership. A good example of this concept is found in the English Game Act of 1671. This act continued to limit not only who could hunt but also who could own guns:

- Only wealthy landowners can hunt,
- Those who can't hunt can't own guns,
- Qualified hunters may search unqualified people for guns and seize them, and
- Qualified hunters may hunt anywhere, including others' private property.

(Charles II, 1670 & 1671: An Act for the better preservation of the Game, and for securing Warrens not inclosed, and the several Fishings of this Realm)

Colonial governments clearly saw the ownership of wildlife differently. This was likely due in part to dislike of the very restrictive European model, and in part to the sheer abundance of game (if somewhat short-lived) in colonial-era North America as compared to lower game populations in more developed Europe. Compare the following law from Virginia, passed in 1632:

> Noe man shall kill any wild swine . . . without leave or lycense from the Governor. But it is thought convenient that any man be permitted to kill deare or other wild beasts or fowle in the common wood . . . that thereby the inhabitants may be trained in the use of theire armes, the Indians kept from our plantations, and the wolves and

other vermine destroyed. (Act XLIX, September 1632, in William W. Hening, The Statutes at Large: A Collection of All of the Laws of Virginia, Vol. 1, at 199)

Returning to the concept of public ownership of fisheries, the US Supreme Court described this public ownership in *Shively v. Bowlby*, 152 U.S. 1 (1894). The Crown held title to fisheries on the people's behalf:

> In England, from the time of Lord Hale, it has been treated as settled that the title in the soil of the sea, or of arms of the sea, below ordinary high-water mark is in the king . . . [and] is held subject to the public right of navigation and fishing. (*Shively*, 152 U.S. at 13)

The Supreme Court and other courts picked up this concept in other cases, such as *Martin v. Lessee of Waddell*, 41 U.S. 367 (1842), clearly defining the public trust in fish (and shellfish):

> [After the American Revolution] the people of each state became themselves sovereign; and in that character held the absolute right to all their navigable waters and the soils under them for their own common use, subject only to the rights surrendered by the Constitution to the general government. (*Martin*, 41 U.S. at 407–408)

Following this series of cases, the Supreme Court first recognized state public trust ownership of wildlife in *Geer v. Connecticut*, 161 U.S. 519 (1896), in which it held that the state could regulate the take and export of wild birds because of the public trust ownership of the birds:

> The ownership being in the people of the state, the repository of the sovereign authority . . . it necessarily results that the legislature, as the representative of the people of the state, may withhold or grant to individuals the right to hunt and kill game or qualify or restrict [the right], as . . . will best subserve the public welfare. (*Geer*, 161 U.S. at 533)

Geer represented the Supreme Court's high-water mark of recognition of state public trust ownership of fish and wildlife. In 1979, the court revisited *Geer* in *Hughes v. Oklahoma*, 441 U.S. 322, 339 (1979), overturning it only to the extent that *Geer* permitted state regulation of fish and wildlife to discriminate against interstate commerce. In other words, the court said that state regulation of fish and wildlife resources can sometimes violate the Commerce Clause of the US Constitution. Congress remedied this decision in 2005, when Senator Harry Reid (Nevada) introduced S. 339, the Reaffirmation of State Regulation of Resident and Nonresident Hunting and Fishing Act of 2005. This bill reaffirmed that states should have primary authority to regulate hunting and fishing, and exempted state and tribal regulation of hunting and fishing from the operation of the Dormant Commerce Clause (S. 339, 109th Cong. § 2(b) [2005]). Congress enacted S. 339 as an amendment to a defense appropriations bill, and it was signed into law as P.L. 109–113. Further, the *Hughes* case made several points indicating the importance of and the survival of state authority to manage fish and wildlife. For example, the court recognized that states' "concerns for conservation and protection of wild animals" are "legitimate local purposes similar to the States' interests in protecting the health and safety of their citizens" (*Hughes*, 441 U.S. at 335–337). Justice Rehnquist, in his dissenting opinion, went further, calling the state's interest in conserving fish and wildlife "substantial," "important," and "special" (*Hughes*, 441 U.S. at 342).

As far as US Supreme Court jurisprudence is concerned, the state public trust ownership of fish and wildlife survives today, although with some limits on it, such as the Commerce Clause, the equal protection clause of the Fourteenth Amendment, the Property Clause, the Treaty Power, and the Supremacy Clause, as other cases have noted. In fact, a 2015 case, *Horne v. Department of Agriculture*, 135 S. Ct. 2419 (2015), mentioned public ownership of oysters and differentiated them from private property.

Common law is, therefore, the basis for state ownership of fish and wildlife in the public trust. From this public trust concept grows the state constitutional and statutory assertion of legal authority over fish and wildlife resources and the concurrent assignment of those resources to state fish and wildlife conservation agencies to manage for the public.

Federal Foundational Statutes Assisting States in Fish and Wildlife Conservation

The Wildlife Restoration Act (passed in 1937, popularly called Pittman–Robertson [PR] after its congressional

sponsors, and also known as the Wildlife Restoration Program) and the Sport Fish Restoration and Recreational Boating Safety Act (passed in 1951 and likewise called Dingell–Johnson [DJ] after its sponsors, or the Sport Fish Restoration Program), along with significant amendments passed in 1984 and commonly known as Wallop–Breaux (WB) or the Sport Fish Restoration and Boating Trust Fund, are seminal federal statutes that provide permanent (not subject to congressional appropriations) and dedicated funds to state fish and wildlife agencies for fish and wildlife conservation to ensure fulfillment of their public trust responsibilities. While another chapter discusses conservation funding in detail, these acts' continuing importance to conservation demands that they receive brief treatment here.

Enacted when state fish and wildlife agencies' duties were growing from basic enforcement of game laws to include conservation and habitat programs, PR and DJ direct revenues from their respective excise taxes on hunting and angling equipment to state fish and wildlife agencies for conservation-related uses. Excise taxes are paid quarterly by industry and collected by federal agencies for apportionment to the states during the following federal fiscal year. WB, as well as subsequent amendments, includes additional tackle and equipment subject to the excise tax and apportions federal excise taxes from fuel attributed to motorboats and small engines to the Sport Fish Restoration and Boating Trust Fund, of which over one-half is apportioned to state agencies for sport fish conservation, boating access, and boating safety activities.

The state agencies determine how these funds are utilized, and federal excise taxes normally fund 75 percent of project costs, with states providing a required nonfederal matching share of 25 percent. States' nonfederal matching share usually comes from state hunting and fishing licenses, permits, and stamp revenues. Appropriate state agencies are the only entities eligible to receive apportioned funds from PR and DJ/WB.

Each state's apportionment of PR funds is derived from a formula that is based on 50 percent of its land and inland and coastal water area and 50 percent of its paid licensed hunters in proportion to the national total (the state provides the number of certified licensed hunters each year). Each state's apportionment of DJ/WB funds from the Sport Fish Restoration Program account is based on a formula that includes 40 percent of its land and inland and coastal water area and 60 percent of its paid licensed freshwater and saltwater anglers in proportion to the national total (the state provides the number of certified licensed anglers each year).

By enacting these laws, Congress recognized and affirmed that state fish and wildlife agencies are the primary managers of fish and wildlife within their borders. At the time of passage, these acts provided (and continue to provide) the majority of funds in the states for wildlife conservation, hunter education, shooting range construction, sport fish conservation, boating safety education, boating access, aquatic resources education, and other programs. These statutes are the foundational elements of fish and wildlife conservation and funding thereof, which other countries admire but have not been able to replicate. It is the funding paradigm that implements the North American Model of Wildlife Conservation. Protecting these funding streams is essential to the future of state-led conservation in the United States.

Federal Authority over Fish and Wildlife Resources

State fish and wildlife agencies have primary legal authority over most fish and wildlife species. Their authority is grounded in the public trust concept that can be traced in common law, as outlined above. Generally speaking, the federal government has authority only over federally protected species, such as federally threatened and endangered species, migratory birds, interjurisdictional fish (i.e., salmonids), and marine mammals—species for which Congress has expressly granted certain federal agencies that authority. Only Congress can authorize a federal agency to preempt state authority over fish and wildlife, and then only for specific conservation purposes. The federal government does not have a comparable public trust responsibility for fish and wildlife vis-à-vis the states. In all cases except marine mammals over which Congress has given federal agencies exclusive authority, Congress has affirmed that state authority is concurrent with federal authority. The federal government also has some authority over nonfederally protected fish and wildlife

on federal lands by virtue of the Property Clause found in Article IV, Section 3 of the US Constitution.

The federal government carries out its fish and wildlife conservation duties via federal agencies such as the USFWS and NMFS. Other federal agencies, such as the BLM, the Department of Defense (DoD), and the Department of Agriculture, and power marketing agencies such as Bonneville Power Administration implement fish and wildlife conservation work as a result of their management of federal lands or programs that directly impact fish, wildlife, and the habitats on which they depend. Following are summaries of some key federal laws regulating fish and wildlife resources with the attendant, often complicated jurisdictional and management challenges that arise between state and federal agencies.

Endangered Species Act (16 U.S.C. §§ 1531–1544)

The Endangered Species Act (ESA) is arguably the strongest federal environmental statute. The USFWS and NMFS share responsibility for administering the ESA. These agencies review and list species that, based on the "best available science," are endangered or threatened with extinction. The ESA carries civil and criminal penalties for "take," which is defined broadly and can include harassment and habitat modification. It also prohibits trafficking in protected species. Federal agencies also must refrain from taking species, and they have consultation obligations with the USFWS or NMFS under the law. The statute allows some methods of permitting incidental take of listed species for otherwise legal activities, whether for the federal government or for private parties. For example, Section 10 (16 U.S.C. § 1539) allows the USFWS and NMFS to issue an "enhancement of survival permit" that authorizes acts otherwise barred by the ESA, such as take. Under this authority, the USFWS created some regulatory mitigation tools such as habitat conservation plans and safe harbor agreements to facilitate conservation of listed species on public and private lands and to protect against prosecution under the ESA for a landowner's future actions. The USFWS also uses the ESA framework to regulate species listed under the Convention on International Trade of Endangered Species (CITES), which are species of international conservation concern predominately impacted by trade. Congress intended the states to play a major role under the ESA, requiring that the federal government "cooperate to the maximum extent practicable with the States" (16 U.S.C. § 1535). The ESA establishes a funding mechanism for states to conduct research, management, and recovery activities on federally listed species, as well as the approval framework for such programs. It also provides a mechanism for states to assume the interior secretary's authority (additive to their own) over listed species if they satisfy certain standards consistent with the ESA, and it specifically allows states to enact consistent or more stringent species protection laws. To date, Section 6 of the ESA has not been used as much more than a vehicle to transfer money to states to allow them to do threatened and endangered species conservation work, but the potential exists for a more robust partnership and recovery tool.

Migratory Bird Treaty Act (16 U.S.C. §§ 703–712)

The Migratory Bird Treaty Act (MBTA) dates from 1918 and enacts four treaties between the United States and Britain (signing for Canada), Mexico, Japan, and Russia, agreeing to protect migratory birds that, during their life cycles, cross international boundaries. The USFWS administers the MBTA and maintains a list of migratory birds. Like the ESA, the MBTA prohibits "take," although the definition is different; it also prohibits trafficking. It has taking exceptions for research and for hunting, as some migratory bird species are hunted, and states mostly regulate migratory bird hunting within season and bag limit frameworks established by the USFWS in cooperation with the states. The USFWS shares with states the regulation of migratory bird (waterfowl, doves, etc.) hunting. Like the ESA, the MBTA expressly allows for more stringent state laws, but state laws may not be inconsistent with the MBTA. There is currently no universal mitigation framework for permitting incidental take, although the issue is currently under evaluation; the USFWS has promulgated a take regulation for DoD activities that relate to military training and is considering one for other federal agencies such as the USFS, the BLM,

and the National Park Service (NPS), as well as private parties. If concluded, it will significantly impact both federal and state management of migratory birds.

Lacey Act (16 U.S.C. §§ 3371–3378)

The Lacey Act was the first federal law to protect wildlife. Its sponsors intended the act to curtail commercial markets for game, to control the movement of invasive species, and to provide a strong federal backstop for state enforcement of game laws in interstate commerce. It now covers many different species of fish, wildlife, and plants and their parts. The modern Lacey Act does two main things: it still backstops federal, state, tribal, and foreign laws pertaining to wildlife by prohibiting violations of those laws (meaning that it requires a predicate offense—first, a defendant must have violated the underlying law), and it prohibits mislabeling or failure to label shipments of fish, wildlife, and plants. Some Lacey Act violations are felonies and carry substantial fines. In many cases, state and federal law enforcement officials work together to bring suspected violators to justice under Lacey Act investigations. The Lacey Act also prohibits the importation of injurious species into the United States and their transport in interstate commerce without a permit from the USFWS. The Lacey Act is an important statute that helps discourage the illegal trade of fish, wildlife, and plants and their parts, which can lead to unsustainable exploitation and species' population declines, as well as human health and safety risks.

Magnuson–Stevens Fishery Conservation and Management Act (16 U.S.C. §§ 1801–1884)

The Magnuson–Stevens Act (MSA) regulates fisheries in federal waters. The general purpose of the law is to conserve fish stocks, carry out international agreements pertaining to fisheries, and conserve fish habitat. It has a number of mechanisms to carry out these purposes, including regional management councils, regional fishery planning, and NMFS oversight. The states sit on councils and are significantly involved in management decisions, although final decision-making authority remains with the NMFS. Like other statutes mentioned here, the MSA is a complex statutory scheme that deserves more treatment than this chapter allows.

National Forest Management Act (16 U.S.C. §§ 1600–1687)

The National Forest Management Act (NFMA) requires the secretary of agriculture to, among other things, undertake a comprehensive planning process for National Forest lands. A prior US Department of Agriculture Forest Service Planning Rule (36 C.F.R. § 219.9 in particular) directed that each national forest be managed "for ecological conditions necessary to contribute to the recovery of federally listed threatened and endangered species, conserve proposed and candidate species and maintain a viable population of each species of conservation concern within the plan area." A species of conservation concern is defined for purposes of this regulation as "a species . . . that is known to occur in the plan area and for which the regional forester has determined that the best available scientific information indicates substantial concern about the species' capability to persist over the long term in the plan area." This single definition of species of conservation concern has several vaguely defined terms, which consequently have attracted significant litigation challenging the meaning of those terms. Phrases such as "best available scientific information," "substantial concern," and "species' capability to exist over the long term" have vague definitions and are thus subject to interpretation by the regional forester and very often subject to litigation where the courts define these terms. Regional foresters have discretion whether to consult with state agencies in interpreting these terms, and efforts continue to strengthen collaboration and improve fish and wildlife conservation outcomes. But the all-too-frequent result of these undefined terms has been the federal courts interpreting them, thereby making or thwarting decisions over which the USFS and state fish and wildlife agencies have jurisdictional authority. Litigation and subsequent court decisions have led to virtual USFS paralysis in concluding National Forest plans in spite of the best efforts by USFS professional staff. For example, litigation over the Northwest Forest Plan and the threatened northern spotted owl complicates efforts to manage Pacific Northwest forests. The USFS has at least three times promulgated changes to the Planning Rule, but litigation on two revisions resulted in federal courts setting aside the revised rule in whole or part and directing the USFS back to the original rule language. Finally in 2015, a court dismissed a challenge to the most recent (2012) revision to the Planning Rule promulgated by the USFS at 36 C.F.R. Part 219, et seq., so the Planning Rule stands. See *Federal Forest Resource Coalition v. Vilsack*, 100 F. Supp. 3d. 21 (D.D.C. 2015). Throughout the litigation process, management of our nation's federal forests suffers along with the fish and wildlife that inhabit them. It is likely that the USFS will continue to be hobbled by litigation, especially under the ESA and in states within the jurisdiction of the US Court of Appeals for the Ninth Circuit. See *Cottonwood Env't Law Ctr. v. U.S. Forest Service*, 789 F.3d 1075 (9th Cir. 2015) (holding that the USFS must reinitiate ESA Section 7 consultation when the USFWS designates new critical habitat; the Supreme Court refused to review the case in October 2016, leaving the decision in place).

National Wildlife Refuge System Administration Act (16 U.S.C. § 668dd)

Congress created a comprehensive organic act for the National Wildlife Refuge System when it passed the National Wildlife Refuge System Improvement Act (NWRSIA) in 1997, which amended the National Wildlife Refuge System Administration Act (NWRSAA). The NWRSAA as amended establishes a mission for the system, establishes statutory purposes for the system, gives direction to the secretary regarding administration of the system, gives direction to the secretary regarding expansion of and acquisitions to the system, establishes terms and conditions for the development of comprehensive conservation plans for each refuge, and gives the secretary direction to cooperate with state fish and wildlife agencies during management of individual refuges in the system and acquisition of new refuges, among other things.

A few excerpts from the NWRSIA exemplify Congress's intention that the secretary closely cooperate with the state fish and wildlife agencies. In 16 U.S.C. § 668dd(a)(4)(E), Congress directs that the secretary, in administering the system, shall "ensure effective coordination, interaction and cooperation with . . . the fish and wildlife agency of the States in which the units of the System are located." And in 16 U.S.C. § 668dd(a)(4)(M), Congress requires the secretary to "ensure timely and effective cooperation and collaboration with Federal agencies and State fish and wildlife agencies during the course of acquiring and managing refuges." Further, in 16 U.S.C. § 668dd(b)(4), the secretary is authorized to "enter into cooperative agreements with State fish and wildlife agencies for the management of programs on a refuge." In the subsection on the development of a comprehensive conservation plan for each refuge (16 U.S.C. § 668dd(e)(3)), Congress directs that the secretary "shall, to the maximum extent practicable . . . (A) consult with . . . affected State conservation agencies; and (B) coordinate the development of the conservation plan or revision with relevant State conservation plans for fish and wildlife and their habitats."

Subsequent to enactment of the NWRSIA, the US-FWS integrated a small, regionally representative team of state fish and wildlife agency senior staff into their process of developing policies to implement the new amendments to the NWRSAA. Under an interagency personnel agreement, five state individuals worked in confidence with USFWS staff at every step from the concept stage to final policy over a period of four to five years in the development of many refuge policies. These state individuals retained their state jobs and workload but had their directors' assent to participate in this unique opportunity to jointly develop federal policy. Although there were some significant disagreements, most were resolved by the state and federal professional participants, and those that were not went to the USFWS director for resolution. In large part, the state fish and wildlife agencies were very pleased with this historic and unprecedented process and now seek to use this model of partnership and cooperation in other aspects of fish and wildlife conservation with the USFWS and other federal agencies.

National Park Service Organic Act (54 U.S.C. §§ 100101–100102, 100301–100303, 100751–100753)

The NPS, chartered in 1915 with the Organic Act, manages National Park lands, National Monuments, and a few other types of lands. The Organic Act, moved in 2014 from Title 16 to Title 54, U.S. Code, directs the NPS to manage these lands to ensure that they will remain "unimpaired for the enjoyment of future generations" (54 U.S.C. § 100101(a)). Management decisions, which generally seek to preserve National Park lands rather than to allow sustainable use of their resources, flow from this concept. For example, the Organic Act allows the NPS to cut timber, but only in order to control insect infestations or to "conserve the scenery" (54 U.S.C. § 100753). Likewise, the secretary may lethally manage wildlife and plants if they interfere with the public's use of a particular land. Hunting is generally not allowed on NPS lands, except where it is expressly provided for in the statute establishing the particular park or preserve. Very few National Parks allow hunting, which can make managing some wildlife populations, diseases, and habitat problematic, but hunting is allowed in many National Seashores or National Preserves. Fishing is allowed in most units of the NPS.

In large part because hunting is prohibited on most NPS lands, the NPS is very reluctant to allow the use of skilled volunteer hunters to remove wildlife when that wildlife needs to be managed, instead usually electing to use professional sharpshooters and sometimes leaving the animals where they lie. For example, on a couple of western National Parks, elk populations have exceeded habitat carrying capacity, and some have chronic wasting disease (CWD). In the past, the NPS has at times cooperated with the state fish and wildlife agency to cull excess elk by allowing the elk to cross onto an adjacent state park or private land, thus allowing hunters to take the elk. Productive uses for the meat were found at food banks or homeless shelters. At other times, concerns about CWD and the NPS's refusal to allow hunters to take excess elk in the park have caused an impasse in discussions about elk management. On many NPS lands, the agency has steadfastly refused to allow wildlife management through the use of volunteer hunters, which can adversely affect the health of herds and habitat.

The NPS and states have also disagreed on regulation and management of waters within NPS lands. States claim ownership of the beds of navigable waters, as well as the waters themselves, but the federal government has a Commerce Clause–based navigational servitude in these waters (meaning that Congress can regulate them). Based on that authority, Congress has passed laws such as the 1976 Park Service Administration and Improvement Act that give the secretary of the interior the power to regulate boating and "other activities relating to waters" on NPS lands. This creates a conflict between federal and state authorities, which has most recently manifested in litigation over a hunter's use of a hovercraft in a river in Alaska; the state of Alaska has intervened to assert its sovereign interest in the river. The case, *Sturgeon v. Frost*, 136 S. Ct. 1061 (2016), made it to the US Supreme Court in 2016. The court gave the hunter a partial win, disagreeing with the lower court's interpretation of the Alaska National Interest Lands Conservation Act, but did not decide whether the Park Service had authority to regulate hovercraft use on the river. Litigation continues in lower courts.

As noted at the beginning of this section, if a fish and wildlife species is not federally protected, generally the states regulate and manage the species. Further, under many of these statutory schemes, whether a directive to manage fish and wildlife or to manage federal land, the states and the federal government work cooperatively to conserve species and their habitats.

Federal Law and Policy: Treatment of State Authority

Many other sources of law, including statutes, rules, and policies, implicitly or explicitly recognize the state public trust in fish and wildlife resources. Just about every state has constitutional provisions, statutes, or both that codify the public trust, and examples of these were provided at the beginning of the chapter. Federal laws also recognize the state public trust.

But before delving into these sources of law, it makes sense to pause and return to high school civics for a moment. This chapter covers concepts of federalism, or the relationship between the federal government and state governments. Because of the Supremacy Clause of the Constitution, federal law, whether a constitutional provision, statute, or rule (executive agency–made law), will sometimes preempt state law:

> This Constitution, and the Laws of the United States which shall be made in pursuance thereof; and all treaties made, or which shall be made, under the authority of the United States, shall be the supreme law of the land; and the judges in every state shall be bound thereby, anything in the constitution or laws of any state to the contrary notwithstanding. (US Constitution, Article VI, Section 2)

The Supremacy Clause sets up the possibility of preemption of state authority, which can happen in several ways: expressly, impliedly (conflict or field), or through agency rulemaking. *Caution to the reader*: preemption is a complex and difficult legal concept that challenges trained lawyers; this paragraph grossly oversimplifies it but should suffice for the purposes of this chapter. In most cases, Congress will pass a law based on a power it has, such as the Commerce Clause, giving the federal executive branch the power to regulate in a certain area. One example is the ESA, which gives the federal government the primary power to regulate the take of threatened and endangered species, yet the ESA has various provisions recognizing state concurrent jurisdiction, and its legislative history is replete with affirmation and recognition of the need for this state concurrent jurisdiction over listed species. For example, Section 6 clearly affirms the concurrent jurisdiction of the states over listed species, and another section gives great deference to the states for management of threatened species. USFWS regulation 50 C.F.R. 17.31(a), extending to threatened species all prohibitions on take of endangered species, seems contrary to some of the ESA's legislative history and, as some argue, the text of ESA Section 4(d). However, in 1991, a court upheld the rule, and the issue was not brought up on appeal. See *Sweet Home Chapter of Communities for a Great Oregon v. Lujan*, 806 F. Supp. 279 (D.D.C. 1991). The states never challenged the administrative policy of applying all take restrictions on endangered species also to threatened species, and so it remains.

The NMFS never promulgated a similar regulation, resulting in inconsistent implementation of the ESA by two federal agencies.

Whether a federal law preempts a state law is a question for a court to decide. What matters for the purpose of state authority to manage fish and wildlife (and this chapter) is whether Congress has acted to give a federal agency authority to manage a certain fish or wildlife resource (for example, the ESA, the Marine Mammal Protection Act at 16 U.S.C. § 1361, et seq., and the Wild, Free-Roaming Horses and Burros Act at 16 U.S.C. § 1331, et seq., the subject of *Kleppe v. New Mexico*, below) or whether the federal government owns land that is affected by fish and wildlife (see *Hunt v. United States*, 278 U.S. 96 [1928], which involved state-owned deer damaging trees on federal land). However, barring the ESA or other such statutes assigning conservation responsibilities (thus authority) over fish and wildlife species themselves to the federal government, the federal government does not own the fish and wildlife and therefore does not have trust responsibilities for species that occupy land it owns. The US Supreme Court stated this plainly in *Kleppe v. New Mexico*, 426 U.S. 529 (1976), a case that ultimately found that Congress designated wild horses and burros as federal property:

> States hold primary jurisdiction over the management of fish and wildlife on federal lands unless Congress explicitly declares otherwise. (*Kleppe*, 426 U.S. at 529)

On the other hand, the Constitution also explicitly states what happens when a power is not given to the federal government. The Tenth Amendment to the US Constitution reserves those powers not granted to the federal government either to the states or to the people:

> The powers not delegated to the United States by the Constitution, nor Prohibited by it to the States, are reserved to the States respectively, or the people. (US Constitution, Amendment X)

Generally, if Congress has not passed a law on a particular topic, or if another constitutional provision does not grant authority over that area of law to the federal government, the federal government may not regulate in that area. Thus, states can choose to regulate in that area, as long as their regulations do not run afoul of the Commerce Clause (as in *Hughes*, 441 U.S.

332) or of other constitutional provisions such as the equal protection clause in the Fourteenth Amendment (see *Takahashi v. California Fish and Game Commission*, 334 U.S. 410 [1948]).

Sometimes, Congress and the federal executive agencies expressly recognize state authority over fish and wildlife. Myriad federal statutes, rules, and policies contain language doing just that.

The ESA and the MBTA both include specific provisions that preserve some state authority over species otherwise assigned to federal authority by the acts and in doing so recognize the concurrent jurisdiction and responsibilities of federal and state governments to conserve species listed under these acts' regulatory schemes:

> (f) Conflicts between Federal and State laws. Any State law or regulation which applies with respect to the importation or exportation of, or interstate or foreign commerce in, endangered species or threatened species is void to the extent that it may effectively (1) permit what is prohibited by this chapter or by any regulation which implements this chapter, or (2) prohibit what is authorized pursuant to an exemption or permit provided for in this chapter or in any regulation which implements this chapter. *This chapter shall not otherwise be construed to void any State law or regulation which is intended to conserve migratory, resident, or introduced fish or wildlife, or to permit or prohibit sale of such fish or wildlife.* Any State law or regulation respecting the taking of an endangered species or threatened species *may be more restrictive* than the exemptions or permits provided for in this chapter or in any regulation which implements this chapter but not less restrictive than the prohibitions so defined. (ESA, 16 U.S.C. § 1535(f); emphasis added)

> Nothing in this subchapter shall be construed to prevent the several States and Territories from making or enforcing laws or regulations *not inconsistent with* the provisions of said conventions or of this subchapter, or from making or enforcing laws or regulations which *shall give further protection to* migratory birds, their nests, and eggs, if such laws or regulations do not extend the open seasons for such birds beyond the dates approved by the President in accordance with section 704 of this title. (MBTA, 16 U.S.C. § 708; emphasis added)

Section 6 of the ESA explicitly sets up a mechanism (the cooperative agreement) whereby the secretary may authorize an eligible state to assume the secretary's authority (additive to the state authority) to conserve listed species, including authorizing take. In practice, this has not been realized by the USFWS and NMFS, and Section 6 has largely become a mechanism to provide federal funds (albeit minimal) to the states. The states have not challenged this for many reasons, including the fact that most states do not have a healthy budget for threatened and endangered species management, and thus this Section 6 assumption of the secretary's authority remains largely unimplemented except in a few states.

The Sikes Act (16 U.S.C. §§ 670–6700), which, very generally, requires the federal government to work with states to manage fish and wildlife on two separate categories of federal lands, military installations and lands administered by the Department of the Interior (DOI), is especially instructive on the interplay of state and federal authorities pertaining to managing fish and wildlife. The act requires for each military installation with significant natural resources that the DoD prepare management plans for fish and wildlife that "reflect the mutual agreement of" the installation, the state fish and wildlife agency, and the USFWS (16 U.S.C. § 670a). This is the highest standard of federal cooperation with states in federal statute. The requirements for DOI lands are not as comprehensive. In both cases, though, Congress recognized the need for state and federal agencies to work together on conserving fish and wildlife but also highlighted the important and primary role of the states.

Other federal statutes also contain clauses that recognize state authority and preserve it; some call these provisions "savings clauses." Consider the following clauses:

> That nothing in this Act shall be construed as authorizing the Secretary concerned to require Federal permits to hunt and fish on public lands or on lands in the National Forest System and adjacent waters or as enlarging or diminishing the responsibility and authority of the States for management of fish and resident wildlife. (Federal Lands Policy and Management Act, 43 U.S.C. § 1732)

> Nothing herein shall be construed as affecting the jurisdiction or responsibilities of the several States with respect to wildlife and fish on the national forests. (Multiple Use-Sustained Yield Act of 1960, 16 U.S.C. § 528)

> Nothing in this Act shall be construed as affecting the authority, jurisdiction, or responsibility of the several States to manage, control, or regulate fish and resident wildlife under State law or regulations in any area within the System. (National Wildlife Refuge System Administration Act of 1966 as amended, 16 U.S.C. § 668dd)

> (a) Fish and wildlife. Nothing in this chapter shall affect the jurisdiction or responsibilities of the States with respect to fish and wildlife. (National Wild and Scenic Rivers Act, 16 U.S.C. § 1284(a))

In each of these clauses, Congress explicitly recognizes and preserves state legal authority to manage fish and wildlife. While one federal appeals court considered the savings clause in the NWRSIA in *Wyoming v. United States*, 279 F.3d 1214 (10th Cir. 2002), and ultimately decided that the clause was not enough to preserve state authority where, as in this case, state and federal regulation and management actually conflict, this was a single federal circuit court, so its decision has limited applicability, and it left untouched many other arenas for state management authority to hold primacy. This decision unfortunately failed to take into account that the NWRSAA is replete with congressional direction to the secretary to cooperate with state agencies in implementing the act, including explicit authority for the USFWS to enter into cooperative agreements with states to manage programs on national wildlife refuges. Further, state authority is preserved in several other federal laws, regulations, and policies, as well as court cases upholding the long-standing premise of states' authority and jurisdiction for management of fish and wildlife on most federal lands.

The Conference Committee Report for the revisions to the Animal Welfare Act as amended by the 2002 Farm Bill affirms the secretary of agriculture's authority and responsibility under emergency situations to manage animal diseases, including in free-roaming fish and wildlife to minimize threats to livestock, but goes on to state, "However, nothing in this section or in this title should be construed as impliedly vesting in the Secretary authority to manage fish and wildlife populations." The report goes on to underscore the importance of consultation: "If fish or wildlife is affected by . . . measures proposed by the Secretary in an extraordinary emergency . . . [Congress] expects that the Secretary will consult with" the state fish and wildlife agency, "as current practice is in those instances" (House Report No. 107-424, Farm Security and Rural Investment Act of 2002, 107th Cong.).

Rules

Turning now to federal agency rules to examine their treatment of state authority to manage fish and wildlife, it is best to explain in the briefest sense how federal agencies write rules and on what basis they carry out work related to fish and wildlife.

Returning to civics, federal agencies are part of the executive branch, headed by the president. Congress writes laws that the executive branch carries out. A law will assign a duty to an agency, and federal agencies are created by law as well. For example, the DOI was created by an act of Congress in 1849. Other, later laws gave it more duties; for example, the ESA, passed in 1973, requires the DOI and the Commerce Department to conserve species they list as threatened or endangered. Generally, the law assigning a duty to an agency will allow the agency to write rules (regulations) to carry out the duties. Regulations have the force of statutory law and can only be final and binding if the agency follows certain procedures, including public notice and comment opportunities. Policies do not have the force of law, but they provide guidance to agencies and indicate to the public how agencies may act.

Following is a list of federal agencies that have significant fish and wildlife portfolios, either by virtue of having direct mandates to manage fish and wildlife or by land and water management or resource management duties:

- Department of Agriculture
  - Forest Service
  - Natural Resources Conservation Service
  - Animal Plant and Health Inspection Service
- Department of Commerce, National Oceanic and Atmospheric Administration
  - National Marine Fisheries Service
  - National Ocean Service
- Department of Defense
  - Army Corps of Engineers
  - Military Installations with "significant natural resources"
- Department of the Interior
  - Bureau of Land Management
  - Bureau of Reclamation
  - Fish and Wildlife Service
  - National Park Service
- Public power marketing agencies (Bonneville Power Association, Southeast Power Association, Southwest Power Association, Western Area Power Association, Tennessee Valley Authority*)
- Environmental Protection Agency

All of these agencies write rules and policies that state how they will manage fish and wildlife or natural resources. Following are excerpts from rules and policies that specifically recognize state agencies' authority to manage fish and wildlife:

> Nothing in the regulations in this part shall be construed as affecting the jurisdiction or responsibility of the several States with respect to wildlife and fish in the National Forests. (US Forest Service regulations, 36 C.F.R. § 293.10)

43 C.F.R. 24.3 General jurisdictional principles.

(a) *In general the States possess broad trustee and police powers over fish and wildlife within their borders, including fish and wildlife found on Federal lands within a State.* Under the Property Clause of the Constitution, Congress is given the power to "make all needful Rules and Regulations respecting the Territory or other Property belonging to the United States." In the exercise of power under the Property Clause, Congress may choose to preempt State management of fish and wildlife on Federal lands and, in circumstances where the exercise of power under the Commerce Clause is available, Congress may choose to establish restrictions on the taking of fish and

---

* Technically not a public power marketing agency, but it operates similarly to others in the list.

wildlife whether or not the activity occurs on Federal lands, as well as to establish restrictions on possessing, transporting, importing, or exporting fish and wildlife. Finally, a third source of Federal constitutional authority for the management of fish and wildlife is the treaty making power. This authority was first recognized in the negotiation of a migratory bird treaty with Great Britain on behalf of Canada in 1916.

(b) The exercise of Congressional power through the enactment of Federal fish and wildlife conservation statutes has generally been associated with the establishment of regulations more restrictive than those of State law. The power of Congress respecting the taking of fish and wildlife has been exercised as a restrictive regulatory power, except in those situations where the taking of these resources is necessary to protect Federal property. With these exceptions, and despite the existence of constitutional power respecting fish and wildlife on Federally-owned lands, *Congress has, in fact, reaffirmed the basic responsibility and authority of the States to manage fish and resident wildlife on Federal lands.*

(c) Congress has charged the Secretary of the Interior with responsibilities for the management of certain fish and wildlife resources, e.g., endangered and threatened species, migratory birds, certain marine mammals, and certain aspects of the management of some anadromous fish. *However, even in these specific instances, with the limited exception of marine mammals, State jurisdiction remains concurrent with Federal authority.* (Department of the Interior Regulations, 43 C.F.R. Part 24.3; emphasis added)

While the several States therefore possess primary authority and responsibility for management of fish and resident wildlife on Bureau of Land Management lands, the Secretary, through the Bureau of Land Management, has custody of the land itself and the habitat upon which fish and resident wildlife are dependent. Management of the habitat is a responsibility of the Federal Government. (Department of the Interior Regulations, 43 C.F.R. Part 24.4(d); emphasis added)

Policies

Below are examples of federal agencies' policies, which serve as guidance and direction to agency employees when implementing a rule:

[The director] must *interact, coordinate, cooperate, and collaborate with the State fish and wildlife agencies . . .* on the acquisition and management of refuges, and appurtenant wilderness areas.

The director must also ensure that Refuge System regulations and management plans are, to the extent *practicable, consistent with State laws, regulations, and management plans. . . .*

[The director] *must provide State fish and wildlife agencies . . .* meaningful opportunities *to participate* in the development and implementation of programs. (US Fish and Wildlife Service, General Overview of Wilderness Stewardship Policy, 610 F.W. 1.11(A) (1), (A)(2), (C); emphasis added)

Both the USFWS and the state fish and wildlife agencies have responsibilities for fish and wildlife conservation on the National Wildlife Refuge System:

7.4 What is the Service's policy on coordination with the States?

A. Effective conservation of fish, wildlife, plants and their habitats depends on the professional relationship between managers at the State and Federal level. *We acknowledge the unique expertise and role of State fish and wildlife agencies in the management of fish and wildlife.*

B. *Both the Service and the State fish and wildlife agencies have authorities and responsibilities for management of fish and wildlife on national wildlife refuges,* as described in 43 CFR 24. Consistent with the National Wildlife Refuge System Administration Act, as amended by the National Wildlife Refuge System Improvement Act:

(1) The Director will:

(a) Interact, coordinate, cooperate, and collaborate with the State fish and wildlife agencies in a timely and effective manner on the acquisition and management of national wildlife refuges; and

(b) Ensure that National Wildlife Refuge System regulations and management plans are, to the extent practicable, consistent with State laws, regulations, and management plans.

(2) Refuge managers, as the designated representatives of the Director at the local level, will also carry out these directives.

(3) We will provide State fish and wildlife agencies timely and meaningful opportunities to participate in the development and implementation of programs

conducted under this policy. This opportunity will most commonly occur through State fish and wildlife agency representation on the CCP planning team; however, we will provide other opportunities for the State fish and wildlife agencies to participate in the development and implementation of program changes that would be made outside of the CCP process. We will continue to provide State fish and wildlife agencies opportunities to discuss and, if necessary, elevate decisions within our hierarchy. (USFWS Manual, National Wildlife Refuge System, 601 FW 7 [2008]; emphasis added)

The USFS and BLM also have cooperative policies for work with the states:

Recognize the role of the States to manage wildlife and fish populations within their jurisdictions. . . . *Recognize the State fish and wildlife agencies as a public agency with management responsibilities for wildlife on the National Forests* and include them as partners in planning and implementation of activities that affect wildlife and fish. (Forest Service Manual, Title 2610.3(1) and (2); emphasis added)

It is the policy of the BLM to maintain *effective coordination and communication with State wildlife management agencies*. . . . The BLM policy requires that each State Office develop and maintain an up-to-date memorandum of understanding with the appropriate State wildlife management agencies that outlines policies and procedures for efficient and effective management of fish and wildlife resources under the jurisdiction of the State.

States have broad trustee and police powers over resident fish and wildlife found on Federal land within their borders. (BLM Interim Management Policy H-8550-1(G) (1); emphasis added)

Conclusion

The US system of fish and wildlife conservation is unique and the envy of the rest of the world. There is an assembly of state and federal laws governing and directing responsibilities and authorities for fish and wildlife conservation, and the bottom line is that success in fish and wildlife conservation requires respect of these respective authorities and rigorous cooperation between state and federal agencies, as well as with private landowners. The state–federal relationship for conserving fish and wildlife has received only minimal adjudication, but court challenges to fish and wildlife conservation laws and regulations can be best precluded by hearty partnerships and open dialogue and cooperation between the state and federal agencies in implementing those laws, regulations, and policies. Our fish and wildlife resources, their habitats, and our citizens deserve nothing less.

4

RONALD J. REGAN AND
STEVEN A. WILLIAMS

Evolution of Funding for State Fish and Wildlife Agencies

Promises Fulfilled, Promises to Keep

State fish and wildlife agencies are on the front lines of conservation delivery. They manage fish, wildlife, and their habitats, in the public trust, with a toolbox that includes law enforcement, education, research, monitoring, and management programs at their disposal. For more than 75 years, state agencies have relied in large measure on hunting and fishing license and permit fees and federal excise taxes on hunting, angling, and recreational shooting equipment, coupled with a motorboat fuel tax, to carry out such work. Indeed, these foundational financial resources have enabled state agencies to develop and maintain a sophisticated, comprehensive, science-based conservation system of unprecedented scale. This successful partnership among sportsmen—the hunting, angling, boating, and shooting sports communities—and government agencies is informally known as the American System of Conservation Funding, a user-pay/public-benefit model (G. Kania, personal communication).

Given the nature of the funding base—license fees and excise taxes—it should come as no surprise that research and management focus has been directed at game and sport fish populations. Since the inception of the Pittman–Robertson Act, or Wildlife Restoration Act, in 1937, the conservation record is replete with conservation success stories ranging from white-tailed deer to wood ducks and from pronghorn to brook trout across the United States.

There are a large number of species under state management authority, especially within taxonomic orders such as reptiles, amphibians, freshwater invertebrates, birds, and small mammals, for which there has not been similar dedicated funding. Broader state agency missions, new mandates, and new constituents have led to some state agencies receiving appropriated general fund dollars for permitting issues (e.g., energy development), broader conservation law enforcement (e.g., motorboat and ATV safety), or specific focal areas (e.g., watershed initiatives). But these funds, while important, have proven to be at risk during economic downturns, such as that of 2008–2009 attests. Resource issues related to energy development, habitat degradation at landscape scales, and Endangered Species Act (ESA) listings and litigation are examples of additional long-term funding needs for state agencies.

Even though a few state agencies have, in fact, secured dedicated sales tax or other revenue to address broader conservation needs, funding for the full suite of fish and wildlife under state authority has proven elusive at best. It is a legacy challenge.

We now focus on *promises fulfilled* by state agencies with a distinctive conservation funding construct and the *promises to keep* for future conservation capacity and success.

Promises Fulfilled

When the Federal Aid in Wildlife Restoration Act (Pittman–Robertson Act, or PR) was passed in 1937, no one could have expected the recent amounts of federal excise tax revenues collected ($300 million to $800 million annually). These large sums were the result of increased firearm and ammunition sales that first occurred in 2009. In addition, archery equipment sales experienced an increase. Compare these annual amounts to the first year of PR funding in 1939, which consisted of a national apportionment of $890,000. The foresight and unselfish commitment of hunters and firearm and ammunition manufacturers, contending with the Great Recession and the launch of World War II in Europe, have paid huge dividends for wildlife conservation for more than 75 years.

Recognizing that huntable wildlife populations in the United States were diminished and unsustainable, political leaders, hunters, and industrialists realized the necessity of a long-term financial commitment to our nation's wildlife. Through the PR Act, an existing manufacturer's excise tax was amended and levied on firearms and ammunition to provide a sustainable funding source for state fish and wildlife agencies. Then and now, these funds can be used to enhance wildlife management for all wild birds and wild mammals, hence the claim of many hunters that hunting and the revenue generated from license sales and excise taxes constitute the primary financial engine that drives conservation. In the past few years, gross license sales and PR funds totaled more than $1.1 billion. This user-pay/public-benefit system undergirds the North American Model of Wildlife Conservation. Indeed, unpublished 2015 survey data of state agencies suggest that sportsperson-derived funding sources compose nearly 60 percent of the average state agency budget.

Similarly, in 1950, Congress, with support from anglers and the fishing industry, passed the Federal Aid in Sport Fish Restoration Act (Dingell–Johnson Act, or DJ). Funds consist of manufacturers' excise taxes on a variety of fishing equipment and gear and a tax on motorboat fuel and motors. The annual apportionment of these excise taxes contributed approximately $350 million to state agencies for sport fish conservation and management. At the start of DJ funding in 1952, these funds totaled just $2.7 million. Like the PR program, the combination of DJ funds and gross license sales totals more than $1 billion annually for state fish and wildlife agencies. However, unlike PR, DJ funds are only eligible for use on sport fish species that have

value for sport or recreation. In practice, DJ restoration and conservation efforts provide substantial and widespread benefits to all aquatic species.

Even though the PR and DJ programs have contributed more than $10 billion to fish and wildlife conservation since their beginnings, the programs are poorly understood by the public and most wildlife professionals. In its current form, PR funding is derived from an 11 percent excise tax on firearms and ammunition, a 10 percent tax on handguns and revolvers, and an 11 percent tax on archery equipment and arrow components (43 cents per arrow shaft). DJ funding is derived from a 10 percent tax on fishing equipment, a 3 percent tax on electric trolling motors, a fuel tax attributable to motorboats, a small engine fuel tax attributable to boat motors, and import duties on tackle, pleasure boats, and yachts. These taxes are collected by the Internal Revenue Service. Tax revenues are deposited in Federal Treasury accounts, in essence, trust funds for fish and wildlife conservation. These federal funding programs are unique because Congress does not appropriate these dollars on an annual basis. They are permanent appropriations for the states. The US Fish and Wildlife Service (USFWS) administers the PR and DJ programs and annually apportions the tax dollars to each of the states and territories of the United States. There is no other user-pay/public-benefit system of this magnitude for any other type of outdoor recreation.

Guidance for Federal Aid in Wildlife and Sport Fish Restoration

The general elements of the apportionment formula for both PR and DJ consist of the number of certified licenses sold in each state and the geographic size of each state. There is a minimum and maximum amount apportioned so that all states and territories receive funding. In order to be eligible for these funds, state and territorial governments must have passed legislation (assent legislation) mandating that hunting and fishing license dollars can be used only for conservation purposes, as opposed to being diverted to fund general activities of the state. Canada does not have a similar federal source of funds; provincial and territorial hunting and fishing license revenues go into the treasury of each respective jurisdiction, and general funds are appropriated to these fish and wildlife agencies for conservation programs.

To ensure that PR and DJ funds are spent appropriately, the USFWS has established regulations consistent with each act that define eligible activities. In addition, approved grant funds are released on a reimbursement basis for up to 75 percent of eligible project costs. For example, if a state agency spends $500,000 for eligible costs to restore a wetland complex, the PR program would reimburse the agency for $375,000. Eligible projects include research, restoration, conservation, management, and enhancement of fish and wildlife and their habitats and providing public benefit from these resources. In general, ineligible activities include public relations, revenue production, commercial purposes to benefit individuals or groups, enforcement of game and fish laws and regulations, publishing and distributing regulations, constructing public facilities not directly related to conservation efforts, and most types of wildlife damage management activities. The PR program does not allow expenditure of funds to support stocking game animals to provide recreation only. To ensure program integrity, internal and external financial and administrative audits are conducted at least every five years to gauge compliance with applicable laws and regulations.

Federal Aid in Wildlife Restoration

Tax revenue deposited in the Federal Treasury is used for a variety of purposes. The great majority is passed on to the states for wildlife conservation activities. Interest earned on the trust fund is transferred to the North American Wetlands Conservation Fund to assist in the management of waterfowl and wetlands. The Multistate Conservation Grant Program receives an annual amount of $3 million. Hunter education and shooting range programs receive $8 million annually. Half of the taxes collected on handguns and archery equipment are apportioned for hunter education programs. The USFWS receives a small percentage of the total fund to administer the act.

The Wildlife Restoration Act has been amended a number of times since 1937. The seven amendments included making the funds permanent and indefinite

Table 4.1 Example of national accomplishments of the Wildlife Restoration Program—major expenditure categories and percentages

Expenditure category	Percentage of expenditures
Habitat improvement	15
Hunter education	9
Land acquisition	9
Operations and maintenance	41
Research—populations	26

(1951), increased the excise tax from 10 percent to 11 percent on firearms and ammunition (1954), included the 10 percent excise taxes from pistols and revolvers and allowed use of these funds for hunter education (1970), created an 11 percent excise tax on archery equipment and allowed the use of these funds for hunter education (1972), changed the tax formula on arrows and arrow components (1997), set aside $8 million for hunter education and shooting range development (2000), and exempted certain small manufacturers from paying excise taxes on firearms (2005).

Apportionments (grants that go directly to the states) for PR in the past five years have ranged from approximately $233 million to $472 million. Average annual apportionments during that time period totaled $324 million.

There are hundreds of PR projects under way across the nation involving wildlife research, habitat management, program administration, hunter education, waterfowl impoundments, planning, shooting range development, land acquisition and easements, and private and public land management. The major expenditure categories for PR are presented in table 4.1.

Federal Aid in Sport Fish Restoration

Just as the Sport Fish Restoration Act's revenue sources are more varied than the Wildlife Restoration Act, so are the programs and distribution of funds supported by the act. The Multistate Conservation Grant Program receives $3 million, the Sport Fishing and Boating Partnership Council receives $400,000 annually, and $800,000 is distributed to four regional Fisheries

Table 4.2 Example of national accomplishments of the Sport Fish Restoration Program—major expenditure categories and percentages

Expenditure category	Percentage of expenditures
Hatchery renovation	19
Aquatic education	10
Hatchery maintenance	22
Operations and maintenance	19
Research—populations	30

Commissions. The USFWS retains a small percentage for administration of the act. Sport Fish Restoration Programs receive 57 percent of the remaining funds. The balance is distributed to coastal, recreational boating, clean vessel, and boating infrastructure grant programs and a national outreach and communication program.

The Sport Fish Restoration Act has been amended five times since its inception in 1950. The Wallop–Breaux Amendment (1984) expanded and captured additional funds from a broad base of fishing and boating items, included motorboat access projects, included marine and freshwater projects, and created the Aquatic Resources Trust Fund. The 1991 amendment added small engine gas taxes to the fund and apportioned a percentage for wetland and coastal wetland conservation. Later amendments authorized funding for outreach and boating infrastructure and safety (1998), reduced or removed excise taxes on a narrow list of products (2004), and established a percentage-based allocation for grant programs (2005).

Apportionments (grants that go directly to the states) for DJ in the past five years have ranged from approximately $291 million to $404 million. Average annual apportionments during that time period totaled $366 million.

As with PR, there are hundreds of DJ projects under way across the nation involving fisheries research, river and stream improvement, program administration, aquatic education, hatchery construction and renovation, planning, fish passage improvements, boating infrastructure development, and reservoir management. The major expenditure categories for 2009 are presented in table 4.2.

The Greatest Story Never Told

For more than a century, hunters and anglers have been a major driver of fish and wildlife conservation funding in the United States. Although these individuals may focus their personal interest on game species, their conservation impact extends to all species. Federal funding associated with excise tax revenues has helped acquire, secure, maintain, and enhance millions of aquatic and wildlife habitat acres across the nation. Fish and wildlife research and advances in biological monitoring, life history, population modeling, habitat management, and the economic impact of fish and wildlife have also been supported by federal excise tax funding through both state fish and wildlife agencies and university research programs such as the Cooperative Fish and Wildlife Research Units. Aquatic and hunter education efforts have introduced millions of Americans over the past 25–30 years to the importance of fish and wildlife and their habitat, conservation, and sustainable and ethical use of these resources. It is not a story well understood by the American public.

As our nation continues to become more urban and our citizens lose their physical connection to our fish and wildlife resources, the financial, social, and political support for sustainable and utilitarian fish and wildlife conservation is at risk. Hunters and anglers contribute not only through license fees and excise taxes but also by their support for conservation organizations such as Ducks Unlimited, National Wild Turkey Federation, Pheasants Forever, Trout Unlimited, and the Rocky Mountain Elk Foundation. In recent years, these five national organizations, supported by hunters, anglers, and corporate donors, have expended approximately $300 million annually in pursuit of their fish and wildlife conservation missions. Other national conservation organizations contribute substantial funds for fish and wildlife conservation, although they pale in comparison to the on-the-ground conservation efforts of the previously mentioned national organizations. Federal taxes support the excellent work of federal agencies such as the USFWS and the US Forest Service. However, these federal agencies do not have authorization or responsibility to manage nonfederal trust species. These species are resident fish and wildlife species within the borders of state and territorial boundaries. They also make up the bulk of species and abundance of fish and wildlife in the nation.

The apparent national decline in hunter and angler numbers and participation rates does not bode well for the future of fish and wildlife conservation or its primary funding source. Hunter and angler recruitment and retention programs are being carried out across the country to sustain and enhance the contributions of these sportsmen and women. The Recreational Boating and Fishing Foundation and the Council to Advance Hunting and the Shooting Sports are focused on recruitment, retention, and reengagement (R3) of hunters and anglers in partnership with state agencies, industry, and conservation organizations. A national R3 plan will lay the groundwork for marketing, customer service, and opportunities for exposure to or direct participation in hunting and recreational shooting activities. Marketing and promotion of nature-based recreational activities, including hunting and fishing, are widespread.

Promises to Keep

During the twentieth century, state agencies produced a remarkable record of fish and wildlife conservation accomplishments in the United States, funded in large measure by license fees and excise taxes (Prukop and Regan 2005). For nearly four decades, however, state agencies and other conservation partners have recognized the need for broader funding to ensure that state agencies have the capacity to address the basic management needs for the full suite of species under state authority and to address new management challenges driven by systemic ecosystem stressors such as climate change, invasive species, development, and fish and wildlife disease. The Missouri Department of Conservation cracked the code, if you will, for sustainable, broad-based funding with dedicated sales tax legislation in 1976. Other states, including Florida, Virginia, Arkansas, Minnesota, and Iowa, have also secured new, dedicated income streams based on sales taxes, real estate transfer taxes, or other stateside general fund sources.

A philosophical and historical underpinning to wildlife conservation in this country suggests that a broader funding model would be desirable. Even though game species and the role of hunting are often featured in discussions about the North American Model of Fish and Wildlife Conservation, all wildlife resources can and should benefit from the public trust doctrine (Regan and Prukop 2008; Organ et al. 2012).

In addition, many Americans turn to fish and wildlife agencies for relief from wildlife damage, new recreational access opportunities, and basic educational resources. Let us also not forget that healthy ecosystems, maintained in part by fisheries and wildlife management initiatives, translate into cleaner water and cleaner air. State agencies have simply not done a good job of telling the story about the value and contributions of their work to the quality of life of all Americans.

The story or quest for broader funding begins with the Fish and Wildlife Conservation Act of 1980. This act, with no appropriated dollars, was intended to create a new federal funding path for nongame species. After a decade of no funding, the Association of Fish and Wildlife Agencies (AFWA) rolled up its sleeves to tackle this issue head-on (i.e., secure appropriated dollars to fund the act). Born on the wings of a 1994 Bridge to the Future Conference in St. Louis, the AFWA formally endorsed a Wildlife Diversity Funding Initiative (Belanger and Kinnane 2002). Under that banner, three focal streams of funding opportunities would emerge over time.

Teaming with Wildlife

The first funding opportunity focused on a new set of excise taxes on outdoor recreational equipment or products with a tangible connection to the enjoyment of wildlife. The "Teaming with Wildlife: A Natural Investment" brand was launched in 1994. A user-pay model, with excise taxes on hunting and angling gear, had set a bar of conservation success, especially for game and sport fish, and logic would suggest the merits of replicating that approach with an excise tax on other recreational wildlife products such as binoculars, bird seed, and field guides. In 2011, 71.8 million people age 16 years old and older, representing 30 percent of the

US population, observed, fed, or photographed wildlife (US Department of the Interior et al. 2011). Hikers, mountain bikers, campers, canoeists, bird-watchers, and others could directly support wildlife conservation, and a national Teaming with Wildlife Coalition made that case and appeal. However, a new tax, even one narrowly proscribed, was a nonstarter in Congress, and the concept never bore monetary fruit.

Conservation and Reinvestment Act

In spite of political reticence for conservation funding grounded in new taxes, there was bipartisan congressional support for the basic goal of securing broader fish and wildlife funding for state agencies. With the tacit and full support of the AFWA, an ambitious legislation package, entitled the Conservation and Reinvestment Act (CARA), would emerge as a vehicle to support stateside fish and wildlife conservation. The specific source of funds consisted of existing offshore oil and gas receipts or royalties. In addition to fish and wildlife programs, the $900 million bill would also fund historic preservation, coastal management, and outdoor recreation needs. In 2000, the CARA passed in the House of Representatives by a large margin, but it failed in the Senate, driven largely by concerns of appropriators for "off-line" dedicated funding.

State Wildlife Grants

In the midst of this conservation funding defeat, new life or hope emerged when Congress crafted the Wildlife Conservation and Restoration Program, with one-year funding ($50 million), and the State Wildlife Grants Program (with initial funding of $50 million), subject to an annual appropriation. The expectation, or hope, was that over time Congress would appropriate the original CARA funding target of $350 million via the State Wildlife Grants Program. As will be noted below, such annual funding would never materialize.

The federally appropriated State Wildlife Grants Program has directed over $1 billion to state agencies for species of greatest conservation need over the past decade, but this remains well short of the estimated $1.3 billion needed *annually* to implement State Wild-

life Action Plans (M. Humpert, personal communication). These plans, a condition of funding, first written in 2005 and now undergoing revisions, have identified over 12,000 species of greatest conservation need. Annual appropriations have varied from $60 million to nearly $100 million annually, and these dollars have become the core funds for state agency efforts to manage wildlife diversity programs. They have become agents of internal organizational change; funded basic inventory, research, management, and monitoring programs; and become the first line of defense in keeping species off of the federal endangered species list.

The Future

The need for dedicated broader wildlife funding for state agencies has not diminished. It is true that all Americans already fund fish and wildlife conservation in that federal taxes support federal agencies that manage such resources. The Land and Water Conservation Fund is an example of another federal investment that delivers habitat conservation and recreational access through land acquisition. In addition, many Americans who do not hunt or fish support conservation advocacy and land conservation through membership dues and donations, respectively. However, state agencies, those managers on the front lines of virtually every fish and wildlife conservation issue in this nation, do not necessarily benefit from those funds. These state agencies need dedicated operational funds to supplement those already provided by hunters, anglers, and trappers for over a century, to get the job done.

In 2015, the AFWA assembled the national Blue Ribbon Panel on Sustaining America's Diverse Fish and Wildlife Resources. This panel includes leaders from the business, energy, industry, conservation, and landowner communities—perhaps for the first time—to address conservation funding for state agencies. The right minds, right talent, and right passion are at the table to think creatively about this important need. The work of the panel is ongoing at the time of this writing, but recommendations for future solutions may very well focus anew on energy-based royalties, including the renewables sector, or some new private/public partnership.

The magnitude and complexity of new issues, such

as federal endangered species listing decisions, over broad geographic areas are creating new collaboration models. For example, the Western Association of Fish and Wildlife Agencies has embarked on a new partnership with oil, gas, private landowners, and others to deliver landscape-scale conservation across a five-state area for lesser prairie-chickens. This partnership has generated over $40 million in conservation funds to date.

It is a new world—one with unprecedented fish and wildlife conservation challenges, threats to historic and base funding, and capacity deficits. New science, new constituents, new organizational values, and leadership change—the average tenure of a state agency director is 3.4 years—create cause for concern. They also create opportunities for passionate conservation leaders to step up and manage to a new destiny, one that is grounded in the sportsmen-based system of funding but embraces an enhanced, diversified financial portfolio for the future.

LITERATURE CITED

Belanger, D. O., and A. Kinnane. 2002. Managing American wildlife. Montrose Press, Rockville, Maryland, USA.

Organ, J. F., V. Geist, S. P. Mahoney, S. Williams, P. R. Krausman, G. R. Batcheller, T. A. Decker, R. Carmichael, P. Nanjappa, R. Regan, R. A. Medellin, R. Cantu, R. E. McCabe, S. Craven, G. M. Vecellio, and D. J. Decker. 2012. The North American Model of Wildlife Conservation. Technical Review 12-04. The Wildlife Society, Bethesda, Maryland, USA.

Prukop, J., and R. J. Regan. 2005. In my opinion: The value of the North American model of wildlife conservation: An International Association of Fish and Wildlife Agencies position. Wildlife Society Bulletin 33:374–377.

Regan, R. J., and J. Prukop. 2008. A view from the trenches—reflections on the North American model of fish and wildlife conservation from a state agency perspective. Transactions of the North American Wildlife and Natural Resources Conference 73:255–264.

US Department of the Interior, US Fish and Wildlife Service, and US Department of Commerce, US Census Bureau. 2011. National survey of fishing, hunting, and wildlife-associated recreation.

5 State Wildlife Law Enforcement

BRIAN NESVIK,
RANDY STARK, AND
DALE CAVENY

This chapter presents a general overview summarizing wildlife law enforcement efforts by state agencies and their relevance to wildlife management. In short, it outlines thoughts and ideas related to the conduct of wildlife law enforcement, but more importantly, it highlights the links between good wildlife law enforcement, public expectations, wildlife law enforcement's place in the North American Model of Wildlife Conservation, public credibility, and how all of these components are critical to an agency's ability to manage wildlife. It is important to note that most wildlife management agencies have at least some responsibilities and duties related to public safety, and in fact, many states have a large role in the safety of their citizens. While these are important components of state wildlife law enforcement programs, we focus primarily on wildlife protection in this chapter.

Wildlife law enforcement officers are given a variety of titles. Most titles have roots in the early wildlife management history of a particular state, jurisdiction, or geographic region. Game warden and conservation officer are the most common titles; however, others include fish and wildlife agent, game ranger, environmental protection officer, or district wildlife manager. We will use these titles interchangeably throughout the chapter.

Historical Emergence and Importance of Conservation Law Enforcement

Understanding the history of conservation law enforcement is aided when viewed through the events and context that gave birth to its existence. In this section, we'll explore some of the key people and events that influenced public thinking, the conservation movement, and the emergence of conservation law enforcement.

Between the 1870s and early 1900s, individual states began hiring officers to enforce fish and game laws. In recognition of the need to improve the training of conservation law enforcement officers, a number of states established academies dedicated to training conservation law enforcement officers beginning in the 1920s. These early wardens, who were hired prior to the advent of the automobile, often traveled by horseback or train to carry out their job responsibilities. They were paid small monthly salaries, some of which were derived from the fines collected from the people they arrested. Many wardens early on were political appointees. Owing to corruption, poor performance, and the institution of the civil service system, wardens were later hired using a competitive civil service process.

Early conservation law enforcement officers faced a significant challenge in that they were confronted with an indifferent public attitude at best and more often than not a hostile attitude from those they contacted in the field. Many wardens were killed or injured in the line of duty. These initial officers had the task of changing public attitude toward wildlife lawbreakers. At the time, it was socially acceptable to shoot wildlife out of season and sell it for commercial gain or for "pot hunters" to illegally shoot wildlife provided they used it to "feed their families." Often when early game wardens apprehended lawbreakers, the courts dismissed cases or levied small fines that were of little deterrent effect. Through the work of these early wardens at the local community level, attitudes began to change. George Bird Grinnell also aided changing attitudes through his articles in *Forest and Stream*. Grinnell wrote articles designed to shift the culture from one of indifference about wildlife violators to piquing the conscience of Americans and shattering their previous faith in the inexhaustibility of wildlife (Reiger 2000).

The Lacey Act was passed in 1900. Introduced into Congress by Representative John F. Lacey, an Iowa Republican, the Lacey Act became the first federal law protecting wildlife and the first clear federal assertion of authority over wildlife. The Lacey Act provided penalties for the illegal trade of wildlife across state lines and prevented the importation of noxious species into the country. This important federal law continues to be critical in the protection of state and federal wildlife resources.

Today laws designed to protect fish and wildlife are embedded in many federal and state statutes and administrative codes in all states across the nation. Modern laws encompass a wider range of issues than those enacted when wildlife conservation was in its infancy. Laws related to technology and methods of take, fair chase, public safety, and wildlife protection all make up the much more inclusive and detailed codes of modern times.

A highly professional cadre of state conservation law enforcement officers is now a part of every state's wildlife management program. Conservation law enforcement officers of the twenty-first century have broad duties within the law enforcement spectrum, from ensuring that licensing requirements are met by hunters, fishermen, and trappers to conducting detailed investigations in order to solve wildlife crimes. Determining the causes of fatal recreational vehicle accidents, investigating environmental crimes, and enforcing regulations designed to prevent the spread

of wildlife disease and invasive species are all included in the duties of today's conservation officer.

Conservation officers also engage in a myriad of public safety responsibilities, such as search and rescue and assisting other local, state, and federal law enforcement agencies. In many states, conservation officers also share other non-law-enforcement roles, including assisting with wildlife management duties such as conducting wildlife surveys or setting hunting seasons. They may assist landowners in finding solutions to wildlife damage problems. They may teach hunter education classes or initiate other public education programs in their local communities highlighting the importance of wildlife management, habitat conservation, sustainable use of our natural resources, and the responsibility everyone has in passing on our natural resources to future generations. The work officers do is very important to everyone's future.

While these officers continue to use some effective traditional strategies from the past, they pride themselves on adapting to continuously shifting new realities brought about through changes in social, technological, environmental, and economic circumstances. While poachers, wildlife traffickers, and other fish and game law violators will always be a focus of officers, new threats to wildlife such as habitat loss, pollution, wildlife disease spread through illegal transportation of live wildlife, or invasive species transported across state lines present new challenges to officers and wildlife populations nationally.

Conservation law enforcement officers are essential in protecting our fish, wildlife, and natural resources and providing for safety during the public's enjoyment of these things. They play a critical role in ensuring that the tenets of the North American Model of Conservation remain in force, particularly the elements ensuring that wildlife is not privatized or illegally commercialized. Without conservation law enforcement officers working to gain compliance with laws designed to protect fish, wildlife, and natural resources and provide public safety, we would not have the abundant wildlife populations we experience today; species such as the turkey, whitetail deer, elk, antelope, and others may not have huntable populations; and the multibillion-dollar contributions to state economies across the country based on hunting, fishing, boating, and other outdoor recreation would not be possible.

Current State Methods to Achieve Compliance and Prosecute Violators
Overt Patrols

Far and away the most common techniques used by state wildlife agencies to achieve compliance start with overt marked patrols by wildlife law enforcement officers. Uniformed presence in marked patrol vehicles provides an outward deterrent to would-be violators simultaneous to providing law-abiding outdoor users with assurances that laws are being enforced. Here we will discuss three primary objectives of high-profile patrols and community engagement: (1) building relationships; (2) officer/sportsman contact; and (3) gathering intelligence, timely investigation of reported crime, and emergency response.

BUILDING RELATIONSHIPS

Most game wardens will tell you that building relationships with constituents and user groups is really where you can "make hay," so to speak. The wildlife version of community policing, unlike in urban areas of the United States, is not a new concept. Conservation officers have been engaged in this type of law enforcement for decades. The wildlife officer who can quickly build rapport with someone she has never previously met stands to improve her ability to achieve desired results more quickly. Achieving respect from and credibility with the public exponentially increases effectiveness and efficiency.

There are a couple of primary reasons this approach to law enforcement has been common and effective for so long in the conservation law enforcement profession. After all else is considered, most warden contacts are with those folks who are trying to enjoy outdoor recreation, whether it be hunting, fishing, boating, ATV riding, or trapping. The conservation officer must be able to do his job with minimal disruption to the user's activity. Unlike a city police officer who oftentimes only contacts those already identified as being in violation, the game warden's contact is more often with a person who hasn't broken any laws. An initial

contact based in a conversational and nonintrusive approach has a much better chance for success, particularly when the contact is with a person who knows that they aren't in violation. Since many wildlife, fishing, watercraft, and recreational laws don't require intent as an element of the crime, many in violation aren't aware that they are in violation until contacted. As an example, the fisherman who lazily failed to read new requirements regarding the size of smallmouth bass allowed in his creel may not realize that he violated the law until he is contacted by the conservation officer who points out his violation. While regulations detailing the types, gender, and sizes of wildlife allowed to be taken are important in managing individual species, a sportsman can inadvertently violate these types of laws, not because of ill intent, but because they fail to know all important regulations. The softer initial approach transitions well into a situation where enforcement action is required for a nonintentional violation.

Secondly, many will assert that they developed as a professional using this public-friendly style out of necessity for survival. When an individual officer's safety is based more largely on their influence on a particular situation versus how fast a backup unit can arrive, there is a subconscious and sometimes conscious tendency to reduce the potential for interpersonal conflict through behaviors and actions. Conservation officers work in remote areas in a majority of situations, and many of the people they contact are armed. Most conservation officers cover large areas (thousands of square miles in many districts) with low potential for quick assistance. Those who are successful and survive are good communicators who seek to gain compliance by positive influence and relationships over force and dictation.

Lastly, over the past 100 years warden jobs have required their dealing with people who live in more rural settings. In many states wardens are required to interact regularly with farmers and ranchers and those who make their living off of the land. Wildlife inhabits areas where there are fewer people, which obviously draws the conservation officer's job duties to less populated areas and smaller towns. A more interpersonal approach tends to work better with those types of people in these types of communities. As is the case when dealing with violators, dealing with landowners who have been negatively impacted by wildlife can be equally as contentious. Conservation officers who have found success in dealing with rural communities and rural people in difficult and uncomfortable situations tend to know the business of personal relationships. If they didn't start their career with those skills, they developed them out of sheer need to accomplish their objectives and do a good job. Dealing with rural America has been part of the wildlife law enforcement officer's job since their inception and probably will continue to be for the long term.

OFFICER/SPORTSMEN CONTACT

A major premise of effective wildlife law enforcement is that those contacts and compliance checks with non-violators serve proactively to influence future compliance. In fact, contact is not always required. Just being seen in the field can have a tremendous deterrent effect for both intentional and nonintentional violators. Subtle hints during contacts with the public validate this point. Statements like "When I saw you checking that boat over there, I started looking for my . . ." or "On opening morning while glassing my favorite hillside, I saw your truck coming and made sure I had my license" are commonly heard by field wardens who make a significant number of contacts.

Aside from the deterrent effect, presence and visibility serve to build public confidence in knowing that laws and regulations are being enforced. As is covered in other parts of this book, public support is absolutely essential to a state agency's ability to manage and protect wildlife for future generations. Seeing, talking with, and discussing issues with the local conservation officer transcends into "knowing" the local conservation officer. A smart game warden once said that it is very difficult for a person to stab you from behind if they are looking you in the eye. This concept has relevance figuratively and literally. There are huge benefits for the game warden being known as something more than a badge and a gun.

Wildlife law enforcement officers employ a variety of techniques to contact users of natural resources and those who recreate in the outdoors. Game wardens take great pride in doing their job in all areas of their assigned district of responsibility. "I have been fishing in this remote area for 20 years and have never seen a game warden," "What are you doing here? I haven't seen another person in this area for days," and "Where did you come from?" are all statements from outdoor users that most conservation officers have heard throughout their career. With large areas of responsibility and the fact that many sportsmen make the trek to remote areas to get away from people, it stands to reason that it is very difficult for the game warden to routinely interact with these folks without a significant field presence. These types of contacts with the public are invaluable and are critical to wildlife protection.

States agencies invest significant funds in providing the necessary resources for wildlife law enforcement officers to go where the wildlife and the user are, no matter how remote and difficult it may be. ATVs as patrol vehicles are common in most states. Eastern Seaboard states provide large offshore patrol boats to ensure compliance with commercial fisheries laws. Northern states like Wisconsin and Minnesota provide snow machines to access remote fishing lakes in the winter. Western states like Idaho and Wyoming provide horses so that wardens are able to access areas 10–40 miles from the nearest roads. Many states provide a variety of patrol boats, ranging from 12-foot aluminum johnboats with 5-horsepower motors to 25-foot (or larger) highly visible lake patrol boats equipped with light bars and emergency equipment. It is now common for officers to use personal watercraft to patrol high-use areas for both fishing and watercraft safety compliance. Alaska state troopers use float planes and other small planes capable of landing on tundra to access remote areas. These are a few examples of how state agencies manage transportation assets to ensure compliance across all areas where users may choose to recreate.

GATHERING INTELLIGENCE, TIMELY
INVESTIGATION OF REPORTED CRIME, AND
EMERGENCY RESPONSE

Another significant advantage to the conduct of overt compliance patrols lies in the wildlife agency's ability to more quickly respond to and investigate reported crimes. Most states have some type of a "tip line" to provide the public a means to report wildlife crimes (more to follow on tip lines below). Public confidence in the agency's priority for law enforcement can be strongly informed by the timely response to wildlife crimes in progress. Oftentimes, some of the most egregious wildlife criminal cases begin with a report from the public. Quicker responses often result in the gathering of perishable evidence, identification of suspects while they are still in the area, and identification of important witnesses before they are no longer available. All of these factors can make the difference between a prosecutable case and an open, unsolved case.

Equally important is the wildlife officer's ability to collect valuable information from the public and users. All officers have experienced the situation where during a routine field contact, a person reports a recent crime or suspicious activity. These violations would often go unreported had the officer not been in the field and face-to-face with the reporting party. Additionally, the officer's proximity to the location of a particular crime when he receives one of these types of reports can add significantly to the potential for successful apprehension and prosecution. Seasoned officers will confirm that making highly visible, overt contacts in the field using marked vehicles has positive consequences far beyond the direct detection of violations from these techniques. The second-order effects are often more significant with regard to the severity of crimes and the intent involved in committing those crimes.

Additionally, swift response to emergency situations such as boat and ATV crashes or hunting accidents is often paramount to developing strong cases and prosecuting those responsible. Conservation officer presence in high-use areas and being available to respond to these situations is important to ensure public safety and, again, build credibility with the public. Investigating stale cases hours after the incident occurred is a reality of the job; however, having officers in the field and in frequent contact with recreational and outdoor users significantly improves the chances that this will not always be the case.

Wildlife Crime Investigation

Spending time in the field with the wildlife resource and the folks who enjoy it is certainly a primary element of good wildlife law enforcement. The conducting of thorough and timely investigations by competent and well-trained officers is a second important component of any wildlife law enforcement program. The fact of the matter is simple: state wildlife management agencies must have the ability to conduct wildlife crime investigations in order to meet their state's charge, protect wildlife, and achieve public credibility. We briefly explore a variety of methods used by state wildlife management agencies to conduct thorough and timely investigations, including (1) wildlife investigative units, (2) field officer investigations, (3) wildlife forensic labs, and (4) social media and electronic evidence.

WILDLIFE INVESTIGATIVE UNITS

Most states employ a wildlife investigative unit; however, most are small (Florida being one exception, with an investigative unit composed of more than 100 officers). Many states, such as Georgia, Arkansas, North Dakota, and South Dakota, only have a small number of full-time investigators and supplement their investigative unit with field officers (National Association of Conservation Law Enforcement Chiefs 2015). Most of these units are staffed, equipped, and structured to focus on larger, more complex cases, often with a federal and/or wildlife commercialization nexus. These teams and work units often focus on both overt and covert work; however, many states employ an undercover unit focused solely on covert cases. Often thought of as the detectives for wildlife agencies, most wildlife crime investigators have opportunities for specialized training and enhanced access to intelligence resources. Regardless of individual state agency structure, wildlife investigative units have proven to be a key component of state wildlife law enforcement programs.

Generally, wildlife investigative units invest significant amounts of time proactively seeking out criminal enterprises and developing adequate evidence to begin an investigation. Most wildlife detectives spend a large amount of time poring over taxidermist, outfitter, and commercial operator records; comparing databases; and analyzing a variety of media to identify potential criminal activity. Unlike the field conservation officer, the wildlife investigator invests significant time protecting wildlife from an office versus in the field. Examples of long-term, complex cases include wildlife trafficking and the illegal sale of wildlife parts, large-scale illegal take of fish and wildlife for profit and moving it across state lines, importation of illegal wildlife, and environmental crime with large negative impacts.

While the wildlife investigator plays a significant role in generating their own casework, they also rely heavily on field officers, other law enforcement agencies, and other intelligence sources to provide the initial impetus to begin an investigation. Some of the best cases started with a minor violation encountered by a game warden in the field that generated some suspicion of other illegal activities. Also fairly common are the cases that begin with information discovered consequential to a nonwildlife crime arrest or the service of a search warrant when nonwildlife enforcement officers find evidence of wildlife crimes. In most cases field agents have strong relationships with local communities and access to critical human intelligence. Many large cases start with the compilation of bits and pieces of intelligence gathered locally by field officers. Conservation officers certainly have an investigatory role, but in the more difficult and complex cases, it is often in the better interest of successful investigations to turn them over to a specialized unit.

A recent Colorado case highlights the importance and effectiveness of long-term investigations by wildlife investigators (from the case files of Bob Thompson, Colorado Department of Parks and Wildlife, 2014).

Lions, Bobcats, Cages, and Traps

Christopher W. Loncarich, 56, of Mack, Colorado, was sentenced on November 20, 2014, in Denver's US District Court to 27 months in prison, followed by three years of probation, for conspiring to violate the Lacey Act, a federal law prohibiting the interstate transportation and sale of any wildlife taken in an illegal manner. Until his probation has been completed, he cannot hunt, pursue, or trap any wildlife and must undergo substance abuse and mental health treatment while on probation. In addition, Loncarich will appear before a Colorado Parks and Wildlife Hearings Officer, where he may receive up to a lifetime ban from hunting and fishing in Colorado, as well as 43 other Interstate Wildlife Violator Compact states.

Loncarich and his assistant, Nicholaus J. Rodgers, 31, of Medford, Oregon, were indicted in January by a grand jury on 17 counts of illegally trapping and maiming mountain lions and bobcats. Loncarich pled guilty to one count of conspiring to violate the Lacey Act in

August of 2014. Rodgers pled guilty to the same charge in July of 2014.

A three-year investigation by Colorado Parks and Wildlife, Utah Division of Wildlife Resources, and the US Fish and Wildlife Service (USFWS) revealed what Colorado Northwest Regional Manager Ron Velarde said was one of the worst examples of poaching he has seen in his 40-plus-year career managing Colorado's wildlife.

According to the indictments, between 2007 and 2010, Loncarich, aided by his daughters, Rodgers, and assistant guide Marvin Ellis, conspired to capture lions and bobcats and then cage them, hold them in leg traps, or shoot them in the foot or stomach. Coordinating by radio communication, they released the hindered cats when their client arrived. The goal was to make the cats easier for their clients to kill during excursions along the rugged Book Cliff Mountains in western Colorado and eastern Utah.

Several cats killed in Utah were illegally transported to Colorado, where Loncarich falsified documents to obtain the required seals for the hides. The outfitter's clients then transported the illegally taken cats back to their home states, in further violation of the Lacey Act.

Loncarich charged 18 clients between $3,500 and $7,500 for each lion hunt and between $700 and $1,500 for each bobcat hunt, sharing his earnings with his assistants. Investigators say that approximately 30 cats were killed in this manner.

In what wildlife officials say was a particularly egregious example of their activities, the group captured a mountain lion and fit it with a radio-tracking collar. Aided by the device, they captured the same lion a year later, immobilizing it overnight with a leg-hold trap. The next day, they placed the lion in a cage and took it to Loncarich's residence in Mack, where it was held for approximately one week while the outfitters waited for their client to arrive from Missouri. They then placed the lion in a box and transported it via snowmobile to a predetermined area, where it was released for the client to kill. Loncarich charged $4,000 for the outing.

One of Loncarich's daughters pled guilty to her role in the scheme and was sentenced on two misdemeanor Lacey Act violations on September 30, 2014. She received one year of probation, a $1,000 fine, and

60 hours of community service, 30 of which must be spent with the Colorado Parks and Wildlife Hunter Education program. The other daughter also pled guilty and was sentenced on a misdemeanor Lacey Act violation, receiving one year of probation, a $500 fine, and 36 hours of community service, half of which must be spent with the Colorado Parks and Wildlife Hunter Education program.

Ellis also pled guilty to a felony on June 3, 2013, and he was sentenced to three years of probation and six months of home detention and ordered to pay a $3,100 fine.

Loncarich's 2008 Ford truck and Ellis's 1995 Dodge truck were seized during the investigation, having been used in the commission of Lacey Act violations. Both vehicles were subsequently forfeited to the federal government. In addition, three of Loncarich's clients have been issued federal, Lacey Act violation notices. Those clients have paid a total of $13,100 in fines.

FIELD OFFICER INVESTIGATIONS

Larger, complex, and long-term investigations are certainly better left to specialized units; however, a majority of wildlife investigations do not fall into this category. A majority of investigations would not be conducted without the field conservation officer. Examples of cases requiring shorter-term and less complex investigation are numerous; these types of cases are often worked on a daily basis by field wardens. Taking wildlife out of season and without licenses, thrill killing (shooting and leaving wildlife with no intention of taking it into possession), license fraud, using illegal methods to take fish and wildlife, and smaller-scale commercial crimes are all examples of cases worked by field agents on a regular basis.

Timeliness has proven to be critical in wildlife investigations, validating the need for an immediate response by field wardens who are trained and know the area. Environmental conditions often degrade evidence quality and even its existence in many scenarios, making field investigations all the more important. Many states have found the use of a comprehensive team approach to crime scene investigation to be most effective. Using a team of officers to thoroughly canvas crime scenes and offer diverse perspectives enhances the probability for successful identification of suspects and establishment of the facts of the case. Most field officers carry and are trained in the use of crime scene investigation kits, which include metal detectors for finding bullets in carcasses, blood and biological evidence collection materials, high-resolution cameras, scalpels for conducting necropsies (comparable to an autopsy but specific to wildlife), and special collection containers for electronics (particularly cell phones).

Once crime scenes are processed and available evidence has been collected, the conservation officer finds herself in the position to conduct interviews of suspects and witnesses. Wildlife officers are trained in a variety of interview and interrogation techniques and are generally known for their keen abilities to use personal interactions to gather admissions and develop important evidence. As is the case in many duties of the game warden, they often find success in leveraging their relationships and their ability to build rapport with people to collect critical evidence necessary to successfully prosecute crimes. Community relationships can be invaluable in putting the finishing touches on an important case or in finding the key piece of information that leads to the identification of the primary suspect.

Data management has become a significant factor in improving the efficiency of short- and long-term investigations. With the development and improvement of databases, wildlife officers are able to quickly access licensing records (even those from other states), harvest reports, commercial operator records, and a myriad of other important information. What once would take days and weeks of poring over records now may take minutes with the advancement of information management. These capabilities enable field officers to much more quickly solve crimes that may have required an investigative unit in the past. The following is an example of good field investigation work coupled with help from the local community (Wyoming Game and Fish Department 2013).

The Cappuccino Caper

Many western Wyoming trophy mule deer buck poaching cases begin with a report from a concerned citizen detailing their finding of a headless deer carcass. The

Cappuccino Caper started in the same typical fashion in late October of 2008. A winter range traveler contacted the department after finding a large-bodied mule deer carcass in the middle of the winter range. The reporter stated that the carcass was fresh and there were several vehicle tracks in the area.

Local wardens responded to the area quickly to begin an investigation. Three game wardens combed over the entire scene, finding tire tracks, shell casings, garbage (including a coffee cup), and a deer carcass they determined had been killed the prior day. Wardens were able to determine by tire tracks that a vehicle had stopped in a crooked position, with the front of the vehicle pointed towards the deer. Identical tracks were found all along the small dirt road, but initially it appeared that the vehicle had traveled the road two times as there were two sets of tracks parallel to one another.

Investigators were able to use the location of two shell casings on the road to determine where the vehicle stopped, pointed it's headlights toward the deer carcass, and allowed garbage to fall off of the floor board on the passenger side of the vehicle. Further investigation over approximately 0.5 miles of the road allowed wardens to determine that the parallel tire tracks were made by a pickup with dual rear wheels. They were also able to follow blood drops from the carcass to a location that was consistent with the back end of a pickup.

Wardens concluded their work at the scene with a few small pieces of important evidence, but no suspect. They knew that the deer most likely was shot in the dark based on the position of the vehicle in relation to the deer. They knew that the shooter was most likely the driver, and they knew that the deer head had been cut off, carried, and loaded on the driver side into the back of a pickup. They also had a small, unique Styrofoam coffee cup.

The wardens examined the cup and determined that there was only one place in the county that served coffee using that type of cup. A visit to the "Mountain Mocha" drive-thru coffee shop confirmed that this establishment served this type of cup. An employee at the establishment examined the cup and was able to determine that it contained cappuccino when it was sold. She also noted that it was very uncommon for any of the normal customers to purchase a small cappuccino. She could remember selling a small cappuccino to a fairly regular customer, but couldn't remember who it was or what they drove. The local warden asked her to keep her eye open for a dual-wheeled pickup where the passenger routinely purchases a small cappuccino. She agreed to do so.

The following morning the coffee shop employee called the local warden and stated that a white dually Dodge pickup had just come through the drive-up and that the passenger ordered a small cappuccino. She also provided a license plate number and stated that there was a scoped rifle on the seat next to the driver.

Wardens quickly mobilized to find the vehicle and were successful by midmorning, finding the vehicle at a local business. Wardens conducted a search of the vehicle and found blood and hair evidence on the driver's side of the truck, as well as rifles, ammunition, and a knife with blood and hair. The owner of the vehicle eventually confessed to poaching the deer and led wardens to a location where he hid the head near his rural residence. The suspect's accomplice was quickly identified and confessed as well. Both suspects described their crime in detail and stated that they were drinking Crown Royal whiskey while driving around in the dark and saw the large buck on the side of the road. Both men were convicted of multiple wildlife violations, paid several thousand dollars in fines, and lost their hunting and fishing privileges for several years.

WILDLIFE FORENSIC LABS

A handful of wildlife forensic labs are spread around the United States and Canada. There are only seven wildlife-specific forensic labs located in US wildlife agencies. In total, there are over 25 laboratories if you include academic laboratories (housed and operating out of a university facility), federal laboratories (working on threatened and endangered species, as well as two laboratories dedicated exclusively to marine animals), and private laboratories (D. Hawk, personal communication). While they are few in number, their importance in wildlife law enforcement is immeasurable.

Many American labs have the ability to extract, analyze, and match wildlife DNA for a number of different species. DNA matching can be absolutely pivotal in matching suspects to crime scenes and specific illegally taken wildlife. DNA analysis has progressively become much more refined over the past couple of decades, to a point where small bits of evidence from hair, blood, and bone can provide enough material to scientifically confirm a match of one evidentiary item to another to determine that two items originated from the same animal (or different animals) with the statistical probability able to withstand judicial scrutiny in the courtroom. In addition to DNA matching, most labs are able to use this same biological evidence to determine species and gender. Some facilities are able to assist with determining cause and time of death. The science is solid, and use of this type of evidence only stands to grow in importance in the future. State wildlife forensic labs in the United States include the Wyoming Game and Fish Department Wildlife Forensics Lab, Laramie; the Idaho Fish and Game Department Wildlife Forensics Program, Caldwell; the Florida Department of Law Enforcement; the New York State Department of Environmental Conservation Wildlife Pathology Unit, Delmar; the California Department of Fish and Game Forensics Laboratory, Rancho Cordova; the Tennessee Wildlife Resource Agency, Big Sandy; and the Texas Parks and Wildlife A. E. Wood Fish Hatchery, San Marcos. In addition to these state labs, the USFWS houses a lab in Ashland, Oregon, with some unique capabilities and resources often used in wildlife law enforcement work. Cases processed at this facility often focus on wildlife trafficking of species such as rhino and elephant. They have likely the largest collection of vouchered wildlife specimens in the world and have capabilities that rival any wildlife forensics lab worldwide (D. Hawk, personal communication).

SOCIAL MEDIA AND ELECTRONIC EVIDENCE

Social media and electronic devices, particularly cell phones, have come to be commonplace as tools used to investigate crimes. Wildlife- and environmental-related criminal activity is no exception. The massive amounts of information available on the Internet and the dozens of available social media sites have served to help wildlife law enforcement officers in their efforts to gather evidence and build solid cases. Consider that as of September 2014, three-fourths of online users are also users of social media sites, with Facebook (71%), Pinterest (28%), Instagram (26%), and Twitter (23%) leading the pack as the most popular sites (Pew Research Center 2014). Nine of 10 US Internet users between the ages of 18 and 29 and 78 percent between the ages of 30 and 49 employ social media. Considering that 70 percent of anglers and 68 percent of hunters are under the age of 55 (US Department of the Interior et al. 2011), it is easy to see how important social media and Internet-related competency is to wildlife law enforcement officers.

Since communication between younger Americans occurs so frequently on the Internet, and considering that communications include information about behavior and observations, it stands to reason that a computer-literate game warden can indulge in a plethora of high-quality information through his laptop computer in his truck or office. In the relatively short existence of the Internet, online evidence has provided vast amounts of information leading to significant wildlife cases. The number of those who are willing to steal wildlife resources from the public and are also willing to brag about their activities on social media is disturbing, but this type of online bragging is helpful to those who are charged with protecting wildlife.

Cell phone technology has emerged nearly simultaneously to Internet development. The Pew Research Center estimates that 90 percent of adult Americans own a cell phone and that 81 percent use their cell phone to send and receive text messages (Pew Research Center 2014). Wildlife enforcement agents are now finding significant benefit in the information that can be obtained from a suspect's cell phone. Text messages, locations, photos, and videos are some of the data forms that wildlife officers often find on mobile devices. Those who are concerned with evidence of their criminal activity residing on their phones have learned ways to quickly delete information on their mobile device remotely to protect it from law enforcement. Game Warden training today includes methods to preserve cell phone evidence following seizure while proper search warrants are obtained.

Public Involvement in Wildlife Law Enforcement

Public involvement in wildlife law enforcement is absolutely paramount to the protection of the resource. First, state wildlife management agencies draw their authority and ability to manage wildlife from those citizens for whom the resource is held in trust. Their involvement in its protection fundamentally makes sense. Second, there are so few game wardens, conservation officers, wildlife agents, and game rangers per unit area of land containing wildlife habitat that additional eyes and ears are essential in order for wildlife protection to be effective. Many officers have areas of responsibility comprising thousands of square miles. Many warden districts are made up of large waterways, remote wilderness, or wild lands and terrain, making access difficult. Engaged citizens who help the local warden by acting as additional observers serve an important role, improving the wildlife officer's chances of successfully prosecuting cases. The following describes how a concerned citizen made a violation report resulting in the prosecution of serious wildlife crime (National Association of Conservation Law Enforcement Chiefs 2015).

Penalties in Snow Goose Case Top $55K

Five have pleaded guilty to combining to take 265 snow geese over the permitted limit. The five people charged with combining to kill 265 snow geese over the permitted limit all have pleaded guilty to charges and together will pay more than $55,000 in fines and replacement costs, the Pennsylvania Game Commission announced today.

The last of the five defendants pleaded guilty Monday to all charges he faced, bringing resolution to the case, which stems from an April 1 incident in Marion Township, Berks County. On that date, Wildlife Conservation Officer (WCO) Brian Sheetz, along with WCO Dave Brockmeier and Deputy WCO Ed Shutter, received information about a lot of shooting in the area of Church Road in Myerstown.

The officers arrived and found evidence that a large number of snow geese had been shot. Through their investigation, it was learned that the five defendants had killed 365 snow geese. The daily bag limit is 25 snow geese per hunter. And since one of the hunters also didn't possess the required migratory bird license, he wasn't permitted to harvest any snow geese.

Between May 19 and June 1, each defendant pleaded guilty to all charges he faced. Magisterial District Judge Gail Greth, of Fleetwood, accepted the guilty pleas. Norman Brubaker, 30, of Bernville, pleaded guilty to one count of hunting without a migratory bird license, one count of a violation involving federal laws, and 73 counts of unlawful taking and possession of snow geese. He also agreed to pay $14,990 in fines and replacement costs. Laverne Frey, 34, of Womelsdorf, pleaded guilty to one count of a violation involving federal laws and 48 counts of unlawful taking and possession of snow geese. He also agreed to pay $10,040 in fines and replacement costs. Nevin Frey, 28, of Myerstown, pleaded guilty to one count of a violation involving federal laws and 48 counts of unlawful taking and possession of snow geese. He also agreed to pay $10,040 in fines and replacement costs. Kenneth Oberholtzer, 26, of Womelsdorf, pleaded guilty to one count of a violation involving federal laws and 48 counts of unlawful taking and possession of snow geese. He also agreed to pay $10,040 in fines and replacement costs. Nelson Sensenig, 25, of Lebanon, pleaded guilty to one count of a violation involving federal laws and 48 counts of unlawful taking and possession of snow geese. He also agreed to pay $10,040 in fines and replacement costs.

While most of the snow geese shot by the hunters were taken illegally, they did not go to waste. After the birds were gathered and evidence was collected, Game Commission officials transported the carcasses to a processor and then donated 288 pounds of goose meat to the Central Pennsylvania Food Bank in Harrisburg. The cost of processing was added to the defendants' penalty.

Game Commission Executive Director R. Matthew Hough said the Game Commission's wildlife conservation officers, through their hard work to enforce Pennsylvania's Game and Wildlife Code, fulfill a vital role in curbing poaching activity, which steals from law-abiding hunters and trappers.

Public Tip Lines

Nearly all states provide some type of "tip line" whereby concerned citizens have the ability to report by phone or text observations and suspicions of illegal wildlife activity. "Operation Game Thief," as it is called in many states (Texas, Maine, and Colorado, to name a few), "Catch a Poacher" in Maryland, "Report all Poachers (RAP)" in North Dakota, "1-800-POACHER" in Ohio, and the "Stop Poaching Hotline" in Wyoming are a few examples from around the nation of the names used for public phone banks where violations can be reported. The percentage of cases that start with a call to a state tip line varies, but it can be substantial, particularly during peak use periods. Most wildlife officers will tell you that their caseload would decrease significantly without poaching reports from the public.

Public Reward Programs

In addition to publically provided opportunities to easily report wildlife crimes, most states have some type of reward program. Most of these programs provide financial rewards for those who provide information that leads to the arrest and conviction of wildlife violators. Many state government rules prevent agencies from resourcing and sponsoring these programs. Consequently, most of these programs are sponsored by nonprofit charitable organizations. Many reward amounts are based on the volume and quality of information provided and the value of the wildlife being reported as illegally taken. These financial incentives have proven extremely effective in motivating people to report serious wildlife crimes.

The Importance of Collaboration across Borders

While the mission of protecting people and natural resources has not changed since the first conservation law enforcement officers were established in the late 1800s, many aspects of society have changed, requiring increased collaboration between state and international conservation law enforcement agencies. In the late 1800s, prior to the existence of the automobile, traveling from one state or country to another via train or horseback was a multiday event. Communication networks were slow, occurring via delivered mail or telegraph. Consequently, most legal and illegal harvesting of wildlife was done more locally, not across state borders or internationally.

Today, society has become very mobile with greatly enhanced highway systems, reliable vehicles, campers, air travel, transoceanic passenger ships, and international train passenger networks. People can travel to and legally or illegally hunt, fish, and commercially harvest and transport wildlife in multiple states—sometimes in a single day or in a matter of hours. Shipments of illegally harvested wildlife can be sent via overnight mail to anywhere in the world. Consequently, in contemporary society it is common practice for people to travel to locations across the country to hunt, fish, boat, commercially harvest wildlife, and engage in other outdoor recreation. The Internet, worldwide cellular and satellite communication networks, and social media have increased the capacity for commercial guiding businesses to advertise their services to people across the world. The same capabilities are availed to illegal wildlife traffickers for use in communicating and coordinating illegal activities. These societal changes have made the illegal trade in wildlife a globalized venture. Where demand exists and supply is low, considerable profits lure illegal wildlife traffickers to engage in illegal international trade in protected wildlife. The convergence of these societal changes requires conservation law enforcement agencies to effectively collaborate across borders in order to effectively protect our wildlife and other natural resources.

The Lacey Act of 1900—a Critically Important Tool in Protecting Wildlife

State conservation officers and federal special agents routinely collaborate with each other on investigations involving the Lacey Act. This act is a very important tool used by state and federal conservation officers to protect wildlife and plants. Stiff penalties associated with violations of the act mete out appropriate justice for the crime, reduce economic incentives to violate wildlife laws and illegally traffic in wildlife, and act as a significant deterrent to others.

Under the Lacey Act, it is unlawful to import, export, sell, acquire, or purchase fish, wildlife, or plants that are taken, possessed, transported, or sold (1) in violation of US or Indian law or (2) in interstate or foreign commerce involving any fish, wildlife, or plants taken, possessed, or sold in violation of state or foreign law. Following are some examples of Lacey Act prosecutions from across the United States.

CALIFORNIA

In 2007, four commercial fisherman involved in the aquarium trade were prosecuted by the US Attorney in San Francisco for illegally catching and selling at least 465 baby leopard sharks. The sharks, worth approximately $20 each, were sold in the United States, United Kingdom, and Netherlands over an 11-year period. A settlement of more than $500,000 was paid to fund habitat restoration in San Francisco Bay.

MISSOURI

In 2008, agents of the Missouri Department of Conservation and the USFWS apprehended an individual engaging in illegal commercial fishing activities on Table Rock Lake. The individual was apprehended while in possession of 78 pounds of extracted paddlefish eggs, taken illegally using commercial methods. He was also found to be in illegal possession of 98 pounds of processed paddlefish caviar at his residence. As the investigation unfolded, officers learned that the individual was a licensed commercial fisherman in Arkansas and had been selling caviar illegally taken from Missouri paddlefish across state lines. To sell illegal caviar on the open market, the individual funneled illegally harvested Missouri paddlefish roe through his Arkansas commercial fishing license. During the month prior to his apprehension, the individual sold approximately 387 pounds of paddlefish caviar to a company in Tennessee for $35,820.

This individual pleaded guilty to Lacey Act violations and was sentenced to one year and one day in federal prison without parole and ordered to pay $30,002 in restitution. He was ordered to forfeit all equipment utilized in commission of the offense (boat, motor, trailer, GPS unit, and assorted equipment).

COLORADO

In a joint investigation between Colorado Parks and Wildlife, Iowa Department of Natural Resources, and the USFWS, evidence collected over a 17-year period showed that six poachers had hunted without proper and valid licenses, hunted out of season, willfully destroyed wildlife, and illegally sold that wildlife. Through the course of investigation, one poacher was identified in the illegal killing of 47 white-tailed deer, 17 elk, and 10 mule deer. This individual received 47 months in federal prison, $40,000 in fines and restitution, and a lifetime suspension from hunting and fishing in the United States for Lacey Act violations.

WISCONSIN

In May 2005, officers in 11 states executed search warrants, served subpoenas, and interviewed dozens of suspects and witnesses. During the investigation, officers seized 32 deer mounts, 14 sets of antlers, 7 firearms, and 10 compound bows. Thousands of pieces of evidence were seized and analyzed. Thirty people were convicted in two separate Wisconsin County Circuit Courts, and six people were convicted in US District Court. The Lacey Act allowed Wisconsin wildlife officers, working closely with federal special agents, to bring these violators to justice in Wisconsin Circuit Court and US District Court.

Interstate Wildlife Violator Compact

The underlying purpose, intent, and concept of the Interstate Wildlife Violator Compact (IWVC) were first developed in the 1980s. It was designed to aid conservation law enforcement agencies dealing with violators who break wildlife and resource laws outside of their home state. The states of Colorado and Nevada worked to draft the first IWVC documents. These documents were passed as legislation in 1989 in Colorado, Nevada, and Oregon and formed the foundation for the current IWVC.

The IWVC was created to promote compliance with the laws, regulations, ordinances, resolutions, and administrative rules that relate to management of wildlife resources in member states. The IWVC achieves this goal through two main strategies:

1. The IWVC establishes a process whereby wildlife law violations committed by a nonresident while in a member state may be handled as if the person were a resident in the state where the violation took place. This allows an officer to accept a personal recognizance promise to appear in the field instead of having to arrest, book, and bond an arrested subject. This process is convenient for citizens of member states, saves taxpayers money owing to lower administrative costs associated with arrest, and increases the efficiency of conservation law enforcement officers.

2. The IWVC includes a reciprocal agreement enabling license privilege suspensions in other member states. Consequently, anyone whose license privileges are suspended in a member state is also suspended in their home state and all other member states. This holds wildlife law violators accountable and provides a deterrent effect promoting nationwide compliance as wildlife law violators' illegal activities in one state can affect their privileges in all participating states. This interstate agreement enhances the ability of all member states to protect and manage wildlife resources for the benefit of law-abiding residents and visitors.

National Association of Conservation Law Enforcement Chiefs

The National Association of Conservation Law Enforcement Chiefs (NACLEC) works collaboratively with state, federal, and international conservation law enforcement agencies and in partnership with other diverse alliances to protect people, sustainably conserve our wildlife and other natural resources, and promote safe, diverse, and enjoyable outdoor experiences. To achieve its mission, the NACLEC focuses on the following overarching goals:

- Fostering collaboration and cooperation among natural resource organizations.
- Supporting a professional and diverse workforce.
- Developing future leaders.
- Creating diverse partnerships.
- Increasing public support through outreach and education.
- Promoting the integral role conservation law enforcement plays in enhancing quality of life.

NACLEC membership is composed of the executive leadership (chiefs and colonels) of state or federal agencies, having primary responsibility for administering and enforcing conservation laws. Having the membership consist of chief executives builds relationships between agencies across borders and enables the direction of agency resources to interstate issues to meet both long-term national strategic needs and short-term interstate and international investigative projects.

The NACLEC collaborates very closely with the USFWS on both strategic and investigative initiatives. As an example of a strategic initiative, the NACLEC collaborates with the USFWS on the National Conservation Law Enforcement Leadership Academy held at the USFWS's National Conservation Training Center in Shepherdstown, West Virginia. The academy is designed to enhance relationships between states and international agencies and to develop the next generation of leadership in conservation law enforcement in North America.

Conclusion

Wildlife law enforcement has deep roots in natural resource conservation history. The founding principles premised in the public trust doctrine hold true today and clearly delineate the importance of wildlife law enforcement and the conservation officer in modern wildlife management. Authority to manage wildlife is granted by those for whom wildlife is managed in trust. Therefore, public involvement, trust, and credibility all play key roles in a state agency's ability to manage wildlife. The conservation law enforcement officer plays a vital role in representing wildlife management as an institution and in establishing strong rapport with the public. Simply put, wildlife management would not be possible without publicly promulgated laws enforced by a professional cadre of officers dedicated to serving people and conserving wildlife.

LITERATURE CITED

National Association of Conservation Law Enforcement Chiefs. 2015. www.naclec.org.

Pew Research Center. 2014. Mobile technology fact sheet. www .pewinternet.org/fact-sheets/mobile-technology-fact-sheet/.

Reiger, J. F. 2000. American sportsmen and the origins of conservation, 3rd edition. Oregon State University Press, Corvallis, Oregon, USA.

US Department of the Interior, US Fish and Wildlife Service, and US Department of Commerce, US Census Bureau. 2011. National survey of fishing, hunting, and wildlife-associated recreation.

Wyoming Game and Fish Department. 2013. Wildlife crime: Stories from Wyoming's wildlife officers, 4th edition. Cheyenne, Wyoming, USA.

6

VERNON C. BLEICH AND
DANIEL J. THOMPSON

State Management of Big Game

Large carnivores and large ungulates have been described as charismatic megafauna. Members of the orders Carnivora and Artiodactyla have been of interest to the American people, have been the objects of extreme efforts to eradicate or increase individual taxa, are thought by many to be among the most desirable of harvestable species, and are, in fact, frequently referred to as trophy species. Indeed, the most recent compilation of North American trophy records (Reneau and Spring 2011) includes black bear (*Ursus americanus*), grizzly and brown bear (*U. arctos* ssp.), polar bear (*U. maritimus*), and two felids, the jaguar (*Panthera onca*) and mountain lion (*Puma concolor*). Also considered trophy species and included in the compilation of Reneau and Spring (2011) are various subspecies of a number of ungulates, including elk (*Cervus elaphus*), mule and black-tailed deer (*Odocoileus hemionus*), white-tailed deer (*O. virginianus*), moose (*Alces alces*), caribou (*Rangifer tarandus*), bighorn sheep (*Ovis canadensis*), thinhorn sheep (*O. dalli*), pronghorn (*Antilocapra americana*), bison (*Bison bison*), mountain goat (*Oreamnos americanus*), and musk ox (*Ovibos moschatus*). Only pronghorn and mountain goat are endemic to North America; genera of the other ungulates have a distribution that includes representatives occurring primarily north of the equator.

The North American Artiodactyla have received attention from conservationists, wildlife managers, and biologists (Krausman and Bleich 2013). Those large mammals have been hunted for food or clothing for millennia, and several of them have declined in number as a result of unregulated harvest (e.g., market hunting), habitat loss, disease, or a combination of these and other factors. Declines spurred early conservationists to embark on programs to prevent the further demise of wildlife (Williamson 1987), resulting in initial efforts to conserve wildlife, the founding of the discipline of wildlife management, and the formulation of the North American Model of Wildlife Conservation (Krausman and Bleich 2013).

The path toward science-based management of carnivores or omnivores such as mountain lions, wolves, and bears has progressed at a slower pace. Large carnivores were generally reduced or extirpated from much of North America owing to factors related to human safety or competition for livestock and livelihood. Changes in perspectives projecting wildlife as a public resource also transcended to carnivorous animals as game species after decades (centuries in some cases) of efforts to reduce their numbers. Bounties were rescinded on bears, wolves, and mountain lions in the latter part of the twentieth century, and the advent of radio telemetry provided further insight into the habits of these animals, allowing managers to more accurately manage those species. The shift from eradication and predator control to conservation and management of mountain lions and black bears has resulted in range expansion of both these species, with black bear management/harvest occurring throughout North America and mountain lion populations expanding into areas of the Midwest and eastern portions of the United States and Canada.

Unfortunately, there is a misperception that no conservation movement existed in the United States until the twentieth century; that movement actually began much earlier (Reiger 1975), evidenced in the passage of legislation that provided at least some protection for species in decline. For example, (1) laws in California were enacted as early as 1852 that provided closed seasons on elk, deer, and pronghorn on a local basis, and in 1854 those seasonal closures were extended statewide (Neasham 1973); (2) in 1872 legislation that protected elk, pronghorn, and mule deer for eight months of the year was enacted; (3) in 1878 additional legislation established a four-year moratorium on take of any elk, pronghorn, female deer, or mountain sheep; and (4) in 1883 additional legislation extended the moratorium on take of any mountain sheep indefinitely (Bleich 2006). During the same period, other states also passed laws providing protection to many trophy species, and as a result, a general shift toward regulated hunting and management occurred throughout the United States. As emphasized by Leopold (1933), regulation of hunting opportunity played an important role in the founding of game management (i.e., wildlife management) as a professional discipline.

Since its inception (Leopold 1933), the discipline of wildlife management has evolved substantially beyond its humble beginnings. This evolution is evidenced not only by the plethora of publications produced by the thousands of scientists or managers now employed in that discipline but also by efforts to protect habitat, efforts to protect or recover endangered taxa, implementation of scientifically based harvest management, and the numerous successful efforts to restore large mammals to historical ranges (Krausman and Bleich 2013). Indeed, early management focused on protection or enforcement of regulations, but conservation objectives changed over time to incorporate emphases on habitat protection and habitat enhancement (Bleich 2006). These efforts resulted in improvements in the status of many taxa and increased interest in wildlife conservation among the public.

Increases in the popularity of conservation biology as a discipline (Soule and Wilcox 1980a) have further stimulated interests in wildlife management.

Unfortunately, as noted by Soule and Wilcox (1980b), "While wildlife management, forestry and resource biologists . . . struggled to buffer the most grievous or economically harmful of human impacts . . . , the large majority of their academic colleagues thought the subject [of conservation biology] was beneath their dignity." Even with the early prejudice against conservation biology as a respectable science aptly described by Soule and Wilcox (1980b), the discipline of wildlife management and goals associated with conservation continued to move forward, and wildlife managers have been successful in their efforts to ensure the futures of numerous taxa, among which are many of the large mammals referenced in this chapter.

Although there are exceptions (e.g., those taxa that receive federal protection), conservation and management responsibilities of the ungulates and carnivores considered in this chapter largely are the responsibilities of the various states. To that end, it is our intent to summarize techniques that state wildlife managers employ to collect and analyze population data for the sustained harvest of ungulates and large carnivores, implement management strategies for ungulates and large carnivores, and discuss the diversity of ways states manage game species based on available wildlife abundance, habitat quality, funding constraints, and public desires. We also emphasize how the best available science is used in management, and in some cases, why it is not possible to do so. While there are a multitude of contributing factors (e.g., wildlife diseases, expansion of invasive species, introduction of exotic species) that managers must take into account when developing management strategies, these topics are covered in further detail elsewhere in this book and will not be fully addressed in this chapter.

There are quite literally dozens of books written about the management of big game, generally specific to the species of interest, for both the laymen and the professional. Our chapter has no illusions of being all-inclusive with regard to information on state management of these large mammals. Nevertheless, we provide information that stresses the major concepts of science-based management of big game and trophy game populations in North America. More detailed understanding of population demography and behavior allows agency personnel to develop management strategies based on the ecology of the animal (Vander-meer and Goldberg 2003; Mills 2007), a much more meaningful approach than trying to manage different species by applying the same methods or harvest regimes—a strategy that can lead to failure.

Methods of Data Acquisition

There are various methods available to collect information used in assessing populations of large mammals (Locke et al. 2012; Pierce et al. 2012a). As a result, data collection and analysis differ considerably across management jurisdictions, ranging from extensive formal sampling constructs to efforts with informal designs and best characterized as convenience sampling (Mason et al. 2006). Although the benefits of standardization of techniques, with an emphasis on mule deer and elk, were clearly explained by Mason et al. (2006), there remains a need for development and application of standardized techniques that would increase comparability of information collected by diverse jurisdictions. The amount of effort expended among those jurisdictions is, however, a function of available resources, demands of user groups, bureaucratic inertia, or some combination of these factors. Nevertheless, statistically valid survey methods and analytical processes clearly will be beneficial in the contexts of assessing populations, setting harvest objectives, or evaluating responses of ungulates to management strategies (Rabe et al. 2002; Mason et al. 2006). The benefits of decades of experience, with both successes and failures, led to the principles of adaptive management of natural resources, wherein managers can assess the results of past actions and acquire and incorporate new strategies (Walters 1986; Organ et al. 2012) to more accurately assess wildlife populations and amend or alter population perturbations through management based on applicable techniques.

It is beyond the scope of this chapter to cover every method of data acquisition or technique currently in use or available for deriving demographic information to be used in managing the species covered here. Multitudes of books, recommendations, and professional papers have been written that describe techniques, provide information on statistical analyses and interpretation, or provide the results of those analyses in the context of either research or management; a helpful summary of such information is available in Silvy (2012). As appropriate, we call attention to some salient works and concentrate on providing examples of the various methods that are applied to species to help students and others gain an appreciation for the variety of techniques available.

Ground Methods

Many state wildlife agencies continue to use a variety of ground-based surveys to assess populations, determine responses to management actions, and generate data necessary to derive harvest recommendations. Data obtained via such surveys have been especially productive for bighorn sheep and have resulted in detailed, long-term population assessments (Wehausen et al. 1987; Rubin et al. 1998; Holl et al. 2004; Wiedmann and Bleich 2014). Likewise, biologists have for many years used ground-based surveys—on foot, on horseback, or in vehicles—to assess demographic characteristics of populations of mule deer (Heffelfinger 2006); in some situations, population density is so low that aerial surveys do not yield sample sizes from which precise estimates can be inferred, and ground-based surveys remain important methods (Thompson and Bleich 1993). Ground-based surveys are not widely used, however, to acquire demographic data useful in setting harvest objectives for many species of ungulates. Nevertheless, some states do set regulations using data obtained by ground observers for bighorn sheep (Wiedmann and Hosek 2013) or mule deer (Heffelfinger 2006), and ground-based surveys have been used effectively to gather demographic information for white-tailed deer (Harwell et al. 1979; McCullough 1979, 1984; Fafarman and DeYoung 1986).

Demographic data for white-tailed deer frequently are obtained using a variety of ground survey methods. This is especially true in the more open habitats occupied by that species. Acquisition of demographic data in densely vegetated areas occupied by some white-tailed deer has a number of limitations, and aerial methods also can be problematic in such situations (Larue et al. 2007). For example, adequate snow cover is necessary during the sampling period, ground cover frequently is adequate to obscure deer from aerial observers, and aerial surveys frequently underestimate the number of deer present (Beringer et al. 1998; Larue et al. 2007). As a result, deer will be missed, and the proportion missed is unknown. Further, appropriate aircraft might not be available when needed to conduct surveys, posing an additional problem for managers.

In some jurisdictions, ground-based surveys are used to obtain demographic data on bison (MSRM 2002; PADEMP 2015) and mountain goats (Schulze et al. 2008). Ground-based surveys, especially track surveys, have been used for carnivores, although difficulties have arisen in attempts to accurately quantify density or minor changes in trend through time, depending on local carnivore densities and road densities (Russell et al. 2012). In many cases that involve transect sampling, the application of distance sampling (Buckland et al. 2001, 2004) is appropriate if the assumptions of the technique can be met.

Learning from shortcomings associated with the technique (Hughson et al. 2010), some jurisdictions have implemented demographic surveys using remote-triggered cameras (camera traps) for mule deer (Marshal et al. 2006) or white-tailed deer (Jacobson et al. 1997; Roberts et al. 2006; Soria-Diaz and Monroy-Vilchis 2015), and methods of obtaining information via this technology are developing rapidly (O'Connell et al. 2010; Locke et al. 2012). This ground-based method has also been applied to bighorn sheep with some success (Jaeger et al. 1991; Perry et al. 2010). In some situations, mark-recapture population estimates based on individually identifiable animals using Lincoln–Peterson-type estimators can be derived, but Larrucea et al. (2007) cautioned that data obtained from camera traps do not always provide unbiased estimates. If timed correctly, placing camera traps along migration corridors and established travel routes can provide indices to local ungulate population size and age/sex ratios. As with all these techniques addressed throughout the chapter, combining data can provide further insight for managers to effectively evaluate harvest strategies for ungulates.

Camera-trap surveys are becoming widely used for cryptic species such as large felids (Karanth and Nichols 1998; Silver et al. 2004; Kelly et al. 2008). Spotted felids such as leopards, jaguars, and tigers are individually discernible based on distinctive pelage markings and are treated as "marked" animals to calculate density and abundance (Karanth and Nichols 1998). These techniques have been efficiently used to document presence or absence; methodologies are still being refined, however, for species such as mountain lions (Kelly et al. 2008). Unique facial and body marking (scarring, leg barring, residual spotting; Fifield et al. 2015) may allow investigators to use such physical characteristics to differentiate individual mountain lions and, when coupled with other information, could prove effective for use in estimating abundance (Kelly et al. 2008). Techniques based on the use of camera traps are continually being refined to systematically survey an area to develop statistically rigorous results for estimating presence, density, and even life history characteristics of cryptic carnivores.

Genetic-Based Estimators

Increased technology and understanding of genetic monitoring methodologies have led to a wide spectrum of techniques for monitoring large carnivores in North America. The monitoring strategies frequently applied to ungulates often are not as efficacious for animals such as grizzly bears, black bears, and mountain lions. Acquiring DNA from individuals allows managers insight into demographics and annual trend assessments through a variety of techniques employed by agencies. Many times these types of estimators are used in addition to harvest data to evaluate population status and develop harvest regimes.

FECAL DNA SAMPLING

Collection of DNA from scat or hair has become a useful method for identifying individual animals in mark-recapture studies and eliminates the need to handle animals. As a result, sample collection is simplified and generally less expensive than otherwise would be the case. Fecal-based genetic techniques are also being improved (Brinkman and Hundertmark 2009; Brinkman et al. 2013) and are used on an increasingly frequent basis to assess the status of populations (Brinkman et al. 2011). The use of fecal DNA in population monitoring is being applied more and more frequently among a variety of species, but, as with all techniques, there are weaknesses. Potential sources of error include those associated with the genotyping process itself and erroneous identification and resultant acquisition of pellets where species overlap (e.g., mule deer and bighorn sheep, or mule deer and white-tailed deer).

Many management jurisdictions employ the use of the technique generally referred to as hair snares, especially for ursids. Hair-snare density estimators have been used in California, Colorado, Utah, Tennessee, and Pennsylvania as an aspect of monitoring strategies in management plans for a variety of carnivores. Bears are driven by their olfactory system, and with the use of nonfood reward lures, managers can acquire hair samples to be used in presence-absence surveys or in mark-recapture techniques to estimate abundance or density of local populations (Kendall et al. 2008). These types of estimators can be used with or without harvest data and generally do not require capture or handling of individuals. Rub stations have been used to acquire genetic material for felids, but the success rate is lower than that for ursids. Systematic ground track surveys can provide DNA through hair obtained by backtracking travel routes of mountain lions (Sawaya et al. 2011) and can result in an estimate of abundance of localized populations using DNA analyses similar to those used for bears.

BIOPSY DARTING

An efficient means to acquire tissue samples from mountain lions is through the use of specialized "biopsy darts," wherein a bayed mountain lion is darted and a tissue sample obtained without immobilization or handling of the animal (Russell et al. 2012). Samples obtained can then be used to acquire DNA from individual animals to develop a genetic database of a sample population. The representative "marks" of the sample population can then be recaptured through either harvest or additional darting at a later period. This technique is used in several western states and has been reported to be successful depending on overall densities and accessibility of the study area. These

techniques can be replicated over time to assess population trend (realizing that there are constraints due to sample size). Ancillary data from genetic analyses also provide insight into movements within or among populations.

SCAT DOGS

A burgeoning DNA-based technique employs the use of trained "scat dogs" to locate species-specific scat for DNA acquisition (Wasser et al. 2004; Thompson et al. 2012; Davidson et al. 2014). This technique has primarily been used with mesocarnivores but appears to be applicable to documenting presence of mountain lions in areas of range expansion. It also may yield estimates of abundance depending on the terrain and population density of felids in a given area. Currently the technique is spatially limited in that extrapolation to larger mountain landscapes could potentially lead to spurious results, a problem that is not unique to this method (Pierce and Bleich 2014; Rinehart et al. 2014).

From DNA to Demography

Once DNA is acquired, there are several methods that can be used to derive population-level assessments of ungulates and carnivores. Many of the population estimators are based on variations of mark-recapture techniques (White and Burnham 1999), where a representative sample of the target population is "marked" genetically and then is resampled either through another sampling effort or through samples obtained from harvested animals (Russell et al. 2012). Similar methodologies are used with camera-trap surveys through a variety of modeling programs that rely on individual identification and resampling to derive density in a given area (Karanth and Nichols 1998; Silver et al. 2004). Appropriate sampling schemes and study designs are critically important to accurately convey what is occurring on the ground and its application to management strategies (Mills 2007; Schwartz and McKelvey 2009). Use of systematic surveys allows for an accurate representation of the area of interest (Pierce and Bleich 2014). To develop accurate population estimates using mark-recapture techniques, managers must determine the detection probability of rare populations or those populations that occur at lower densities on the landscape (White and Burnham 1999).

Aerial Methods

FIXED-WING AIRCRAFT

Despite increased dangers associated with aerial monitoring (Bleich 1983; Sasse 2003), techniques involving the use of manned aircraft largely have superseded ground-based methods. In part, this has been the result of the inefficiency of data collection combined with the need to obtain adequate sample sizes from which to draw meaningful inferences. These factors, when combined with the remoteness of the geographic areas occupied by many species of large mammals or the ruggedness of the terrain in which those ungulates occur, dictate that increased efficiency is an important factor in data acquisition. Norton-Griffiths (1978) described and discussed the basics of acquiring demographic data using aerial methods.

Aerial techniques are used extensively to obtain demographic information for use in assessing population trends, estimating population size, determining population sex or age structure, and, ultimately, developing harvest objectives. Early application of aircraft to the census of large mammals occurred in Africa, and fixed-wing aircraft now have been used for decades in North America to obtain demographic information for pronghorn (Johnson et al. 1991; Pojar 2004), elk (Buechner et al. 1951), moose (Timmerman 1993), caribou (Bergerud 1963; Siniff and Skoog 1964), and musk ox (Gunn and Adamczewski 2003). Despite the rugged terrain in which they occur, biologists also use fixed-wing aircraft to assess some populations of bighorn sheep and mountain goat (Johnson 1983; McDonald et al. 1990). Krausman and Hervert (1983) and Krausman et al. (1986) examined the responses by mule deer and bighorn sheep, respectively, to fixed-wing aircraft flying at low elevations. Mule deer responded much less severely than did bighorn sheep, with few individuals changing habitats, compared with more than 41 percent of bighorn sheep that did so in a similar investigation. Individuals using aerial techniques must be mindful of the responses of wildlife to the disturbance associated with the technique (Bleich et al. 1990, 1994).

Regardless of the generally reclusive nature and elusiveness of some large carnivores, aerial surveys have application in monitoring species such as grizzly bears in the more open alpine habitats of the Greater Yellowstone Ecosystem (O'Brien and Lindzey 1998) and brown bears in Alaska and Canada, with potential application in mountain lion snow-track surveys during winter (Van Sickle and Lindzey 1991). Wolves are readily visible from the air during the winter months in western North America, and aerial monitoring of wolves or their tracks can be used effectively to estimate annual populations depending on population size and terrain, both of which are important variables (WGFD et al. 2014).

ROTARY-WING AIRCRAFT

Jurisdictions queried by Rabe et al. (2002) indicated that none used an exclusive method to conduct surveys of any one species. During recent times, however, the use of helicopters for evaluating populations of large mammals has become widespread. Many jurisdictions currently use helicopter surveys to determine the demographics of mule deer, moose, elk, and, in particular, mountain goat, bighorn sheep, and Dall's sheep (MSRM 2002; Rabe et al. 2002). Reactions of bighorn sheep, Dall's sheep, and mountain goats to helicopters appear to be especially strong (i.e., changes in behavior, flight, or distribution; Bleich et al. 1990, 1994; Frid 2000a, 2003; Cote et al. 2013) when compared to the reactions of other North American artiodactyls.

Observers in helicopters have advantages over those in fixed-wing aircraft in that helicopters can be flown at slow speeds, can hover for extended periods to allow a more thorough assessment of animals being observed, and can get closer to those individuals of particular interest (Johnson 1983). Nevertheless, helicopters are apt to cause animals to move away from the source of disturbance (Bleich et al. 1990, 1994; Frid 2000b, 2003), thereby increasing the probability of an animal or animals not being seen by the observers, or of double-counting some individuals. Helicopters are not necessarily as efficient as other methods in all situations. For example, Weckerly and Kovacs (1998) determined that acquisition of demographic information by ground observers was superior to the use of a helicopter when working with elk in a heavily vegetated part of northwestern California.

Although there are advantages, there are also limitations to the use of rotary-winged aircraft. For example, for survey results to be comparable from year to year, it is essential that survey intensity be standardized (Wehausen and Bleich 2007), and variation in annual agency budgets frequently affects the ability to do so. Further, bighorn sheep respond strongly to helicopters during aerial surveys (Bleich et al. 1990, 1994; Frid 2000a, 2000b, 2003), and those responses can result in violations of assumptions inherent in mark-recapture sampling (Bleich et al. 1994). Bleich et al. (1990) described these responses as the "Bo-Peep Effect" because those ungulates moved long distances as—and possibly before—the aircraft entered the survey polygon; as a result, observers could not be certain that observations of individual animals were not replicated or missed as a result of the disturbance associated with the aircraft. Mountain goats also have been reported to respond strongly to disturbance from helicopters, perhaps even more so than bighorn sheep. Indeed, Cote (1996) reported that goats responded very strongly to helicopters and emphasized that managers must take those responses under consideration when planning aerial censuses.

Some investigators have attempted to develop methods using a combination of fixed-wing and helicopter surveys. In such situations, observers in a fixed-wing aircraft have "marked" groups of animals by location, and after exiting the survey area, a second group of observers in a helicopter located "marked" groups and unmarked groups that had not been recorded during the initial flight. The Bo-Peep Effect (Bleich et al. 1990), however, makes it nearly impossible to distinguish between previously marked (i.e., groups that were seen) and unmarked groups of animals, thereby violating one of the basic tenets of mark-recapture sampling and limiting the utility of the technique. Nevertheless, there are exceptions where animals that are uniquely identifiable can be incorporated into the survey design (McClintock et al. 2006).

An analogous technique involving two fixed-wing aircraft rather than helicopters has been used to successfully "mark" and "resight" tule elk (*C. e. nannodes*) in open environments (Bleich et al. 2001), and a method involving observations made by ground-based observers and aerial observers has been used successfully to monitor the status of a bighorn sheep population for several decades (Holl et al. 2004). This and similar methods are often referred to as a simultaneous double-count. Observers must perform independently of each other, but the technique can involve observers in the same aircraft (Graham and Bell 1989), in separate aircraft (Bleich et al. 2001), or on the ground and in an aircraft (Bodie et al. 1995; Holl et al. 2004).

THERMAL IMAGING

Mammals are homeothermic and, as such, generate heat that is released in the form of infrared radiation (IR). Sensors designed specifically to detect IR have been developed and used in the acquisition of demographic data through the technique of thermal imaging (Naugle et al. 1996; Wakeling et al. 1999; Haroldson et al. 2003; Drake et al. 2005; Potvin and Breton 2005), which has been applied in ground-based and aerial demographic research. Much of this work has involved the use of Forward Looking Infrared Radar (FLIR), which has been used primarily with aerial platforms, albeit with limited success, particularly as it relates to age or sex composition (Keegan et al. 2011). Various types of thermal imaging or the use of FLIR can be implemented with either helicopters or fixed-wing aircraft, and there has been some application using ground-based surveys. Collier et al. (2007) noted the inability of thermal imaging to distinguish between age and sex classes of white-tailed deer.

UNMANNED AERIAL VEHICLES

Unmanned aerial vehicles (UAVs, or drones) have the potential to assist biologists with the acquisition of demographic information (Horcher and Visser 2004; Jones et al. 2006; Watts et al. 2010). References to the practical application of this technology to wildlife science are beginning to appear in the professional literature on a frequent basis (Rango et al. 2006; Martin et al. 2012; Vermeulen et al. 2013). As technology improves in terms of operational capabilities, optical resolution, and battery life, it is likely that more and more demographic data will be obtained using this method.

This method, however, is not without controversy as to applicability or regulations of its use, and it will likely be subject to numerous regulations.

Radio Telemetry and Monitoring of Ungulates and Large Carnivores

One of the most common methods used to acquire information on large mammals involves the use of radio telemetry. Two very important books on the subject (White and Garrott 1990; Millspaugh and Marzluff 2001) provide insight into monitoring through telemetry. Additionally, much information on the use of aerial telemetry has been compiled by others (Fuller et al. 2005). Despite the financial burden of maintaining a representative sample of the population, the technique of capture and tracking of animals through very high frequency (VHF) and global positioning system (GPS) technology is widely used by managers for ungulates and large carnivores. In instances of federally threatened or endangered species or populations, a high level of precision to evaluate status is especially necessary, and actively tracking marked animals provides the fine-scale data necessary to understand overall demographics that likely could not be attained through less rigorous techniques. More importantly, data can be used to develop baseline information and then combined with some of the standard monitoring techniques described earlier (Johnson et al. 2010; Pierce and Bleich 2014) to move forward with more applicable and logistically feasible methods to monitor population status. Data provided from research and monitoring have provided the foundation for harvest management of all game species.

Synthesis and Integrated Population Data Management

There is no single "black box" technique that can be administered to evaluate populations of any species; more importantly, techniques that may be the most effective manner to efficiently and accurately monitor a population are variable by taxon and by region. Managers do not always have the luxury to accurately derive estimates of the total number of animals in a specific area, or to adequately determine the structure of a popula-

tion; therefore, they use a combination of these and other techniques to appropriately gather information needed to develop appropriate management strategies based on available data.

For many species a combination of aerial, ground survey, and harvest data is used to determine recruitment rates and trajectory of a population. Use of historic trend data allows managers to reflect on changes in techniques and results over several generations of populations, thereby facilitating adaptive management (Walters 1986) incorporating new methodologies to maintain science-based conservation strategies and harvest regimes. Empirical data on populations are used in multiple formats and modeling techniques to project demographic data, such as population size, population growth, survival rates, reproduction, and recruitment, and the baseline data required to understand population dynamics (i.e., birth, immigration, death, and immigration; Mills 2007).

Regardless of the method used to acquire demographic information, it is essential that the data be gathered in a manner that allows the information to be used in statistically robust analyses. Among the many methods used to acquire data, there are well-established techniques such as quadrat sampling (moose, Evans et al. 1966; Gasaway et al. 1986), strip sampling (mule deer, Keegan et al. 2011), transect sampling (mule deer, White et al. 1989; pronghorn, Guenzel 1997; Whittaker et al. 2003), sightability correction methods (elk, Samuel et al. 1987; Bleich et al. 2001; mule deer, Ackerman 1988; bighorn sheep, Bodie et al. 1995; moose, Anderson and Lindzey 1996), mark-resight (bighorn sheep, Neal et al. 1993), and distance sampling (mule deer, Koenen et al. 2002; Cobb 2014; white-tailed deer, Larue et al. 2007). These methods frequently are used to gather information for use in estimating total numbers or population density, but data on the age and sex structure of populations can be gathered simultaneously to the extent practical (Gasaway et al. 1986). For mountain lions and black bears the sex and age ratios of harvested animals are examined to assess population trajectory and evaluate harvest strategies. The application of these various techniques was assessed for mule deer, and Keegan et al. (2011) have provided a very thorough description and evaluation of each of these methods.

Volumes have been written on the application of survey results and other demographic data and their statistical treatment and interpretation. Sampling bias is a constant problem associated with the acquisition of demographic information. The species addressed in this chapter are all large, sexually dimorphic ruminants or large carnivores. The ungulates segregate by sex for much of the year (Bowyer 2004), and the potential for the acquisition of misleading information is enhanced if the timing of data acquisition does not account for this factor (Bleich et al. 1997; Rubin and Bleich 2005). Thus, it is essential that life history traits be considered when gathering demographic data and that any surveys conducted during periods of sexual segregation encompass large areas and include habitats used by males and females during that period to decrease the possibility of biased demographic data (Schaller and Junrang 1988).

The application and interpretation of demographic data and, in turn, their applicability to setting management goals or harvest regulations can be complex. Many models have been developed and are used to formulate meaningful inferences from data gathered using the methods described above. Very recent treatments include the material summarized by Pierce et al. (2012a), but such details are beyond the scope of this chapter. What is exceedingly clear is that accuracy and precision, as well as the ways that data can be affected by biases associated with differing methods of data acquisition, are important factors that students, researchers, and managers must consider. As a result, sampling design, discussed in detail by Garton et al. (2012), is an extremely important consideration.

Harvest Management: Two Case Histories

Regulated hunting continues to be the primary tool applied to the management of carnivores and ungulates in North America. For many sportsmen, hunting of black bears is a time-honored tradition that provides opportunity, food, and sport while affording for close ties to the natural world (Hurst et al. 2012). For others, the pursuit of large ungulates, especially wild sheep, is the ultimate hunting experience. Harvest evaluation generally equates to assessing the sex and age structure of harvested animals, and those data then provide

insight into what was occurring on the landscape for target populations and what was removed through hunting activities. Space limitations dictate that we restrict our discussion of harvest management. To do so, we discuss two species, black bears and bighorn sheep, both of which are of interest to hunters and the public in general; harvest programs for both of these species have generated considerable controversy. Although this discussion of harvest management centers on two taxa, the information presented is relevant in one way or another to the management and conservation of all species discussed in this chapter.

Black Bears

The low fecundity of ursids results in many agencies relying on estimates of abundance or density in relation to population trend data from harvest to drive management regimes (Miller 1990; Garshelis and Hristienko 2006). Currently all agencies and provinces that harvest black bear rely on some form of annual monitoring tool, with several using harvest data to evaluate population status. Data such as sex and age ratios of harvested animals, density of mortality, and others including hunter success and harvest rate are further used to evaluate efficacy of management strategies. Hristienko and McDonald (2007) suggested that objectives of black bear management should center around balancing the goals of maintaining viable black bear populations, safeguarding human safety and livelihood, and satisfying the needs of various stakeholders in a cost-effective manner. Management and harvest strategies also depend on local black bear population densities. For example, many areas in the northeastern United States are able to achieve the harvest they desire through a short-duration fall hunt season; the high density of bears and hunters allows for this type of scenario.

For black bears, the primary methods for hunting include the use of trained dogs, baiting, and—to a lesser extent—spot and stalk practices. Although not without controversy (Hristienko and McDonald 2007), the use of baits or dogs to bay or tree bears can allow for additional selectivity and protection of females with dependent young. Many agencies use spring hunting seasons as a strategy to take black bears (Hristienko and McDonald 2007; Hurst et al. 2012). Timing of spring hunting seasons is developed with sex-specific den emergence in mind, wherein females with young are generally the last cohort to leave dens and, therefore, are less susceptible to harvest depending on season dates. Spring hunting seasons are more effective at stabilizing or reducing populations when compared to fall-only hunting scenarios (Garshelis and Hristienko 2006; Hristienko and McDonald 2007). Hristienko and McDonald (2007) provide an encompassing review of trends and controversies associated with managing black bears into the twenty-first century. Because of the lower fecundity and potential for overharvest, many states critically evaluate the proportion of female black bears harvested within their respective jurisdictions and examine ages of harvested animals, in conjunction with other monitoring criteria, to evaluate success of harvest objectives.

To regulate population viability and take of black bears, agencies employ the use of quotas, mortality limits, timing of season, or season length to manage a population toward specific population objectives. As with ungulates, the use of general versus limited quota permits also may be applied to reach harvest objectives. Hunt areas are devised based on local habitat, topography, and bear densities. Management by hunt area allows managers to direct harvest to applicable areas on a localized level within larger bear management units; this technique is applicable in many western states with large tracts of public lands and low bear densities when compared to eastern North America. Efficacy of management strategies is determined using monitoring techniques such as mandatory checks of harvested bears to accurately portray specific sex and age data from harvested animals.

To further quantify success of harvest and management, many states use harvest surveys to evaluate the quality of the hunt, hunter success, days hunted, and additional relevant ancillary data. Systematic surveys allow managers to evaluate the intangibles from successful and nonsuccessful hunters by providing information on hunter effort, dates hunted, and wounding loss. In addition to hunter surveys, valuable data can be obtained from the general public and nonhunting community about wildlife and wildlife management to gauge sentiments toward regional wildlife activities

and population status (discussed in chap. 13). These data are not the quantitative data acquired from monitoring, but they allow managers to assess public attitudes of a wide range of stakeholders about multiple issues that impact wildlife management—an important component of wildlife conservation.

Many factors have the potential to affect black bear harvest regulations, such as comparative prey populations and potential for human conflicts (Hurley et al. 2011). Targeted control of offending individuals continues to be the most efficacious method for removing problem individuals associated with human conflicts (e.g., livestock depredations, human safety; Bodenchuck 2011). While increased harvest or removal of mountain lions and black bears may have site-specific benefits to localized prey, the long-term effects of increased harvest of carnivores to positively benefit prey populations are negligible (Hurley et al. 2011; Pierce et al. 2012b). More importantly, for any type of predator reduction or control program to have positive effects, managers must first determine whether predation is a factor limiting prey populations (Hurley et al. 2011; Pierce et al. 2012b).

Bighorn Sheep

Conservation successes over the past half century (Krausman and Bleich 2013; Hurley et al. 2016) have yielded increasing opportunities to harvest bighorn sheep, an iconic species of western North America. Bighorn sheep are the only ungulate managed almost exclusively for trophy hunting, although a few exceptions exist. Bighorn sheep occur across the West in 14 of the contiguous United States, each of which provides for the harvest of adult males (Wild Sheep Foundation Professional Biologist Meeting Attendees 2008b) on a near annual basis. A total of three of those jurisdictions, however, do provide very limited opportunities for the harvest of females (Wild Sheep Foundation Professional Biologist Meeting Attendees 2008a; NDOW 2016). Primary season structures include limited entry seasons for archery or rifle, and some jurisdictions include opportunities for hunting with primitive weapons (i.e., muzzle-loading firearms) or handguns (Larkins 2010).

Limited entry or quota strategies derive a total number of tags to be distributed to hunters based on the number of bighorn sheep to be harvested in an area determined by regional conservation objectives. Unlimited entry hunts are almost unheard of, but they have been in place in specific parts of Montana for many years (Alt 1998; MFWP 2010). In general, bighorn sheep harvest quotas are extremely restrictive. The extremely low harvest rate is consistent with the status of wild sheep as the quintessential trophy ungulate in North America (Monteith et al. 2013). Conservative harvest rates, combined with restrictions on minimum horn size or minimum age used to define legal animals, are designed to produce males with large horns (Wild Sheep Foundation Professional Biologist Meeting Attendees 2008b). The combined annual harvest of Rocky Mountain bighorn sheep and desert bighorn sheep consists almost exclusively of adult males, averaging among hunted populations across the western United States 2.5 males (range 1.3–3.5) per 100 individuals, which represents 7–12 percent of all males but only about 2.5 percent of entire populations (Wild Sheep Foundation Professional Biologist Meeting Attendees 2008b). Nevertheless, some states have adopted regulations allowing the take of "any" male bighorn sheep (NDOW 2016).

Montana has long been an exception to male-only harvest regulations (Alt 1998; MFWP 2010), but several other states have also incorporated female harvest into their management strategies (Larkins 2010; NDOW 2016). Harvest of female bighorn sheep can be a controversial topic and remains unacceptable to many people despite the substantial female harvest among numerous other North American ungulates whose populations continue to thrive (Monteith et al. 2013). Despite these demonstrated successes, harvest of female bighorn sheep is seldom employed as a population management tool (Wild Sheep Foundation Professional Biologist Meeting Attendees 2008a; Monteith et al. 2014), perhaps because management agencies or regulatory entities are reluctant to face outright unacceptance or the social animosity potentially associated with such a strategy (Monteith et al. 2013). Ewe harvests are restricted to 15 of 48 hunted herds in Montana and 7 of 51 hunted herds in Colorado (Wild Sheep Foundation Professional Biologist Meeting Attendees 2008a); Nevada now provides limited opportunities to harvest female bighorn sheep in 4 of 40 desert bighorn sheep hunt zones open to the harvest of male bighorn sheep and 1 of 12 hunt zones open to the harvest of "California" bighorn sheep (NDOW 2016).

Harvest of female bighorn sheep has been employed to increase hunter opportunity as well as to reduce population size to levels below carrying capacity, often in conjunction with translocation projects (Wild Sheep Foundation Professional Biologist Meeting Attendees 2008a). In addition to being a means by which population densities can be lowered, female harvest also is a strategy to be employed if disease concerns, logistics of capture operations, or lack of suitable release sites limit translocation options (Brewer et al. 2014). However, it is ironic that translocation of female bighorn sheep to establish populations in historical habitat is widely supported, but harvest of female bighorn sheep stirs such controversy. Bighorn sheep are large herbivores, populations of which are regulated by density-dependent processes. Thus, it is paradoxical that those specialized ungulates might not respond to female harvest in a manner similar to that demonstrated by numerous North American ruminants, or that they are somehow not subject to the effects of resource limitation and density-dependent feedbacks (Monteith et al. 2013). Female harvest, thus, can be an effective tool to increase per capita availability of nutrients by decreasing population density (Jorgenson et al. 1993; Monteith et al. 2014), and may well gain greater support for managing populations for production of trophy males as it grows in acceptance (Monteith et al. 2013).

With few exceptions, bighorn sheep hunting regulations are based on lotteries (Larkins 2010), with high application rates for the very few permits that generally are available. For example, in California the mean rate at which applicants are successful in drawing a sheep tag is 0.22 percent (± 0.053 SD) over the 29 years that bighorn sheep hunting has occurred in that state. In 2016, available data (Epps et al. 2003; CDFW 2016; CFGC 2016) indicate that 10,233 individuals applied for the 17 opportunities available, yielding a success rate of <0.17 percent. The odds of drawing a tag to harvest a female bighorn sheep in Colorado (~50%) or Montana (~70%) are many times greater than those of drawing a tag for a male, the odds of which in some cases are as great as 4,000:1 (Wild Sheep Foundation

Professional Biologist Meeting Attendees 2008a; Larkins 2010).

Despite the overall very limited harvest of male bighorn sheep, controversy regarding the effects of taking the largest, or fastest-growing, males is intense (Monteith et al. 2013). Much of the controversy has been centered on information obtained from a small, isolated population of bighorn sheep that had been subjected to virtually unlimited hunter opportunity for many years (Coltman et al. 2003). Those results became the subject of considerable criticism and concern over inadequate consideration of environmental effects on horn size (Heimer 2004; Festa-Bianchet et al. 2006; Coltman 2008; Traill et al. 2014). A subsequent reanalysis of the Ram Mountain data clearly indicated that severe and selective harvest was consistent with a genetically based reduction in horn size, but it also demonstrated that the effect was lessened with a change in harvest regulations that restricted a legal male to those having "full curl" horns (Pigeon et al. 2016). The role of genetics, selective harvest, and subsequent evolutionary consequences has been—and likely will continue to be—kept alive by the popular press. Additionally, other critics (e.g., Darimont et al. 2009; Simon 2016) likely will ensure that managers of big game populations will continue to contend with these issues well into the future.

An additional area of contention regarding the harvest of bighorn sheep and several other North American artiodactyls has been the availability of fund-raising tags sold specifically to generate monies for use in state wildlife management and conservation programs. Recreational hunters have paid large sums of money to purchase difficult-to-obtain but highly coveted permits (Erickson 1988; Whitfield 2003; Festa-Bianchet and Lee 2009; Festa-Bianchet 2012; Palazy et al. 2012). The majority of fund-raising tags are sold at auction to the highest bidder, and many millions of dollars have been raised to augment agency budgets for specific management programs. Auctions of permits to fund conservation programs have yielded millions of dollars, having produced more than $400,000 for a single bighorn sheep permit (Landers 2013), $390,000 for a mule deer permit (Prettyman 2015), $315,000 for a Dall's sheep permit (CBC 2008), and $385,000 for a Rocky Mountain elk permit (Wagner 2013). Bighorn sheep—and

Dall's sheep—likely represent the quintessential North American trophies, and harvest regulations for both species emphasize age or horn size as defining criteria for a legal animal throughout most of North America. Unfortunately, however, success often breeds controversy, and such has occurred with fund-raising permits even though the proceeds directly benefit wildlife conservation (Damm 2008; Festa-Bianchet 2012). This controversy results from the belief of some (e.g., Simon 2016) that these types of high-bidder licenses not only represent the total commercialization of wildlife but also severely limit hunting opportunities for those other than the extremely wealthy.

Regulatory Processes

Once biologists have acquired and analyzed demographic data, the resultant information is used to formulate regulations that authorize the species discussed in this chapter. Among animals listed as endangered or threatened, such as wolves and Sierra Nevada bighorn sheep, federal regulations generally supersede state regulations. In other cases, state listings have precluded harvest, as with desert bighorn sheep in New Mexico (Rominger et al. 2006). As the result of an effective conservation program, however, desert bighorn sheep have been removed from that state's list of endangered species (Goldstein and Rominger 2013). In other jurisdictions, regulations or legislation preclude the harvest of certain categories of game species. For example, California law preludes the harvest of immature male mule deer (i.e., "spike" bucks and spotted fawns) and mountain lions (Bleich and Pierce 2005) and limits the harvest of bighorn sheep to "mature rams" (Bleich 2006). In situations where certain activities or actions are banned by legislation, the regulatory process is moot. Nevertheless, the regulatory process is the normal means by which hunting seasons, open or closed areas, bag limits, and total allowable harvests are set.

The governors of many, if not most, states have the authority to appoint members to what is commonly referred to as the "fish and game commission." In general, the commissioners serve at the pleasure of the governor and are appointed for various periods of time. In many instances, commissioners come from a diversity of backgrounds and typically represent equally distributed political affiliations. State wildlife agencies provide recommendations to commissions based on demographic information obtained from the various species for which harvest is contemplated. Commissioners also receive recommendations from members of the public and consider input from other agencies and the general population. Upon consideration of input received, commissioners then establish regulations for the upcoming hunting seasons.

In some states hunting regulations are issued by proclamation. For example, biologists from the North Dakota Game and Fish Department acquire demographic information on various species, analyze those data, and make recommendations to the director of the department. The director can also receive input from the general public and from other interested parties, such as nongovernmental organizations, local sporting or conservation clubs, and seven different county advisory boards—the members of which are appointed by the governor. Upon considering input from professional staff and others, the director formulates recommendations that are forwarded to the office of the governor, who then issues a proclamation covering regulations for specific hunting seasons. Although regulatory processes vary from jurisdiction to jurisdiction, regulations are established and then enforced to ensure that wildlife populations are not compromised through sport harvest.

From Recovery to Management— State Responsibilities for Recovering Threatened and Endangered Species

Recovery of imperiled big game species or populations would not have occurred without the devotion of state agency wildlife management personnel and coordinated collaboration with multiple jurisdictions and public or private stakeholders. In recent years, successful recovery of species like wolves and black bears led to these animals being removed from federal protection and transferred to management by the various states. In other situations, bighorn sheep listed as endangered by the federal government have approached, and in some places have even far exceeded, numbers recorded prior to listing.

Unfortunately, the science and biology that are used as the foundation for any type of forward progress often get caught up in litigation and public emotion, both of which somewhat taint the success and coordination necessary to recover a species or population from the brink of extinction. The principles behind management and monitoring strategies for wolves are the same as they are for white-tailed deer, but the controversy is greatly elevated with respect to wolves. Perhaps more than any other game animal, wolves engender an inherent interest and controversy, where public input and consternation distract from the basic science used to effectively manage the species. Litigation and legal interpretations do very little to assist in long-term conservation efforts for that species and many others, further complicating efforts to ensure the viability of several taxa of large mammals.

Managing Conflict

The authors of chapter 11 provide an in-depth look into managing conflicts with wildlife, but the importance of conflict management, especially for large carnivores, warrants discussion. Maintaining healthy and abundant wildlife populations often leads to conflicts with humans. Maintaining human tolerance for wildlife is critically important for viable populations and continued implementation of the North American Model of Wildlife Conservation (chap. 2). This is particularly true for large carnivores, where conflicts impact or impede livelihoods (Thirgood et al. 2005) and may directly influence human safety (Quigley and Herrero 2005). Similarly, conflicts occur through depredation by ungulates on agricultural endeavors (Messmer 2009), or because of pathogens that could be contracted from domestic stock (WSWG 2012) and sometimes have residual impacts (Ryder et al. 1994). Additionally, there is the potential for wildlife to transmit pathogens to livestock (Scurlock and Edwards 2010).

Professionals that deal with wildlife conflicts realize that controversy is inherent in management, and many state agencies have protocols and guidelines in place to deal directly with human–wildlife and human–carnivore conflicts in the form of Wildlife Human Attack Response Teams. These trained personnel allow an immediate and efficient response to issues involving wildlife attacks on humans. Dealing with wildlife conflicts has a proactive and reactive approach, where information and education efforts are in place to reduce the likelihood of conflicts. When they do occur, however, managers must respond to reduce or eliminate the potential for such conflicts recurring. Public perspectives and opinion are omnipresent when dealing with wildlife conflicts, and maintaining the highest level of consistency and professionalism when dealing with all wildlife–human conflicts is essential.

Future Trends and Adaptive Changes in Management and Conservation

It is impossible to predict what the future holds in the world of wildlife management. The advent of the Internet and real-time updates through social media are likely to completely change the realm of wildlife management as science and society "move forward." Despite the need for robust biological information in order to manage wildlife, much of what is done in the wildlife field is called into question by the public on a daily basis. We are fortunate to work with something as a vocation that has inherent public interest, despite the controversy, and having information regarding wildlife conservation and management literally at everyone's fingertips increases the scrutiny on the profession. This situation further emphasizes the importance of integrity and ethics within the wildlife profession, as well as the need to adhere to the principles of the scientific method. Conversely, sharing information with the larger public as to how wildlife populations are managed fosters public ownership and augments understanding of the role that wildlife agencies have in developing data-driven management scenarios to maintain healthy, robust populations of all wildlife for future generations.

We talk at great lengths of carrying capacity for wildlife, both biological and cultural (Caughley 1977; Vandermeer and Goldberg 2003; Mills 2007), but as humans increase in abundance, habitats for wildlife are being evermore fragmented or diminished. Suitable habitats are the bedrock of all wildlife populations, and as areas suitable for wildlife decrease, so does the amount of wildlife those habitats can sustain. The interlacing life histories of predator and prey populations are built on thousands of years of evolution

(Caughley 1977). We currently are seeing instances of interactions and impacts between intact carnivore and prey guilds not observed for hundreds of years in some areas of North America, as populations adapt to natural changes on the landscape and the ways that human perturbations influence wildlife as a function of the life history traits of each particular species (e.g., whether the species are generalists or specialists—or predators or prey).

As wildlife evolve and adapt, it is critically important that wildlife managers adapt and remain on the cutting edge of technology and monitoring techniques. Doing so will not only help maintain wildlife populations, as well as the land ethic and hunting heritage that allow humans to relate to their place in the system, but also help maintain respect for wildlife and its role on the landscape. Human population growth, limited funding, changing climates, and other challenges will continue to force managers to reevaluate and alter management strategies for big game; with the triad of biological understanding of populations, knowledge of the importance of habitat, and human considerations, however, large, wild mammals in North America have the potential to continue to thrive (Krausman and Bleich 2013).

To ensure that potential in perpetuity, future biologists and managers must have effective modeling skills, a firm understanding of statistics and their application to management, a willingness to implement the concept of adaptive management (Walters 1986), and an understanding of structured decision-making (Martin et al. 2009). Nevertheless, Peek (1989:364) emphasized that "courses in basic biology and ecology should not be eschewed for courses in applications, which logically build from basics." Indeed, it is essential that those responsible for the management and conservation of large mammals must also have a firm knowledge of the life history characteristics of the species for which they are responsible, as well as a firm foundation in evolutionary biology (Bleich and Oehler 2000).

Courses in natural history and evolutionary biology have become increasingly less common in the curricula offered by many universities (Bleich and Oehler 2000), and some recently have again called attention to the relevance of those disciplines to the future of wildlife conservation (Bleich 2014; Hutchins et al.

2014). The concept of evolution is common to all aspects of science related to living resources, and as such, evolutionary theory provides a common link between those interested in or responsible for conservation and habitat management (Bleich and Oehler 2000). "Evolution is our employer" (Murie 1954), but, as noted more than 25 years ago by Gavin (1989), most wildlife biologists lack adequate training in evolutionary biology. An understanding of why populations perform the way they do is requisite to making management decisions that affect more than proximate situations, a concept that has its very foundations in evolutionary biology and underlies the many challenges with which wildlife managers will be faced in the future.

LITERATURE CITED

Ackerman, B. B. 1988. Visibility bias of mule deer census procedures in southeast Idaho. PhD diss., University of Idaho, Moscow, Idaho, USA.

Alt, K. L. 1998. To be limited or not, that is the question. Biennial Symposium of the Northern Wild Sheep and Goat Council Proceedings 11:160.

Anderson, C. R., and F. G. Lindzey. 1996. Moose sightability model developed from helicopter surveys. Wildlife Society Bulletin 24:247–259.

Bergerud, A. T. 1963. Aerial winter census of caribou. Journal of Wildlife Management 27:438–449.

Beringer, J., L. P. Hansen, and O. Sexton. 1998. Detection rates of white-tailed deer with helicopter over snow. Wildlife Society Bulletin 26:24–28.

Bleich, V. C. 1983. Comments on helicopter use by wildlife agencies. Wildlife Society Bulletin 11:304–306.

———. 2006. Mountain sheep in California: Perspectives on the past, and prospects for the future. Biennial Symposium of the Northern Wild Sheep and Goat Council Proceedings 15:1–13.

———. 2014. The relevance of evolutionary biology. Wildlife Professional 8(3):8–9.

Bleich, V. C., R. T. Bowyer, A. M. Pauli, M. C. Nicholson, and R. W. Anthes. 1994. Responses of mountain sheep Ovis canadensis to helicopter surveys: Ramifications for the conservation of large mammals. Biological Conservation 45:1–7.

Bleich, V. C., R. T. Bowyer, A. M. Pauli, R. L. Vernoy, and R. W. Anthes. 1990. Responses of mountain sheep to aerial sampling using helicopters. California Fish and Game 76:197–204.

Bleich, V. C., R. T. Bowyer, and J. D. Wehausen. 1997. Sexual segregation in mountain sheep: Resources or predation? Wildlife Monographs 134:1–50.

Bleich, V. C., C. S. Y. Chun, R. W. Anthes, T. E. Evans, and J. K. Fischer. 2001. Visibility bias and development of a sightability model for tule elk. Alces 37:315–327.

Bleich, V. C., and M. W. Oehler. 2000. Wildlife education in the United States: Thoughts from agency biologists. Wildlife Society Bulletin 28:542–545.

Bleich, V. C., and B. M. Pierce. 2005. Management of mountain lions in California. Pages 63–69 in E. L. Buckner and J. Reneau, editors. Records of North American big game, 12th edition. Boone and Crockett Club, Missoula, Montana, USA.

Bodenchuck, M. J. 2011. Population management: Depredation. Pages 135–143 in J. A. Jenks, editor. Managing cougars in North America. Jack H. Berryman Institute, Utah State University, Logan, Utah, USA.

Bodie, W. L., E. O. Garton, E. R. Taylor, and M. McCoy. 1995. A sightability model for bighorn sheep in canyon habitats. Journal of Wildlife Management 59:832–840.

Bowyer, R. T. 2004. Sexual segregation in ruminants: Definitions, hypotheses, and implications for conservation and management. Journal of Mammalogy 85:1039–1052.

Brewer, C. E., V. C. Bleich, J. A. Foster, T. Hosch-Hebdon, D. E. McWhirter, E. M. Rominger, M. W. Wagner, and B. P. Wiedmann. 2014. Bighorn sheep: Conservation challenges and management strategies for the 21st century. Western Association of Fish and Wildlife Agencies, Cheyenne, Wyoming, USA.

Brinkman, T. J., and K. J. Hundertmark. 2009. Sex identification of northern ungulates using low quality and quantity DNA. Conservation Genetics 10:1189–1193.

Brinkman, T. J., D. K. Person, F. Stuart Chapin, III, W. Smith, and K. Hundertmark. 2011. Estimating abundance of Sitka black-tailed deer using DNA from fecal pellets. Journal of Wildlife Management 75:232–242.

Brinkman, T. J., D. K. Person, W. Smith, F. Stuart Chapin, III, K. McCoy, M. Leonawicz, and K. Hundertmark. 2013. Using DNA to test the utility of pellet-group counts as an index of deer counts. Wildlife Society Bulletin 37:444–450.

Buckland, S. T., D. R. Anderson, K. P. Burnham, J. L. Laake, D. L. Borchers, and L. Thomas. 2001. Introduction to distance sampling: Estimating abundance of biological populations. Oxford University Press, Oxford, UK.

———. 2004. Advanced distance sampling. Oxford University Press, Oxford, UK.

Buechner, H. K., I. O. Buss, and H. F. Bryan. 1951. Censusing elk by airplane in the Blue Mountains of Washington. Journal of Wildlife Management 15:81–87.

Caughley, G. 1977. Analysis of vertebrate populations. John Wiley and Sons, New York, New York, USA.

CBC (Canadian Broadcasting Corporation). 2008. Winner pays $315K to hunt Kluane trophy sheep. Canadian Broadcasting Corporation, Toronto, Ontario, Canada. www.cbc.ca/news/canada/north/winner-pays-315k-to-hunt-kluane-trophy-sheep-1.771330.

CDFW (California Department of Fish and Wildlife). 2016. Big game hunting drawing statistics. California Department of Fish and Wildlife, Sacramento, USA. www.wildlife.ca.gov /Licensing/Statistics/Big-Game-Drawing.

CFGC (California Fish and Game Commission). 2016. Mammal hunting regulations 2016–2017. California Fish and Game Commission, Sacramento, USA. www.fgc.ca.gov/regulations /current/mammalregs.aspx.

Cobb, M. A. 2014. Using mark-recapture distance sampling to estimate Sitka blacktailed deer densities in non-forested habitats of Kodiak Island, Alaska. Refuge Report 2014.5. US Fish and Wildlife Service, Kodiak National Wildlife Refuge, Kodiak, Alaska, USA.

Collier, B. A., S. S. Ditchkoff, J. B. Raglin, and J. M. Smith. 2007. Detection probability and sources of variation in white-tailed deer spotlight surveys. Journal of Wildlife Management 71:277–281.

Coltman, D. W. 2008. Molecular ecological approaches to studying the evolutionary impact of selective harvesting in wildlife. Molecular Ecology 17:221–235.

Coltman, D. W., P. O'Donoghue, J. T. Jorgenson, J. T. Hogg, C. Strobeck, and M. Festa-Bianchet. 2003. Undesirable evolutionary consequences of trophy hunting. Nature 426:655–658.

Cote, S. D. 1996. Mountain goat responses to helicopter disturbance. Wildlife Society Bulletin 24:681–685.

Cote, S. D., S. Hamel, A. St-Louis, and J. Maknguy. 2013. Do mountain goats habituate to helicopter disturbance? Journal of Wildlife Management 77:1244–1248.

Damm, G. R. 2008. Recreational trophy hunting: "What do we know and what should we do?" Pages 5–11 in R. D. Baldus, G. R. Damm, and K. Wollscheid, editors. Best practices in sustainable hunting—a guide to best practices from around the world. CIC—International Council for Game and Wildlife Conservation, Budakeszi, Hungary.

Darimont, C. T., S. M. Carlson, M. T. Kinnison, P. C. Paquet, T. E. Reimchen, and C. C. Wilmers. 2009. Human predators outpace other agents of trait change in the wild. Proceedings of the National Academy of Sciences 106:952–954.

Davidson, G. A., D. A. Clark, B. K. Johnson, L. P. Waits, and J. R. Adams. 2014. Estimating cougar densities in northeast Oregon using conservation detection dogs. Journal of Wildlife Management 78:1104–1114.

Drake, D., C. Aquila, and G. Huntington. 2005. Counting a suburban deer population using Forward-Looking Infrared radar and road counts. Wildlife Society Bulletin 33:656–661.

Epps, C. W., V. C. Bleich, J. D. Wehausen, and S. G. Torres. 2003. Status of bighorn sheep in California. Desert Bighorn Council Transactions 47:20–35.

Erickson, G. L. 1988. Permit auction: The good, the bad and the ugly. Biennial Symposium of the Northern Wild Sheep and Goat Council Proceedings 6:47–53.

Evans, C. D., W. A. Troyer, and C. J. Lensink. 1966. Aerial census of moose by quadrat sampling units. Journal of Wildlife Management 30:767–776.

Fafarman, K. R., and C. A. DeYoung. 1986. Evaluation of spotlight counts of deer in South Texas. Wildlife Society Bulletin 14:180–185.

Festa-Bianchet, M. 2012. Rarity, willingness to pay and conservation. Animal Conservation 15:12–13.

Festa-Bianchet, M., J. T. Jorgenson, D. W. Coltman, and J. T. Hogg. 2006. Feared negative effects of publishing data: A rejoinder to Heimer et al. Biennial Symposium of the Northern Wild Sheep and Goat Council Proceedings 15:213–219.

Festa-Bianchet, M., and R. Lee. 2009. Guns, sheep, and genes: When and why trophy hunting may be a selective pressure. Pages 94–107 in B. Dickson, J. Hutton, and W. M. Adams, editors. Recreational hunting, conservation and rural livelihoods: Science and practice. Wiley-Blackwell, Oxford, UK.

Fifield, V. L., A. J. Rossi, and E. E. Boydston. 2015. Documentation of mountain lions in Marin County, California, 2010–2013. California Fish and Game 101:66–71.

Frid, A. 2000a. Behavioral responses by Dall's sheep to overflights by fixed-wing aircraft. Biennial Symposium of the Northern Wild Sheep and Goat Council Proceedings 12:153–169.

———. 2000b. Fleeing decisions by Dall's sheep exposed to helicopter overflights. Biennial Symposium of the Northern Wild Sheep and Goat Council Proceedings 12:153–169.

———. 2003. Dall's sheep responses to overflights by helicopter and fixed-wing aircraft. Biological Conservation 110:387–399.

Fuller, M. R., J. J. Millspaugh, K. E. Church, and R. E. Kenward. 2005. Wildlife radiotelemetry. Pages 377–417 in C. E. Braun, editor. Techniques for wildlife investigation and management, 6th edition. The Wildlife Society, Bethesda, Maryland, USA.

Garshelis, D., and H. Hristienko. 2006. State and provincial estimates of American black bear numbers versus assessments of population trend. Ursus 17:1–7.

Garton, E. O., J. S. Horne, J. L. Aycrigg, and J. T. Ratti. 2012. Research and experimental design. Pages 1–40 in N. J. Silvy, editor. The wildlife techniques manual, 7th edition. Vol. 2. Johns Hopkins University Press, Baltimore, Maryland, USA.

Gasaway, W. C., S. D. DuBois, D. J. Reed, and S. J. Harbo. 1986. Estimating moose population parameters from aerial surveys. Biological Papers 22. University of Alaska, Fairbanks, Alaska, USA.

Gavin, T. A. 1989. What's wrong with the questions we ask in wildlife research? Wildlife Society Bulletin 17:345–350.

Goldstein, E. J., and E. M. Rominger. 2013. Status of desert bighorn sheep in New Mexico, 2011–2012. Desert Bighorn Council Transactions 52:32–34.

Graham, A., and R. Bell. 1989. Investigating observer bias in aerial survey by simultaneous double-counts. Journal of Wildlife Management 53:1009–1016.

Guenzel, R. J. 1997. Estimating pronghorn abundance using aerial line transect sampling. Wyoming Game and Fish Department, Cheyenne, Wyoming, USA.

Gunn, A., and J. Adamczewski. 2003. Muskox. Pages 1076–1094 in G. Feldhamer, B. C. Thompson, and J. A. Chapman, editors. Wild mammals of North America, 2nd edition. Johns Hopkins University Press, Baltimore, Maryland, USA.

Haroldson, B. S., E. P. Wiggers, J. Beringer, L. P. Hansen, and J. B. McAninch. 2003. Evaluation of aerial thermal imaging for detecting white-tailed deer in a deciduous forest environment. Wildlife Society Bulletin 311188-1197.

Harwell, F., R. L. Cook, and J. C. Barron. 1979. Spotlight count method for surveying white-tailed deer in Texas. Texas Parks and Wildlife Department, Austin, Texas, USA.

Heffelfinger, J. 2006. Deer of the southwest. Texas A&M University Press, College Station, Texas, USA.

Heimer, W. E. 2004. Inferred negative effect of "trophy hunting" in Alberta: The great Ram Mountain / Nature controversy. Biennial Symposium of the Northern Wild Sheep and Goat Council Proceedings 14:193–210.

Holl, S. A., V. C. Bleich, and S. G. Torres. 2004. Population dynamics of bighorn sheep in the San Gabriel Mountains, California, 1967–2002. Wildlife Society Bulletin 32:412–426.

Horcher, A., and R. J. M. Visser. 2004. Unmanned aerial vehicles: Applications for natural resource management and monitoring. Proceedings of the 2004 Council on Forest Engineering Annual Meeting. Council on Forest Engineering, Morgantown, West Virginia, USA.

Hristienko, H., and J. E. McDonald, Jr. 2007. Going into the 21st century: A perspective on trends and controversies in the management of the American black bear. Ursus 18:72–88.

Hughson, D. L., N. W. Darby, and J. D. Dungan. 2010. Comparison of motion-activated cameras for wildlife investigations. California Fish and Game 96:101–109.

Hurley, K., C. Brewer, and G. Thornton. 2016. The role of hunters in conservation, restoration, and management of North American wild sheep. International Journal of Environmental Studies 72:784–796.

Hurley, M. A., J. W. Unsworth, P. Zager, M. Hebblewhite, E. O. Garton, D. M. Montgomery, J. R. Skalski, and C. L. Maycock. 2011. Demographic response of mule deer to experimental reduction of coyotes and mountain lions in southeastern Idaho. Wildlife Monographs 178:1–33.

Hurst, J. E., C. W. Ryan, C. P. Carpenter, and L. Sajecki, editors. 2012. An evaluation of black bear management options. Northeast Black Bear Technical Committee. www.state.nj.us/dep/fgw/pdf/bear/mgtoptionseval_nebbtc.pdf.

Hutchins, M., V. Geist, J. Organ, and H. Salwasser. 2014. Evolutionary biology: Applications for wildlife management and conservation. Wildlife Professional 8(2):54–58.

Jacobson, H. A., C. Kroll, R. W. Browning, B. H. Koerth, and M. H. Conway. 1997. Infrared triggered cameras for censusing white-tailed deer. Wildlife Society Bulletin 25:547–556.

Jaeger, J. R., J. D. Wehausen, and V. C. Bleich. 1991. Evaluation of time-lapse photography to estimate population parameters. Desert Bighorn Council Transactions 35:5–8.

Johnson, B. K., F. G. Lindzey, and R. Guenzel. 1991. Use of aerial line transect surveys to estimate pronghorn populations in Wyoming. Wildlife Society Bulletin 19:315–321.

Johnson, H. E., L. S. Mills, J. D. Wehausen, and T. R. Stephenson. 2010. Combining ground count, telemetry, and mark-resight data to infer population dynamics in an endangered species. Journal of Applied Ecology 47:1083–1093.

Johnson, R. L. 1983. Mountain goats and mountain sheep of Washington. Biological Bulletin 181–196. Washington State Game Department, Olympia, Washington, USA.

Jones, G. P., IV, L. G. Pearlstine, and H. F. Percival. 2006. An assessment of small unmanned aerial vehicles for wildlife research. Wildlife Society Bulletin 34:750–758.

Jorgenson, J. T., M. Festa-Bianchet, and W. D. Wishart. 1993. Harvesting bighorn ewes: Consequences for population size and trophy ram production. Journal of Wildlife Management 57:429–435.

Karanth, K. U., and J. D. Nichols. 1998. Estimation of tiger densities in India using photographic captures and recaptures. Ecology 79:2852–2862.

Keegan, T. W., B. B. Ackerman, A. N. Aoude, L. C. Bender, T. Boudreau, L. H. Carpenter, B. B. Compton, M. Elmer, J. R. Heffelfinger, D. W. Lutz, B. D. Trindle, B. F. Wakeling, and B. E. Watkins. 2011. Methods for monitoring mule deer populations. Western Association of Fish and Wildlife Agencies, Cheyenne, Wyoming, USA.

Kelly, M. J., A. J. Noss, M. S. DiBitetti, L. Maffei, R. L. Arispe, A. Paviolo, C. D. De Angelo, and Y. E. Di Blanco. 2008. Estimating puma densities from camera trapping across three study sites: Bolivia, Argentina, and Belize. Journal of Mammalogy 89:408–418.

Kendall, K. C., J. B. Stetz, D. A. Roon, L. P. Waits, J. B. Boulanger, and D. Paetkau. 2008. Grizzly bear density in Glacier National Park, Montana. Journal of Wildlife Management 72:1693–1705.

Koenen, K. G., S. DeStafano, and P. R. Krausman. 2002. Using distance sampling to estimate seasonal densities of desert mule deer in semidesert grassland. Wildlife Society Bulletin 30:53–63.

Krausman, P. R., and V. C. Bleich. 2013. Conservation and management of ungulates in North America. International Journal of Environmental Studies 70:372–382.

Krausman, P. R., and J. Hervert. 1983. Mountain sheep responses to aerial surveys. Wildlife Society Bulletin 11:372–375.

Krausman, P. R., B. D. Leopold, and D. L. Scarborough. 1986. Desert mule deer responses to aircraft. Wildlife Society Bulletin 14:68–70.

Landers, R. 2013. Record $480k bid for Montana bighorn tag. Spokesman-Review, Spokane, Washington, USA. www.spokesman.com/blogs/outdoors/2013/feb/15/record-480k-bid-montana-bighorn-tag/.

Larkins, A. 2010. Wild sheep status and management in western North America: Summary of state, province, and territory status report surveys. Biennial Symposium of the Northern Wild Sheep and Goat Council Proceedings 17: 8–28.

Larrucea, E. S., P. F. Brussard, M. N. Jaeger, and R. H. Barrett. 2007. Cameras, coyotes, and the assumption of equal detectability. Journal of Wildlife Management 71:1682–1689.

Larue, M. A., C. K. Nielsen, and M. D. Grund. 2007. Using distance sampling to estimate densities of white-tailed deer in south-central Minnesota. Prairie Naturalist 39:57–68.

Leopold, A. 1933. Game management. Charles Scribner's Sons, New York, New York, USA.

Locke, S. L., I. D. Parker, and R. R. Lopez. 2012. Use of remote cameras in wildlife ecology. Pages 311–318 in N. J. Silvy, editor. The wildlife techniques manual, 7th edition. Vol. 1. Johns Hopkins University Press, Baltimore, Maryland, USA.

Marshal, J. P., L. M. Lesicka, V. C. Bleich, P. R. Krausman, G. P. Mulcahy, and N. G. Andrew. 2006. Demography of desert mule deer in southeastern California. California Fish and Game 92:55–66.

Martin, J., H. H. Edwards, M. A. Burgess, H. F. Percival, and D. E. Fagan. 2012. Estimating distribution of hidden objects with drones: From tennis balls to manatees. PLoS One 7.6:e38882.

Martin, J., M. C. Runge, J. D. Nichols, B. C. Lubow, and W. L. Kendall. 2009. Structured decision making as a conceptual framework to identify thresholds for conservation and management. Ecological Applications 19:1079–1090.

Mason, R., L. H. Carpenter, M. Cox, J. C. deVos, J. Fairchild, D. J. Freddy, J. R. Heffelfinger, R. H. Kahn, S. M. McCorquodale, D. F. Pac, D. Summers, G. C. White, and B. K. Williams. 2006. A case for standardized ungulate surveys and data management in the western United States. Wildlife Society Bulletin 34:1238–1242.

McClintock, B. T., G. C. White, and K. P. Burnham. 2006. A robust design mark-resight abundance estimator allowing heterogeneity in resighting probabilities. Journal of Agricultural, Biological, and Environmental Statistics 11:231–248.

McCullough, D. R. 1979. The George Reserve deer herd: Population ecology of a K-selected species. University of Michigan Press, Ann Arbor, Michigan, USA.

———. 1984. Lessons from the George Reserve, Michigan. Pages 211–242 in L. K. Halls, R. E. McCabe, and L. R. Jahn, editors. White-tailed deer: Ecology and management. Stackpole Books, Harrisburg, Pennsylvania, USA.

McDonald, L. L., H. B. Harvey, F. J. Mauer, and A. W. Brackney. 1990. Design of aerial surveys for Dall sheep in the Arctic National Wildlife Refuge, Alaska. Biennial Symposium of the Northern Wild Sheep and Goat Council Proceedings 7:176–193.

Messmer, T. A. 2009. Human–wildlife conflicts: Emerging challenges and opportunities. Human–Wildlife Conflicts 3:10–17.

MFWP (Montana Fish Wildlife and Parks). 2010. Montana bighorn sheep conservation strategy. Wildlife Division, Montana Fish Wildlife and Parks, Helena, USA.

Miller, S. D. 1990. Population management of bears in North America. International Conference on Bear Research and Management 8:357–373.

Mills, L. S. 2007. Conservation of wildlife populations: Demography, genetics, and management. Blackwell, Malden, Massachusetts, USA.

Millspaugh, J. J., and J. M. Marzluff. 2001. Radio tracking and animal populations. Academic Press, San Diego, California, USA.

Monteith, K. L., V. C. Bleich, T. R. Stephenson, B. M. Pierce, M. M. Conner, J. G. Kie, and R. T. Bowyer. 2014. Life-history characteristics of mule deer: Effects of nutrition in a variable environment. Wildlife Monographs 186:1–62.

Monteith, K. L., R. A. Long, V. C. Bleich, J. R. Heffelfinger, P. R. Krausman, and R. T. Bowyer. 2013. Effects of harvest, culture, and climate on trends in size of horn-like structures in trophy ungulates. Wildlife Monographs 183:1–28.

MSRM (Ministry of Sustainable Resource Management). 2002. Aerial-based inventory methods for selected ungulates: Bison, mountain goat, mountain sheep, moose, elk, deer and caribou. Standards for Components of British Columbia's Biodiversity 32. Ministry of Sustainable Resource Management, Victoria, British Columbia, Canada.

Murie, O. J. 1954. Ethics in wildlife management. Journal of Wildlife Management 18:289–293.

Naugle, D. E., J. A. Jenks, and B. J. Kerhohan. 1996. The use of thermal infrared sensing to estimate density of white-tailed deer. Wildlife Society Bulletin 24:37–43.

NDOW (Nevada Department of Wildlife). 2016. Big game seasons and applications. J. F. Griffin, Williamstown, Massachusetts, USA. www.eregulations.com/nevada/big-game/.

Neal, A. K., G. C. White, R. B. Gill, D. F. Reed, and J. H. Olterman. 1993. Evaluation of mark-resight model assumptions for estimating mountain sheep numbers. Journal of Wildlife Management 57:436–450.

Neasham, V. A. 1973. Wild legacy. Howell-North Books, Berkeley, California, USA.

Norton-Griffiths, M. 1978. Counting animals, 2nd edition. African Wildlife Leadership Foundation, Nairobi, Kenya.

O'Brien, S. L., and F. G. Lindzey. 1998. Aerial sightability and classification of grizzly bears at moth aggregation sites in the Absaroka Mountains, Wyoming. Ursus 10:427–435.

O'Connell, A. F., J. D. Nichols, and K. U. Karanth, editors. 2010. Camera traps in animal ecology: Methods and analyses. Springer Science and Business Media, New York, New York, USA.

Organ, J. F., D. J. Decker, S. J. Riley, J. E. McDonald, Jr., and S. P. Mahoney. 2012. Adaptive management in wildlife conservation. Pages 43–54 in N. J. Silvy, editor. The wildlife techniques manual, 7th edition. Vol. 2. Johns Hopkins University Press, Baltimore, Maryland, USA.

PADEMP (Peace-Athabasca Delta Ecological Monitoring Program). 2015. Wood bison monitoring. http://pademp.com/research-and-monitoring/research-and-monitoringwood-bison-monitoring-%e2%80%8e/.

Palazy, L., C. Bonenfant, J. M. Gaillard, and F. Courchamp. 2012. Rarity, trophy hunting and ungulates. Animal Conservation 15:4–11.

Peek, J. M. 1989. A look at wildlife education in the United States. Wildlife Society Bulletin 17:361–365.

Perry, T. W., T. Newman, and K. M. Thibault. 2010. Evaluation of methods used to estimate size of a population of desert bighorn sheep (*Ovis canadensis mexicana*) in New Mexico. Southwestern Naturalist 55:517–524.

Pierce, B. L., R. R. Lopez, and N. J. Silvy. 2012a. Estimating animal abundance. Pages 284–310 in N. J. Silvy, editor. The wildlife techniques manual, 7th edition. Vol. 1. Johns Hopkins University Press, Baltimore, Maryland, USA.

Pierce, B. M., and V. C. Bleich. 2014. Enumerating mountain lions: A comparison of two indices. California Fish and Game 100:527–537.

Pierce, B. M., V. C. Bleich, K. L. Monteith, and R. T. Bowyer. 2012b. Top-down versus bottom-up forcing: Evidence from mountain lions and mule deer. Journal of Mammalogy 93:977–988.

Pigeon, G., M. Festa-Bianchet, D. W. Coltman, and F. Pelletier. 2016. Intense selective hunting leads to artificial evolution in horn size. Evolutionary Applications 9:521–530.

Pojar, T. M. 2004. Survey methods to estimate population. Pages 631–644 in B. W. O'Gara and J. D. Yoakum, editors. Pronghorn ecology and management. University Press of Colorado, Boulder, Colorado, USA.

Potvin, F., and L. Breton. 2005. From the field: Testing 2 aerial survey techniques on deer in fenced enclosures—visual double-counts and thermal infrared sensing. Wildlife Society Bulletin 33:317–325.

Prettyman, B. 2015. Antelope Island trophy mule deer tag draws world record bid of $390,000. Where does all that money go? Salt Lake Tribune, Salt Lake City, Utah, USA. www.sltrib.com/home/2212820-155/antelope-island-trophy-mule-deer-tag.

Quigley, H., and S. Herrero. 2005. Characterization and prevention of attacks on humans. Pages 27–48 in R. Woodroffe, S. Thirgood, and A. Rabinowitz, editors. People and wildlife: Conflict or coexistence? Cambridge University Press, New York, New York, USA.

Rabe, M. J., S. Rosenstock, and J. C. deVos, Jr. 2002. Game survey methods used by western state wildlife agencies: A review. Wildlife Society Bulletin 30:46–52.

Rango, A., A. Laliberte, C. Steele, J. E. Herrick, B. Bestelmeyer, T. Schmugge, A. Roanhorse, and V. Jenkins. 2006. Using unmanned aerial vehicles for rangelands: Current applications and future potentials. Environmental Practice 8:159–168.

Reiger, J. F. 1975. American sportsmen and the origins of conservation. Winchester Press, New York, New York, USA.

Reneau, J., and J. E. Spring, editors. 2011. Records of North American big game, 13th edition. Boone and Crockett Club, Missoula, Montana, USA.

Rinehart, K. A., L. M. Elbroch, and H. U. Wittmer. 2014. Common biases in density estimation based on home range overlap with reference to pumas in Patagonia. Wildlife Biology 20:19–26.

Roberts, C. W., B. L. Pierce, A. W. Braden, R. R. Lopez, N. J. Silvy, P. A. Frank, and D. Ransom, Jr. 2006. Comparison of camera and road survey estimates for white-tailed deer. Journal of Wildlife Management 70:263–267.

Rominger, E. M., V. C. Bleich, and E. J. Goldstein. 2006. Bighorn sheep, mountain lions, and the ethics of conservation. Conservation Biology 20:1041.

Rubin, E. S., and V. C. Bleich. 2005. Sexual segregation: A necessary consideration in wildlife conservation. Pages 379–391 in K. E. Ruckstuhl and P. Neuhaus, editors. Sexual segregation in vertebrates: Ecology of the two sexes. Cambridge University Press, Cambridge, UK.

Rubin, E. R., W. M. Boyce, M. C. Jorgensen, S. G. Torres, C. L. Hayes, C. S. O'Brien, and D. A. Jessup. 1998. Distribution and abundance of bighorn sheep in the Peninsular Ranges, California. Wildlife Society Bulletin 26:539–551.

Russell, R. E., J. A. Royle, R. Desimone, M. K. Schwartz, V. L. Edwards, K. P. Pilgrim, and K. S. McKelvey. 2012. Estimating abundance of mountain lions from unstructured spatial sampling. Journal of Wildlife Management 76:1551–1561.

Ryder, T. J., E. S. Williams, and S. L. Anderson. 1994. Residual effects of pneumonia on the bighorn sheep of Whiskey Mountain, Wyoming. Biennial Symposium of the Northern Wild Sheep and Goat Council Proceedings 9:15–19.

Samuel, M. D., E. O. Garton, M. W. Schlegel, and R. G. Carson. 1987. Visibility bias during aerial surveys of elk in northcentral Idaho. Journal of Wildlife Management 51:622–630.

Sasse, G. B. 2003. Job-related mortality of wildlife workers in the United States, 1937–2000. Wildlife Society Bulletin 31:1015–1020.

Sawaya, M. A., T. K. Ruth, S. Creel, J. J. Rotella, H. B. Quigley, J. B. Stetz, and S. T. Kalinowski. 2011. Evaluation of noninvasive genetic sampling methods for cougars using a radio-collared population in Yellowstone National Park. Journal of Wildlife Management 75:612–622.

Schaller, G. B., and B. Junrang. 1988. Effects of a snowstorm on Tibetan antelope. Journal of Mammalogy 69:631–634.

Schulze, D. R., R. F. Schulze, and S. I. Zeveloff. 2008. A ground-based paintball mark re-sight survey of mountain goats. Biennial Symposium of the Northern Wild Sheep and Goat Council Proceedings 16:68–77.

Schwartz, M. K., and K. S. McKelvey. 2009. Why sampling scheme matters: The effect of sampling scheme on landscape genetic results. Conservation Genetics 10:441–452.

Scurlock, B. M., and W. H. Edwards. 2010. Status of brucellosis in free-ranging elk and bison in Wyoming. Journal of Wildlife Diseases 46:442–449.

Silver, S. C., L. E. T. Ostro, L. K. Marsh, L. Maffei, A. J. Noss, M. J. Kelly, R. B. Wallace, H. Gomex, and G. Ayala. 2004. The use of camera traps for estimating jaguar *Panthera onca* abundance and density using capture/recapture analysis. Oryx 38:1–7.

Silvy, N. J., editor. 2012. The wildlife techniques manual, 7th edition. 2 vols. Johns Hopkins University Press, Baltimore, Maryland, USA.

Simon, A. 2016. Against trophy hunting: A Marxian-Leopoldian critique. Monthly Review—an Independent Socialist Magazine 68(4):17–31.

Siniff, D. B., and R. O. Skoog. 1964. Aerial censusing of caribou using stratified random sampling. Journal of Wildlife Management 28:391–401.

Soria-Diaz, L., and O. Monroy-Vilchis. 2015. Monitoring population density and activity patterns of white-tailed deer (*Odocoileus virginianus*) in central Mexico, using camera trapping. Mammalia 79:43–50.

Soule, M. E., and B. A. Wilcox, editors. 1980a. Conservation biology. Sunderland Associates, Sunderland, Massachusetts, USA.

———. 1980b. Conservation biology: Its scope and its challenge. Pages 1–8 *in* M. E. Soule and B. A. Wilcox, editors. Conservation biology. Sunderland Associates, Sunderland, Massachusetts, USA.

Thirgood, S., R. Woodroffe, and A. Rabinowitz. 2005. The impact of human wildlife conflict on human lives and livelihoods. Pages 13–26 *in* R. Woodroffe, S. Thirgood, and A. Rabinowitz, editors. People and wildlife: Conflict or coexistence? Cambridge University Press, New York, New York, USA.

Thompson, C. M., J. A. Royle, and J. D. Gardner. 2012. A framework for inference about carnivore density from unstructured spatial sampling of scat using detector dogs. Journal of Wildlife Management 76:863–871.

Thompson, J. R., and V. C. Bleich. 1993. A comparison of mule deer survey techniques in the Sonoran Desert of California. California Fish and Game 79:70–75.

Timmerman, H. R. 1993. Use of aerial surveys for estimating and monitoring moose populations: A review. Alces 29:35–46.

Traill, L. W., S. Schindler, and T. Coulson. 2014. Demography, not inheritance, drives phenotypic change in hunted bighorn sheep. Proceedings of the National Academy of Sciences 111:13223–13228.

Vandermeer, J. H, and D. E. Goldberg. 2003. Population ecology: First principles. Princeton University Press, Princeton, New Jersey, USA.

Van Sickle, W. D., and F. G. Lindzey. 1991. Evaluation of a cougar population estimator based on probability sampling. Journal of Wildlife Management 55:738–743.

Vermeulen, C., P. Lejeune, J. Lisein, P. Sawadogo, and P. Bouche. 2013. Unmanned aerial survey of elephants. PLoS One 8.2:e54700.

Wagner, S. 2013. Photos: Nosing ahead—Elk Camp 2013. Outdoor Hub, Carbon Media Group, Bingham Farms, Michigan, USA. www.outdoorhub.com/stories/2013/03/11/photos-nosing-ahead-elk-camp-2013/.

Wakeling, B. F., D. N. Cagle, and J. H. Witham. 1999. Performance of aerial forward-looking infrared surveys on cattle, elk, and turkey in northern Arizona. Proceedings of the Biennial Conference of Research on the Colorado Plateau 4:77–88.

Walters, C. 1986. Adaptive management of renewable resources. MacMillan, New York, New York, USA.

Wasser, S. K., B. Davenport, E. R. Ramage, K. E. Hunt, M. Parker, C. Clark, and G. Stenhouse. 2004. Scat detection dogs in wildlife research and management: Application to grizzly and black bears in the Yellowhead Ecosystem, Alberta, Canada. Canadian Journal of Zoology 82:474–492.

Watts, A. C., J. H. Perry, S. E. Smith, M. A. Burgess, and B. E. Wilkinson. 2010. Small unmanned aircraft systems for low-altitude aerial surveys. Journal of Wildlife Management 74:1614–1619.

Weckerly, F. W., and K. E. Kovacs. 1998. Use of military helicopters to survey an elk population in north coastal California. California Fish and Game 84:44–47.

Wehausen, J. D., and V. C. Bleich. 2007. Influence of aerial search time on survey results. Desert Bighorn Council Transactions 49:23–29.

Wehausen, J. D., V. C. Bleich, B. Blong, and T. L. Russi. 1987. Recruitment dynamics in a southern California mountain sheep population. Journal of Wildlife Management 51:86–98.

WGFD (Wyoming Game and Fish Department), US Fish and Wildlife Service, National Park Service, USDA-APHIS-Wildlife Services, and Eastern Shoshone and Northern Arapahoe Tribal Fish and Game Department. 2014. 2013 Wyoming gray wolf population monitoring and management annual report. K. J. Mills and R. F. Trebelcock, editors. Wyoming Game and Fish Department, Cheyenne, Wyoming, USA.

White, G. C., R. M. Bartmann, L. H. Carpenter, and R. A. Garrott. 1989. Evaluation of aerial line transects for estimating mule deer densities. Journal of Wildlife Management 53:625–635.

White, G. C., and K. P. Burnham. 1999. Program MARK: Survival estimation from populations of marked animals. Bird Study 46:S120–139.

White, G. C., and R. A. Garrott. 1990. Analysis of wildlife radio-tracking data. Academic Press, San Diego, California, USA.

Whitfield, J. 2003. Sheep horns downsized by hunters' taste for trophies. Nature 426:595.

Whittaker, D. G., W. A. VanDyke, and S. L. Love. 2003. Evaluation of aerial line transect for estimating pronghorn antelope abundance in low-density populations. Wildlife Society Bulletin 31:443–453.

Wiedmann, B., and V. C. Bleich. 2014. Responses of bighorn sheep to recreational activities: A trial of a trail. Wildlife Society Bulletin 38:773–782.

Wiedmann, B., and B. Hosek. 2013. North Dakota bighorn sheep management plan (2013–2023). Wildlife Division Report A-213. North Dakota Game and Fish Department, Bismarck, North Dakota, USA.

Wild Sheep Foundation Professional Biologist Meeting Attendees. 2008a. Ewe harvest strategies for western states and provinces—2007. Biennial Symposium of the Northern Wild Sheep and Goat Council Proceedings 16:99–102.

———. 2008b. Ram harvest strategies for western states and provinces—2007. Biennial Symposium of the Northern Wild Sheep and Goat Council Proceedings 16:92–98.

Williamson, L. L. 1987. Evolution of a landmark law. Pages 1–7 in H. Kallman, C. P. Agee, W. R. Goforth, and J. P. Linduska, editors. Restoring America's wildlife 1937–1987. US Department of the Interior, Washington, DC, USA.

WSWG (Wild Sheep Working Group). 2012. Recommendations for domestic sheep and goat management in wild sheep habitat. Western Association of Fish and Wildlife Agencies, Cheyenne, Wyoming, USA.

7

David K. Dahlgren,
Michael A. Schroeder,
and Billy Dukes

State Management of Upland and Small Game

Upland and small game are important wildlife resources managed by the states. This chapter addresses both upland (i.e., game birds) and small (i.e., small mammals that are harvested) game because these species are usually managed under the same program within state wildlife agencies. Herein, upland game is specifically defined as avian species within the order Galliformes. We consider small game to be furred game animals that are not regulated as furbearers (see chap. 8). A list of native and introduced upland game can be found in table 7.1. All native upland game are currently hunted except lesser prairie-chicken, Gunnison sage-grouse, masked bobwhite (*C. v. ridgwayi*; a subspecies of northern bobwhite only found in southern Arizona), and Attwater's prairie-chicken (*T. c. attwateri*; a subspecies of greater prairie-chicken only found on the Texas coastal plain). Prairie grouse (i.e., sharp-tailed grouse and prairie-chickens), sage-grouse (both spp.), and masked bobwhite are currently considered species of conservation concern. Small game species are all native and include rabbits (*Sylvilagus* spp.), hares (*Lepus* spp.; jackrabbit and snowshoe), and squirrels (*Sciurus* spp.). Most small game species are currently hunted.

Though the ecological and biodiversity aspects of conserving upland/small game species are important topics and becoming increasingly relevant to public concerns, e.g., conservation and federal listing issues with our native species like grouse (Tetraoninae), historically it was primarily the hunting of these species that drove management and regulations within state

agencies. Harvest remains the leading concern of upland/small game programs; however, conservation of declining species and their habitats has gradually taken a more prominent role. Regulation of harvest was the obvious situation where states could actively manage impacts of human activities. With the development and maturation of state agencies and the passing of the Federal Aid in Wildlife Restoration Act (Pittman–Robertson Act of 1937, or PR), which provided state agencies with funding through a federal excise tax on hunting equipment, states have played an increasing role in provision of habitats necessary to support upland/small game.

In addition to their role in ecosystems, upland/small game provide important public resources and services such as hunting, hunter recruitment, economic and agency revenue, wildlife viewing, biodiversity, umbrella-species conservation, and more. Some species have been well studied (e.g., ring-necked pheasant, greater sage-grouse, northern bobwhite), and significant management and conservation have been implemented or are currently under way. Other species (e.g., dusky grouse, mountain quail, scaled quail) have received comparatively little research and management attention. Interestingly, small game accounts for slightly more hunters and hunting days per year than upland game, but small game tend to receive less research, management, and conservation consideration from most state agencies (US Fish and Wildlife Service 2010). In some state agencies, turkey are managed as

Table 7.1 A list of upland game bird species that currently have wild-propagated populations with distributions in North America

Common name	Scientific name	Origin	Currently hunted	Notes
Ruffed grouse	*Bonasa umbellus*	Native	Yes	
Spruce grouse	*Falcipennis canadensis*	Native	Yes	Sometimes referred to as Franklin's grouse
Dusky grouse	*Dendragapus obscurus*	Native	Yes	Recently (2004) split from blue grouse; see Barrowclaugh et al. (2004)
Sooty grouse	*Dendragapus fuliginosus*	Native	Yes	Recently (2004) split from blue grouse; see Barrowclaugh et al. (2004)
Sharp-tailed grouse	*Tympanuchus phasianellus*	Native	Yes	
Greater prairie-chicken	*Tympanuchus cupido*	Native	Yes	
Lesser prairie-chicken	*Tympanuchus pallidicinctus*	Native	No	
Gunnison sage-grouse	*Centrocercus minimus*	Native	No	
Greater sage-grouse	*Centrocercus urophasianus*	Native	Yes	
White-tailed ptarmigan	*Lagopus leucura*	Native	Yes	
Rock ptarmigan	*Lagopus muta*	Native	Yes	
Willow ptarmigan	*Lagopus lagopus*	Native	Yes	
Mountain quail	*Oreortyx pictus*	Native	Yes	
California quail	*Callipepla californica*	Native	Yes	Sometimes referred to as valley quail
Gamble's quail	*Callipepla gambelii*	Native	Yes	
Scaled quail	*Callipepla squamata*	Native	Yes	Sometimes referred to as blue quail or cotton tops
Northern bobwhite	*Colinus virginianus*	Native	Yes	
Montezuma quail	*Cyrtonix montezumae*	Native	Yes	Sometimes referred to as Mearns's quail
Wild turkey	*Meleagris gallopavo*	Native	Yes	Turkey have been transplanted to many non-endemic areas
Chachalacas	*Ortalis vetula*	Native	Yes	Only found in southern Texas in the United States
Ring-necked pheasant	*Phasianus colchicus*	Exotic	Yes	Native to China
Grey partridge	*Perdix perdix*	Exotic	Yes	Native to Asia; sometimes referred to as Hungarian partridge
Chukar partridge	*Alectoris chukar*	Exotic	Yes	Native to Asia; multiple subspecies have been transplanted to the United States
Himalayan snowcock	*Tetraogallus himalayensis*	Exotic	Yes	Native to Asia (Himalayan Mountains); only found in Nevada in the United States

stand-alone programs or part of big game programs, while in other states turkeys are covered by the upland/small game programs. Though not as common, some states also manage webless migratory game birds (e.g., doves, woodcock, rail) within upland/small game programs. Webless migratory game birds are discussed in chapter 9.

Typically, state agencies have an upland or small game program coordinator who is tasked with interpreting survey data, recommending management actions, and communicating with appropriate user groups. Depending on the state, individuals in this position may focus on management related to population objectives or may also be involved in habitat management and/or research. Other personnel are typically involved in management of upland/small game,

including assistant game coordinators, regional habitat managers, statewide and regional specialists (researchers or managers) who focus on specific species (e.g., wild turkeys) or groups of species, enforcement agents (see chap. 5), and many others.

Sportsman dollars, which come from license sales and PR revenue, provide the majority of funding for management of game species. The overall economic benefit of game management and hunting is much larger than hunter contributions to state agencies. Nationwide approximately 4.5 million upland/small game hunters spend 51 million days hunting each year and spend $2.6 billion per year to pursue these species (US Department of the Interior et al. 2011). In comparison, annually there are 11.6 million and 2.6 million big game and migratory bird hunters, respectively, in the United States (note: numbers of hunters by category are not mutually exclusive, as many hunters participate in multiple types of hunting; US Department of the Interior et al. 2011). From 2001 to 2011, numbers of hunters increased by 5 percent and days spent hunting by 23 percent; however, upland/small game hunters decreased by 17 percent and days spent hunting by 15 percent. Though a concerning trend, the economic benefit of upland/small game hunting cannot be underestimated. Hunting in general has been a multibillion-dollar economic endeavor, and upland/small game hunting plays a significant role in overall economic benefit (IAFWA 2002; US Department of the Interior et al. 2011). Upland game bird and small game hunting generates a minimum of $1.85 billion in retail sales and $1.18 billion in salaries and wages per year. Jobs created as a result of upland game hunting, as well as fuel sales and local, state, and federal income taxes, often benefit rural economies, where the majority of upland and small game hunting takes place (IAFWA 2002).

Native versus Imported Upland Game

While native upland/small game species are important state and national resources, non-native imported upland game play a substantial role in upland/small game programs and interest within this country. The importance of native versus non-native species within upland/small game programs varies geographically. For example, bobwhite quail are the primary species of upland game in the Southeast, and no large-scale populations of non-native upland game species occur. Similarly, ruffed grouse in the Northeast are the main focus of upland game programs and hunter interest. However, ring-necked pheasants have undeniably captured the hearts and opened the wallets of American upland game hunters. Pheasants are the single most hunted species of upland/small game in the United States (US Department of the Interior et al. 2011). Simply put, pheasants garner more interest, revenue, and attention than any other upland game species in North America. In fact, pheasants have largely shaped the history of upland game management and programs within the United States.

In its native country, ring-necked pheasants adapted to coexist and thrive with human agriculture. Judge Denny, in 1881, arranged the import of wild ring-necked pheasants, trapped near Shanghai, China, and released them into the agricultural region of Oregon's Willamette Valley. This was not the first effort to establish pheasants in North America, but it was the first to succeed and result in a wild reproductive population. In fact, within a decade the entire Willamette Valley had filled with pheasants from a relatively few wild-trapped and released individuals. The success of this effort, combined with the birds' beauty, captured imaginations of North American hunters and triggered a momentous effort to establish more pheasants and other non-native species throughout the United States. Many state upland game programs focused on establishing pheasants within their borders. State-owned game farms were a common initial feature of upland game programs nationwide. Pheasant egg delivery and chick-rearing programs were a regular practice in many rural communities. However, the vast majority of these artificial stocking programs failed at establishing breeding populations. Biologists soon learned that providing and managing quality habitat and then transplanting wild-propagated pheasants into these habitats were the most successful and wisest uses of resources. It was only a matter of time before pheasants were established in agricultural regions of the Northeast, Midwest, Northern Plains, Intermountain West, and Pacific states.

By the mid-1900s, thriving wild pheasant populations, and the associated hunting opportunities and economic benefits that followed, fueled state and federal wildlife programs to seek importation and establishment of other non-native upland species. In 1948, the federal Foreign Game Investigation Program began within the Bureau of Sport Fisheries and Wildlife in the US Fish and Wildlife Service (USFWS) and was funded by PR reverted monies (i.e., PR funds not used by state agencies). Staff biologists visited many areas around the world and conducted field observations of exotic upland game species to assess their potential for translocation to the United States. Under this program, 26 foreign species were recommended for trial releases. Many species showed promise, including Japanese green pheasant (*P. versicolor*), black francolin (*Francolinus francolinus*), gray francolin (*F. pondicerianus*), Chinese bamboo partridge (*Bambusicola thoracicus*), and several species of tinamou (Tinamidae). State agency directors and upland game biologists coordinated with federal biologists to set up transfer of trapped individuals and release sites for stocking trials. The program was discontinued in 1972 owing to lack of funding, potential disease transmission, concern over the potentially negative impact of exotic species on native species and habitats, and lack of success with introduction trials. Only chukar (Great Basin), gray partridge (northern parts of the Midwest and West), and Himalayan snowcock (Ruby Mountains of Nevada) established stable wild populations in North America. Notable exceptions occurred in Hawaii, where wild populations of several exotic upland game species were established and are currently hunted.

Leopold (1938:3), when commenting on efforts to establish chukar partridge and other non-native species throughout the country, stated that "few sportsmen ever become immune to the idea that foreign game birds are the answer to the 'more game' problem." He postulated that stocking non-native game birds distracted sportsmen and upland game programs from conserving habitat and populations of native upland game species. Though this is an important criticism to consider, currently non-native species tend to fill niches where native upland game species do not persist and little direct interspecific competition exists. There is a possibility of spreading disease from non-native to native species, but this has rarely been studied. Potentially, the greatest competition between native and non-native species is simply the attention of sportsmen, agencies, and resulting conservation efforts. Conversely, it is likely that our past, present, and future cohort of upland game hunters who support conservation may not be as large if pheasant and other non-native species were not so popular. In reality, the line between native and non-native upland game conservation and subsequent resource allocation is largely indistinguishable. For example, revenue generated from license sales and resulting PR funds from the hunting of non-native species is commonly used for conservation activities for native upland game species.

Generally, stocking programs no longer exist to establish or augment breeding populations of non-native upland/small game. However, for the purpose of increasing fall hunting opportunities, many states continue pen-reared game bird stocking of non-native species, primarily pheasants. Pennsylvania maintains one of the largest state-owned pheasant rearing and stocking programs in the United States (www.pgc.pa .gov/Wildlife/WildlifeSpecies/Ring-NeckedPheasant /Pages/PheasantManagement.aspx). During the mid-1900s, Pennsylvania rivaled popular Midwest pheasant states (e.g., South Dakota, Iowa, and Kansas) for the number of pheasant hunters and total harvest of wild pheasant. Even in 2001, Pennsylvania reportedly had more upland game hunters than any other state (IAFWA 2002). However, in the late 1900s, as agricultural practices and habitat changed in Pennsylvania, wild pheasant populations declined, and a stocking program was implemented to meet hunter demand. Other states (e.g., Idaho) stock pheasants on wildlife management areas, and hunters buy a special harvest permit (https://fishandgame.idaho.gov/public/hunt/?getPage =275). Utah recently began a program where residents adopt and raise pheasant chicks to supplement state-purchased pen-reared pheasants released in designated areas during the hunting season (https://wildlife .utah.gov/rules-regulations/941-r657–4—possession-of-live-game-birds.html). Multiple research projects have shown that stocking programs may increase numbers of birds available for harvest, but pen-reared game birds either do not survive or reproduce at rates too low to maintain or increase breeding population levels (Leif 1994). Although relatively unstudied to date, an additional possibility is that pen-reared birds may reduce viability of wild birds (e.g., genetic dilution, increase in number of predators, and higher risk of disease) in the areas where they are released. While it is important for state agencies to provide hunting opportunities and maintain/increase pheasant or other game bird populations, the quality and quantity of habitat should remain the focus of state agency resources.

Historically, native upland game have also been translocated within and across state boundaries, often to establish populations in areas with potentially suitable habitat but no endemic populations. Some efforts have been unsuccessful, such as prairie-chicken releases in Hawaii and sharp-tailed grouse in Massachusetts and New Mexico. Conversely, some translocations have been successful. Ruffed grouse populations were established through translocations in northwestern

Nevada and south-central Alaska. Additionally, white-tailed ptarmigan populations have been established in Utah and California. The most widely successful effort has been with wild turkeys, which have been translocated and for which populations have been established within their historic range and many non-endemic areas (Kennamer et al. 1992; Tapley et al. 2001).

Harvest Management

Setting harvest regulations for state-trust game species is the only management authority solely held by states, except for management of state-owned wildlife areas. Most, if not all, state agencies have a politically appointed wildlife board or commission. Wildlife boards have a long history of being involved in wildlife management decisions. Currently, wildlife managers within state agencies do not set hunting guidelines, but rather recommend rules and regulations to wildlife boards that, usually after public input, provide the final decision on any proposed rule changes.

Currently, upland/small game programs, with a few exceptions, do not use population data to set and adjust harvest regulations regularly. Regulations are set annually, or for a certain period of years, primarily based on previous regulations. If harvest regulations change, it is usually political/social pressure driving modifications, not population data. Rather, states' approaches to harvest management of upland/small game have generally been through trial and error. A trial-and-error approach to harvest management seems haphazard, but in the case of upland/small game it might be considered relatively cost-effective. Population surveys are expensive to conduct, and license revenue from upland/small game hunting is rarely sufficient to develop and implement complex harvest strategies. At the time of state agency founding in the late nineteenth and early twentieth centuries, most species of upland/small game had few or no limits to harvest. A typical response to declining populations has been to eliminate harvest. This is the most extreme trial-and-error response, but one that has been fairly typical in the history of state agencies (Connelly et al. 2000). Eventually, most state agencies have attempted to "settle" on harvest limits that provide abundant harvest opportunities while also appearing to leave populations of upland and small game that are

somewhat stable in both abundance and distribution. Some of this approach appears to be benefited by portions of game populations that are inaccessible owing to weather, topography, and/or trespass restrictions on private land (Small et al. 1991). Moreover, most harvest regulations have been kept in check by the conservative nature of most wildlife managers, not necessarily owing to good scientific information (Caughley 1985).

Harvest strategies for upland and small game can be placed into three general categories: (1) high-profile species with large constituencies and relatively high revenue potential; (2) species with declining populations, smaller distributions, and/or increasing conservation concerns; and (3) all of the other species. It is possible for species to be in multiple categories, depending on the state. For example, ruffed grouse and bobwhite quail are considered high-profile species in eastern and midwestern states, and conservation concerns exist in many areas for both species. Consequently, the harvest regulations reflect this by being variable, site specific, and restrictive in length. In contrast, in the West, for the purposes of harvest, ruffed grouse in Idaho, Montana, Utah, California, and Alaska are lumped together with other species of grouse such as dusky, sooty, and spruce grouse as an aggregate forest grouse bag. In addition, the forest grouse season in the West typically begins in early September (mid-August in Alaska), midwestern states' seasons open in mid-September, while eastern seasons typically start in October. In contrast to the regional variation in hunter interest and harvest management approach for certain upland game species, some small game populations may be considered underutilized, such as gray squirrels in the East or cottontail in the West.

Although over the past several decades changes in harvest management have not generally been based on population-level data, our current scientific understanding of the effects of harvest on upland/small game has improved considerably. Upland/small game harvest principles have undergone a significant paradigm shift over the past couple decades (Connelly et al. 2005). During the mid-twentieth century, the effects of harvest on upland/small game were reportedly minimal (Errington and Hamerstrom 1935; Errington 1945). It seemed that upland/small game populations regularly produced enough animals to exceed carrying capacity.

Thus, when harvested populations reached a winter bottleneck, harvest mortality was considered compensatory to overwinter mortality, where a harvested individual was considered one that would have died otherwise as a result of natural causes. This theory was termed "doomed surplus." Because of this early body of research, the contemporary prevailing approach to upland/small game harvest included an assumption that harvest had no impact on upland/small game populations. Regulations, including bag limits and season length, became more liberal and resources devoted to monitoring populations became lower priority, because if harvest has no impact, there is no need to monitor changes in relation to harvest (Connelly et al. 2005).

During the late twentieth century, new research began to demonstrate that harvest mortality may actually be additive or at least partially additive to natural mortality. Issues of immigration, emigration, lack of winter bottlenecks, individual heterogeneity, timing of harvest, and high overwinter survival began to surface as factors that earlier studies did not consider. Total compensation may have been a premature conclusion of early studies on the effects of harvest on upland/small game populations. However, well-designed research has shown that harvest mortality can be at least partially compensated by decreases in other types of mortality in many populations of upland and small game (Ellison 1991; Roseberry and Klimstra 1992; Hudson et al. 1997; Sedinger et al. 2010). Compensation has also been observed in the form of increases in productivity during the following reproductive cycle (Boyce et al. 1999), dispersal from areas with lower harvest rates (Small et al. 1991; Smith and Willebrand 1999), and exclusion of surplus animals from breeding populations through territoriality (Jenkins et al. 1964). However, other sound research has shown that some mortality associated with harvest can be additive (Small et al. 1991; Pack et al. 1999; Connelly et al. 2000; Guthery et al. 2000; Alpizar-Jara et al. 2001; Humberg et al. 2009). Harvest of upland and small game is likely both partially compensatory and partially additive (Smith and Willebrand 1999; Sandercock et al. 2011). The fact that states actively manage their upland and small game harvest with licenses, tags, seasons, bag limits, and methods of take illustrates that they have a core philosophical approach to management that mor-

tality associated with harvest (at least at some elevated levels) is not completely compensated.

Matching harvest regulations to species life history is a critical feature of appropriate harvest management (Connelly et al. 2005). Research has demonstrated differing life history strategies for upland game species. While all galliform populations require annual influxes through reproduction, the life history strategies of species can be compared relative to each other. For example, most quail species are characterized by high annual mortality and high reproductive effort (clutch size to body size ratios), while in comparison most grouse are relatively long-lived with comparatively lower reproductive effort. Upland/small game species that exhibit high rates of population turnover (high reproductive output and high adult mortality) have more potential to rebound from additive impacts. Additionally, recent research has shown that harvest may be additive at relatively small scales compared to statewide boundaries. Both historically and currently, states most often set harvest regulations at statewide scales, and critics have called for harvest regulations that match the spatial scale of impacts (Williams et al. 2004). Pheasant and spring turkey hunting is focused on male-only harvest, while for most other species harvest is not sex specific. We currently have little understanding of the differences in sex-specific and non-sex-specific harvest. For example, consider the potential of higher risk to harvest for successful grouse brood hens when associated with chicks early in the hunting season. The results of more recent research have not generally been incorporated into state harvest regulations at this time. Recommendations from scientific studies often have a lag time before integration into management.

However, there is a notable example of harvest regulations being tailored to current scientific and life history information for an upland game species. Greater sage-grouse historically had similar regulations to other species with larger bag limits, longer seasons, and statewide hunting areas (table 7.2). Currently, statewide harvest bag/possession limits no longer exist for sage-grouse, and individual state regulations vary from closed statewide to open only in specific areas, usually where populations are larger and more stable (table 7.2). Some states have implemented adaptive harvest management principles for sage-grouse hunting regulations (table 7.2). For example, in Utah only four large (>500 breeding adults) populations are hunted. Lek counts are used each year to estimate a spring breeding population using assumptions of sex ratio, observer detection probability, and male lek attendance rates (Dahlgren 2010). Then further assumptions of productivity and survival are employed to estimate a fall population for each hunted area. Lastly, after accounting for harvest success rates from previous years, Utah issues a number of tags that will result in no more than 10 percent harvest of the fall population estimate for each population. Oregon currently has a similar adaptive harvest management approach for hunting sage-grouse but manages for a population harvest rate of 5 percent or less. Idaho has a different approach based on past research, where the previous spring lek counts are used to adjust season length (including season closure as an option) and bag limits based on thresholds for low- and high-elevation populations (Connelly et al. 2003a). Though not as rigorous as adaptive harvest management models in waterfowl, current sage-grouse harvest regulations provide the first example of using more scientific data-driven principles to adapt appropriate harvest regulations for an upland game species. We anticipate adaptive harvest management principles becoming more popular in upland/small game harvest regulations. Adaptive harvest management is a more defensible approach, as American landscapes continue to experience broad habitat degradation and loss, resulting population declines, and a public that is increasingly critical of purposeful killing of wildlife. Notably, implementation of adaptive harvest management requires regular population-level monitoring and commitment to provide the necessary resources for this purpose.

It is important to note that just because harvest mortality may be additive at some level does not mean that harvest is inherently harmful and should be discontinued. Upland game populations can continue to exist and even show stability with continued harvest (Connelly et al. 2005). It is up to wildlife managers and state wildlife boards to work responsibly to maintain harvested populations into the future. Hunter interest and concern for the very game species they hunt have funded and supported the abundant conservation efforts of the past century and a half (Trefethen 1975).

While hunters are considered consumptive users of wildlife resources, there are many nonconsumptive users such as wildlife watchers, hikers, and others who are also concerned about the conservation and future of wildlife. Finding ways for nonconsumptive users to have a seat at the conservation funding and management table, while maintaining the rich tradition of sportsman funding and involvement, will be one of the principal challenges of future wildlife professionals.

Few species of upland/small game have the high profile and economic impact of wild turkey. They can be legally hunted in every state except Alaska. Wild turkeys are the only upland game species that is hunted during the spring breeding season. Only bearded (males, commonly referred to as Toms; rarely females) turkeys are allowed to be harvested during the spring. Usually a carcass tag is issued for one or more turkeys. Many states also have a fall hunting season for turkeys, where either males or females (i.e., bearded and nonbearded) can be taken. Even when managed by upland game programs, turkey harvest management resembles big game regulations compared to the majority of other upland game species. The demand for turkeys is so high that states have introduced turkeys in numerous locations where they were not native, particularly the different subspecies (Eastern [*M. g. silvestris*], Florida [*M. g. osceola*], Rio Grande [*M. g. intermedia*], and Merriam's [*M. g. merriami*]) in the same state. Hunters have responded by creating their own challenges, such as the Turkey Grand Slam, which is accomplished by harvesting all four subspecies. The result of this interest is that wild turkeys receive substantial license and tag fees from hunters and abundant attention from state agencies and nongovernmental organizations (NGOs). On the flip side, states sometimes have to use their resources to reduce turkey abundance in areas where they have become pests.

Population Surveys

Understanding changes in wildlife populations and trends is an important function of state wildlife agencies. Monitoring populations is especially important given the state's responsibility to conserve all state-trust species and regulate harvest for game species. Aldo Leopold wrote, "Finding out how many there are

Table 7.2 Historical changes in greater sage-grouse hunting regulations for Utah, Montana, Idaho, Oregon, Wyoming, and Washington demonstrating the application of more conservative and science-based adaptive harvest management principles for an upland game species

State	Years	Daily	Days	Opening day	Area	Other information
Utah	1951–72	2–4	2–7	Early September	Statewide; except four areas with population estimates >500	Two permits/tags per hunter with draw system; lek counts are used with assumptions of survival and production for number of tags at 10% of fall population for each area
	2008–15	2 (tags required)	21	Last Saturday in September		
Montana	1938–57[a]	3–4	1–8	Early September	Statewide	Season length progressively increased from 1958 to 1970
	1958–75	4	8–65	Early September	Statewide	
	2005–13	2	91	September 1	Closed west of Continental Divide	Montana closed parts of the range that have historically been open owing to population declines
	2014–15	2	31	September 1	Two open areas in north-central and southwest	
Idaho	1903–9	12–18	108–137	Mid-July and mid-August	Statewide	
	1939–42 and 1944–47	0	0	NA	NA	Season closed
	1957–78	2–4	4.5–23	3rd Saturday in September	Specific areas; except 1977–78	In 1977–78 the entire state was open
	2011–15	1 or 2	6–7	3rd Saturday in September	Specific areas; high- and low-elevation populations considered separately	Beginning in 2008 season length and bag limits can be adapted annually based on spring lek counts and other population information
Oregon	1949–52	2–4	5–13	Early September	Statewide	The season was closed 1954, 1957, 1960, 1965, 1967, 1976–81, and 1985–88
	1966–88	2	2–3	Early September	Statewide	
	1989–2009	2 (season limit is 2)	2–9	Early September	East-central and southeast	Permit is required and based on an application draw; permit numbers are adjusted based on lek counts each year; only areas with >100 males on leks will be hunted
	2010–15		9	1st or 2nd Saturday in September	13 specific population areas	
Wyoming	1991–94	3	30–31	1st of September	Statewide	Season length was area specific
	1995–2001	3	14–16	Mid- to late September	Statewide	
	2002–7	2	9–11	Late September	Statewide	
	2008–15	2	3–13	Late September	Two open areas in northeast and central to southwest	
Washington	1950–1987	1–2	2, 8–10	Early September	Seasons and bag limit designated by county	The number of open areas increased from the 1950s to the 1970s
	1988–2015					Season closed

[a] Season closed 1945–1951.

left is the least of the purpose of game census. Measuring the response of game populations to changes—deliberate or accidental—in their environment is the big purpose. Continuous census is the yardstick of success or failure in conservation" (Leopold 1933:169–170).

Techniques for surveying wildlife populations have evolved since Aldo Leopold's time with the benefit of scientific evaluation. Population surveys for upland/small game species vary greatly, with greatest distinction between avian and mammalian species. Because a true census is rarely possible, indices have become the norm, with detection probability an important issue (Buckland et al. 1993). Often the scale of sampling and that of the population are not in sync, or matching spatial scales is simply infeasible. In many states historical data rarely influenced harvest decisions and therefore became ineffectual and of lower priority. Because of these issues, and owing to the expense and competing workloads of budgets and biologists, many state wildlife agencies have discontinued some portion of their historic upland and small game population surveys. Nevertheless, most states continue to conduct population surveys of some kind, especially for the more popular and intensively managed upland game species. States often maintain long-term (>20 years) data sets for upland and small game populations at large spatial scales, though robust statistical methodologies to analyze such data are often lacking (Hefley 2014). Upland hunting forecasts are frequently provided by upland game programs for more popular species, such as ring-necked pheasant, and hunters and media outlets seek this information. Forecasts based on population surveys are obviously preferred, but in many cases forecasts are more of a best guess by upland game biologists than a data-driven process.

Throughout the Midwest, pheasants provide the bulk of upland game interest and revenue. Most Midwest states with significant pheasant populations (Kansas, North Dakota, South Dakota, and Iowa) continue to conduct late-summer brood surveys to aid in forecasting populations for the upcoming hunting season. Because pheasants use roadways to obtain grit for digestion, early morning road (usually graded gravel) surveys are conducted from vehicles. The mileage, location, number, and classification (age and sex) of observed pheasants are recorded. New technology is being employed in some states (South Dakota, Kansas, and Iowa) where data are recorded with an application on a handheld device (GPS, smartphone, tablet, etc.) that provides geo-referenced data (see www.cybertracker.org for an example of freeware currently being used). Sample efforts across large spatial scales are needed to detect population changes, but spring crow counts and brood surveys have been shown to predict harvest at statewide, but not necessarily local, scales (Rice 2003). Crow and brood counts have been used to assess habitat selection for pheasants (Eggebo et al. 2003). Additionally, these survey methods have provided critical information for large-scale impacts of national agricultural policy on wildlife populations (Nusser et al. 2004).

In many states with lekking grouse, spring lek counts are completed annually. For prairie grouse (sharp-tailed grouse and prairie-chickens) linear routes are surveyed early in the morning, and the numbers of leks and birds per lek are recorded. For sage-grouse, individual leks are usually in the same location each year and counted consistently (annually or periodically) and usually within lek complexes where individual leks are spatially associated (Connelly et al. 2003b). When counting male grouse, there is an assumption that the number of males is representative of the female portion of the population, and recent research has suggested that this may be a safe assumption (Dahlgren et al. 2016a). To relate lek counts to population levels, assumptions are needed such as detection rates of leks and individual birds on a lek, lek attendance rates, and sex ratios within populations. Walsh et al. (2004) found sage-grouse daily lek attendance rates of only 42 percent, 4 percent, and 19 percent for adult males, females, and yearling males, respectively. However, others have found much higher daily lek attendance rates for male sage-grouse (Emmons and Braun 1984; Dahlgren 2010). Male prairie-chickens and sharp-tailed grouse have been shown to have relatively high (>94%) daily lek attendance rates (Schroeder and Braun 1992; Drummer et al. 2011). An important consideration in daily lek attendance rates for male grouse is the impact of weather and time of year. Often studies, such as Walsh et al. (2004), measure lek attendance rates over the course of an entire spring lekking season, with varying weather conditions on given mornings.

Lek attendance rates decline with more severe (e.g., cloudy, windy, precipitation) weather (Emmons and Braun 1984; Connelly et al. 2003b). Lek survey protocols call for sampling efforts to only occur on mornings with clear, calm weather conditions. Therefore, if lek counts are going to be related to population dynamics, lek attendance rates on mornings where weather meets protocol standards are the most important estimates to use.

In the Midwest and Northeast, ruffed grouse drumming surveys are completed along driving routes with specified stops (listening stations), and the number of drumming grouse is recorded within a given time period. Drumming survey data can be combined with other methods such as occupancy modeling and resource selection functions to monitor the status, trends, and habitat associations, including spatial scale, of ruffed grouse populations (Hansen et al. 2011; Mehls et al. 2014). However, historically in the Intermountain West and on the West Coast little to no spring surveys for forest grouse (ruffed, spruce, dusky, and sooty grouse) have been regularly completed, at least at larger statewide scales, even when these species provide a major portion of upland bird hunting interest. Oregon and Alaska have recently begun statewide breeding surveys for sooty grouse. In some areas in the West, access to breeding habitat is precluded by snowpack and road closures, making spring surveys difficult. Another common issue for many states is the workload of state agency field biologists who are required to conduct breeding surveys for multiple game (big and small) species during the same time of year, often with similar weather restrictions (low wind, low percent cloud cover, etc.) within survey protocols.

Few statewide population surveys are conducted for tree squirrels or rabbits across the eastern range of these species. South Carolina has used a sighting survey to document distribution of southern fox squirrels throughout the state, but the survey does not yield density estimates or even comparative population estimates from year to year. Eastern gray squirrels are typically considered abundant in areas of suitable habitat, and since harvest is considered conservative and compensatory, field surveys are typically not.conducted. Line transect surveys using distance sampling could prove useful for estimating squirrel densities at various

scales (Buckland et al. 1993). In addition, since gray squirrel fecundity and survival have been positively correlated with mast production, it is possible that systematic mast surveys could be used as a predictor of the subsequent year's squirrel population.

Many southeastern and southern Great Plains states conduct spring bobwhite call surveys along routes with specified stops, as well as fall covey count surveys prior to the hunting season. As of 2012, 24 of 25 states within the range of northern bobwhites were still using breeding season auditory surveys as one measure of bobwhite population trends. Important assumptions for bobwhite call counts include individual call rate, the effect of weather on calling rates, and call rates based on time of day and year. Breeding whistle counts usually occur in June, with fall covey counts happening in October. Covey density may have the greatest influence on fall calling rates (Wellendorf et al. 2004). Crosby et al. (2015) used spring whistle surveys to demonstrate that bobwhites can be used as an umbrella species for other avian species, some of which were species of conservation concern. New efforts for bobwhite monitoring, led by the National Bobwhite Technical Committee through their National Bobwhite Conservation Initiative, stress the principles of strategic habitat conservation and prescribe methods for a coordinated monitoring approach across state boundaries that utilizes breeding season call counts, density estimates using fall covey counts, and habitat monitoring (Morgan et al. 2014). Hefley et al. (2013) used whistle count data within state-space models to estimate extinction thresholds, which can help managers make resource allocation decisions to maintain persistence of populations. This technique is likely applicable to other state-owned long-term data sets for many species of upland and small game.

In many midwestern and eastern states, wildlife agencies contract with the US Postal Service to conduct rural mail carrier surveys (RMCSs). RMCSs are completed by mail carriers who count upland/small game species as they drive their regular rural mail routes. This cooperative effort began in the 1960s in many states and continues to provide long-term indices for upland/small game populations. Data are usually recorded by county and often summarized by sightings per 100 miles. These data have been used to evaluate trends in populations, help forecast hunting seasons, and conduct large-scale wildlife population analyses (Williams et al. 2003). While this is a somewhat crude measure of wildlife populations, the large spatial and temporal scales the survey covers are primary strengths.

Hunter activity and harvest surveys are the most consistent method for monitoring upland/small game harvest for most state agencies. Following cessation of hunting season, state agencies use their database of license holder information to contact upland/small game hunters to complete a harvest survey. Historically, mail-in and/or phone surveys were commonly used; however, recently e-mail contact and web-based surveys have become standard. Because most upland/small game species can be hunted with a general hunting license and do not require special permits or tags, a large sample of general license holders is usually needed to have enough respondents to get reliable survey information. Survey questions include the number of individual species harvested, days hunted, areas (usually county level information) hunted, and satisfaction questions. Sometimes a time period within the season is obtained for hunter day and harvest information. While nonresponse bias can result in overestimation and is often a concern when conducting questionnaire surveys, and although some states incorporate a correction factor for nonresponse bias into their harvest estimates, some studies have shown that large biases due to nonresponse did not occur in harvest surveys (Barker 1991; Pendleton 1992). Data are used to estimate an overall statewide or area-specific harvest for each species. Research has demonstrated usefulness of harvest surveys and how they relate to population ecology (Fedy and Doherty 2011).

Some states use an upland bird stamp (California, Indiana, Nebraska, Nevada, Oregon, and Texas) to generate program-specific funding and target upland hunters for surveys. Special permits are often used when hunting species of conservation concern, and hunters who obtain these can be contacted directly. For example, a special permit at a nominal fee is required to hunt prairie-chickens and Columbian sharp-tailed grouse (*T. p. columbianus*) in Kansas and Idaho, respectively. These species have relatively low numbers of hunters each year and are of special conservation concern within these states. Upland game program man-

agers can obtain reliable harvest information by way of these special permits when low numbers of hunters are involved, and fees can be used directly for upland/small game or species-specific management. Oregon has used their sage-grouse permit holders to obtain wing samples from harvested birds and used wing samples to test for West Nile Virus and obtain a large genetic database across their populations. We predict that special permits will become more common in the future, especially with species of conservation concern.

The most common surveys for wild turkeys are harvest surveys and brood surveys. Harvest survey trends are often used as a surrogate for population trends, and brood surveys are used to estimate productivity as a ratio of poults per hen. Combined data from harvest surveys and brood surveys can be used in population reconstruction models to estimate populations of wild turkey and other species. In midwestern states turkeys may be incorporated into brood surveys for other upland game species. Additionally, RMCSs include turkeys with other upland/small game.

Wings of most upland game can be used to classify species, sex, and age class (juvenile, yearling, and adult). Juveniles are birds within their hatch year prior to their first breeding season. Yearlings are in their second year or first breeding year. Adults are birds in any year after their second breeding season. Wing data can provide information concerning the age and sex structure of a population, including productivity indices such as juvenile:adult ratios. Most gallinaceous birds begin to molt during the late spring or early summer. Primary and secondary feathers on the wing are replaced in a pattern from 1 to 10, called a molt pattern. Juveniles hatch with only primaries 1–8 and start to grow primaries 9 and 10 at three to four weeks of age. During their first molt in the late summer and early fall of hatch year, they replace primaries 1–8 but retain their outer two primaries (p9 and p10), and they do not replace p9 and p10 until the fall (usually October) of their second year. Juvenile p9 and p10 have pointed tips, compared to being round for adults. Therefore, in September and early October juveniles can be distinguished from yearlings and adults. Juveniles will have pointed p9 and p10 feathers that are only slightly faded and worn compared to p1–8. Yearlings can sometimes be distinguished from adults and juveniles. In the early

Sex

MALE primaries 20-30 mm longer than FEMALE primaries for each age class, region, and sage-grouse species (sample midpoint cutoffs shown to right – measurements with an asterisk have overlap).

State	P10	P9	P7	P6	P5	P1
Oregon	155*	194	212	218	207	
Washington	158*	207*	222	226	216	
Idaho	158	193		210	206	
Montana	160	205*	228	232	229	
Colorado-GUSG	157	190				140
Colorado-GRSG	160	200				140

Age

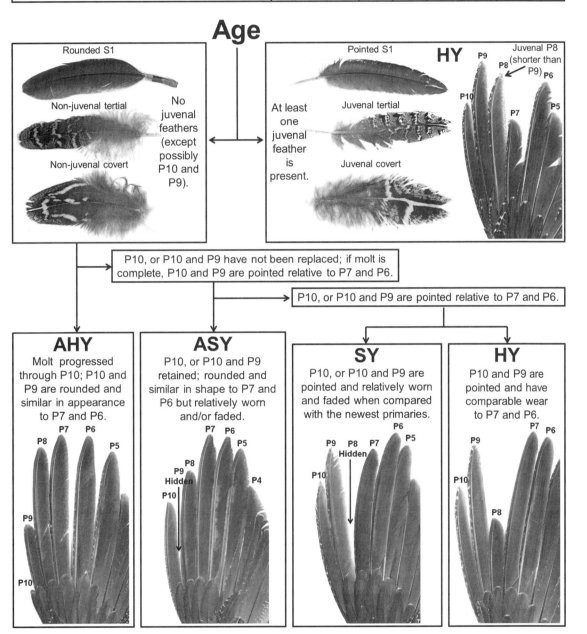

Example of a flowchart for identification of sex and age of sage-grouse based on appearance and measurements of wings (Braun et al. 2015). The primary and secondary molt patterns in sage-grouse are consistent for most upland game birds, including the shape and wear of the primaries. However, the shape of secondary 1 and color patterns and shape of tertial and covert feathers may vary by species. Also, in most species there is no significant difference in the length of primaries between sexes. GUSG refers to Gunnison sage-grouse, and otherwise, if not stated, the reference is to greater sage-grouse.

fall yearlings will have p9 and p10 feathers that have not been replaced yet and are pointed, faded, and worn compared to p1–8 and are even more faded and worn in relation to p1–8 than juvenile p9 and p10. Adults will have round p9 and p10 feathers. However, there is more uncertainty when distinguishing yearlings from adults because the beginning of molt can vary by sex, age, and nesting behavior and some yearlings may have already replaced their p9 and p10 and would be classified as adults. Although wing molt patterns described above are consistent for most upland game species, regional variation has been noted in turkeys (Dickson 1992), and even in greater sage-grouse (Braun and Schroeder 2015).

For many upland game species there are other distinguishing feather characteristics to assess age and sex. As examples, juvenile bobwhite quail have buff-colored tips on their upper primary coverts; juvenile dusky and sooty grouse have pointed tertiary feathers; juvenile sage-grouse often have a pointed first secondary feather; gray partridge males have a more rusty color on their upper wing coverts compared to females; female dusky and sooty grouse have more mottled buffy and dark coloring on their upper wing coverts, compared to males with blue gray drab upper wing coverts; and sage-grouse males of all age classes have longer primary feathers than females of the same age class. For many species, such as ruffed grouse, most quail species, sharp-tailed grouse, and prairie-chickens, there are no distinctive features to separate sexes using wings only; rather, other feather characteristics are needed. For example, ruffed grouse males have longer and differently patterned central retrices (tail feathers) compared to females; female prairie-chickens (all species) have barring on their outer retrices, while male outer retrices are solid black; sharp-tailed grouse females have horizontal barring the full length of their two central retrices, while males have more white and vertical barring comparatively; bobwhite quail males have a white throat patch, while females have a buffy patch; and scaled quail females have faint dark streaking on their throat, while males have no streaking.

A "wing bee" is the common term used for a gathering of biologists collecting data from a harvested wing sample. Wings are usually gathered in barrels from specific hunting areas, or mail-in envelopes are provided

to hunters, similar to waterfowl protocols. There are assumptions for using wing data to assess upland game populations. The first assumption is that harvest occurs randomly across sex and age classes such that the harvest sample characterizes the population accurately (Roseberry and Klimstra 1992; Flanders-Wanner et al. 2004). Another assumption is that enough wings are harvested to provide representative data (Hagen and Loughin 2008). Hagen and Loughin (2008) used asymptotic normal approximations (large sample sizes) of wing data. Based on these analyses, they suggested that a sample size of more than 300 wings would be necessary from a population to obtain the necessary precision when comparing productivity between two areas when assessing conservation actions. However, this number of harvested birds may be unattainable or not recommended for populations where a limited harvest is the objective. When sample sizes are small, bootstrapping techniques may be more appropriate (Manly 1998). As an example of using wings to examine populations, Braun et al. (2015) used over 67,000 sage-grouse wing samples from Oregon and Colorado to analyze fall population structure, including breeding success and survival rates. Another useful method is statistical population reconstruction using these types of data (Broms et al. 2010).

Conservation and Management

Upland/small game programs face many diverse challenges in the coming years. Declining populations will necessitate an increased role for state agencies in delivery of landscape-level habitat conservation. Many factors contribute to the high risk for habitat loss, including urban sprawl, conversion of grasslands to agricultural cropland to meet human food demands, invasive plant conversion, energy development, or some combination of factors. While some upland/small game species (pheasants, rabbits, and gray partridge) are better adapted to inhabiting fragmented or small-scale habitats often associated with agriculture or other development, most upland/small game species need large-scale connected habitats to persist and thrive. Even species like bobwhite quail, which were formerly thought to be a small-scale edge species where "token" management was sufficient, have been shown to need large connected landscapes for population stability (Williams et al. 2004). Certainly, prairie grouse and sage-grouse are notable examples of the need for large landscapes of steppe or shrub-steppe habitat (Knick and Connelly 2011; Van Pelt et al. 2013). It is these landscape species that are proving to be of most conservation concern in the twenty-first century. Landscape species of upland game are often useful as umbrella species for their ecosystems and habitats (Knick and Connelly 2011; Crosby et al. 2015). Thus, by states putting significant resources into conservation for these species, they are indirectly conserving other, often nongame, species and the habitats that support them.

During the mid-twentieth century in the Southeast, declining populations of popular "farm game" species like northern bobwhites, rabbits, and squirrels resulted from decreased habitat spatial scale and quality related to natural forest succession. Conversion to forested landscapes resulted from large-scale farm abandonment, which began in the 1930s and continued throughout the 1950s. The problem was exacerbated by farm consolidation and increased mechanization aided by post–World War II production of larger, more powerful, and more efficient agricultural equipment that allowed cultivation and harvest of larger acreages with less waste grain and uncultivated acres. Resulting population declines were especially acute among species like northern bobwhites, ruffed grouse, and rabbits that are dependent on early seral stages or "early-succession" habitat types. Management efforts historically focused on artificially restocking of bobwhite. Small game programs concentrated on evaluation and development of plant materials and cultivars utilized by small game species. These programs often provided free seed or seedlings to landowners for planting as potential means to increase populations of small game species. As the science surrounding limiting factors of small game species progressed, the token management mentality of the 1960s and 1970s was gradually replaced with more holistic and comprehensive habitat management during the late 1900s. It was a gradual transition to a more habitat-based, landscape-level recovery effort.

States have increased their involvement in habitat and land management over time using a variety of approaches, but particularly with the acquisition of habitat and the use of education, collaboration, and incentives to improve wildlife habitat on other lands (both public and private). Technical assistance, in the form of direct consultation with landowners and the development of customized site-specific management plans, became standard practice. For example, in most states the US Farm Service Agency (FSA) and the Natural Resource Conservation Service (NRCS) offer the Conservation Reserve Program (CRP), which is authorized by the "Farm Bill." The CRP was initiated in the mid-1980s to provide incentives to landowners to put highly erodible land with a cropping history into permanent vegetation cover, usually grasslands, primarily to conserve soil. Biologists quickly realized the benefits to wildlife, especially upland/small game, and became close partners with NRCS and FSA in promoting and providing technical support for managing CRP habitat. The CRP has been shown to benefit many upland/small game and other species (Eggebo et al. 2003; Rodgers and Hoffman 2005; Nielson et al. 2008; Riffell and Burger 2008; Schroeder and Vander Haegen 2011; Dahlgren et al. 2016b). In the late 1990s biologists worked with US Department of Agriculture (USDA) personnel to implement mid-contract management to improve habitat quality and ensure that seed mixes included forbs for CRP establishment and management practices, primarily to benefit upland/small game. This model of technical assistance for private lands, coupled with financial incentives available through USDA Farm Bill programs, continues in many states today. Partnerships between state agencies, USDA, NGOs, and landowner groups are key to the success of this approach. In particular, identifying the key motivations and incentives that convince private landowners to manage for wildlife habitat will remain a key challenge to upland/small game programs.

State agencies have used PR funds, licenses, state funds, and other sources to manage wildlife habitat and acquire or lease land. In the latter case, the benefits to the management of upland and small game have been substantial. Acquisitions and easements are tools for providing opportunities for recreation as well as for improved management of populations. Every state owns and manages areas that have been set aside for wildlife habitat, recreation, and/or protection, with Alaska having the most at 12,800 km^2. Because of the abundance of private land in many states, these public lands offer some of the most important recreational opportunities for upland and small game. Although conservation easements do not result in state ownership of property, they often result in beneficial management and recreational opportunities. Despite the large quantities of land in public ownership, private land provides the most habitat for upland and small game in the majority of states. Even on private land, state wildlife agencies can play a major role in management through planning, research, monitoring, surveying, and education. Additionally, because wildlife are considered publicly owned even when they occur on private land, the state takes a leadership role in the recommendation, establishment, and enforcement of wildlife laws. As a result of the importance of private lands to the conservation of small game species, many eastern states' programs formerly known as small game programs have transitioned to "private lands programs," with emphasis on technical assistance to private landowners. Despite the apparent de-emphasis of small game species in this model, much of the technical assistance delivered under these programs is focused on recovery of upland game birds and other small game species. Though land ownership patterns and proportion of lands in private versus public ownership vary greatly across the United States, in the eastern part of the country land is predominantly privately owned. By providing technical assistance for management of upland/small game to private landowners, state agencies are attempting to instill and foster a stewardship ethic among private landowners that will result in greater conservation of wildlife populations, critical habitat, and other natural resources such as soil and water.

The Association of Fish and Wildlife Agencies (AFWA) is an organization with representation from wildlife agencies in every state. AFWA is divided into regional groups, including Western AFWA (WAFWA), Midwestern AFWA (MAFWA), Southeastern AFWA (SEAFWA), and Northeast AFWA (NEAFWA). The Resident Game Bird Working Group within AFWA oversees issues regarding upland game across all states. Other working groups and technical committees are part of regional AFWA organizations and include the Midwest Pheasant Study Group (MAFWA), Sage and Columbian Sharp-tailed Grouse Technical Committee

(WAFWA), Interagency Sage-Grouse Conservation Team (WAFWA), Lesser Prairie-Chicken Interstate Working Group (WAFWA), National Bobwhite Technical Committee (SEAFWA and MAFWA), and Western Quail Working Group (WAFWA). The Prairie Grouse Technical Council (PGTC) is a professional association of researchers and biologists focused on prairie grouse research and management, and state agencies have historically participated in and hosted the biennial conference, but PGTC is not currently sponsored by an AFWA organization. These AFWA groups often take the lead in upland game conservation in the United States.

Every state agency is involved in planning efforts. Plans are written to establish the agency's vision, as well as providing general goals and strategies. Most often, upland/small game plans that cover interstate boundaries are sponsored by a working group or technical committee within AFWA. In some cases, states collaborate with each other and with nongovernmental groups to produce species-specific plans across multiple jurisdictions (table 7.3). Plans involve collaborative efforts to map habitat, coordinate management strategies across state and jurisdictional boundaries, and implement restoration strategies such as translocations from one state to another (Reese and Connelly 1997; Snyder et al. 1999; Bouzat et al. 2009; World Pheasant Association and IUCN/SSC Re-introduction Specialist Group 2009; Seidel et al. 2013).

These plans have played significant roles in nationwide conservation efforts and decisions. It was the National Bobwhite Technical Committee that developed the first northern bobwhite multistate plan, which eventually led to the National Bobwhite Conservation Initiative (NBCI; http://bringbackbobwhites.org/). US-FWS relied heavily on WAFWA's Lesser Prairie-Chicken Rangewide Plan in their Endangered Species Act (ESA) listing decision, essentially partnering with the five state agencies containing lesser prairie-chicken populations for future conservation of the species. For greater sage-grouse, in September 2015, Interior Secretary Sally Jewell cited the collaboration of conservation efforts, especially within management plans, between state and federal agencies and private partners as the primary reason for an unwarranted ESA listing decision.

NGOs also contribute significantly to upland/small game conservation. A dedicated cadre of small game

Table 7.3 List of upland and small game species covered by regional management, conservation, and/or recovery plans

Species	Reference
Ruffed grouse	Dessecker et al. (2006)[a]
Spruce grouse	Williamson et al. (2008)[b]
Greater sage-grouse	Stiver et al. (2006)
Gunnison sage-grouse	Gunnison Sage-Grouse Rangewide Steering Committee (2005)
Prairie grouse	Vodehnal and Haufler (2008)[c]
Lesser prairie-chicken	Van Pelt et al. (2013)
Attwater's prairie-chicken	US Fish and Wildlife Service (2010)
North American wild turkey management plan	Currently being developed[d]
Ring-necked pheasant	Midwest Pheasant Study Group (2012)[e]
Western quail	Zornes and Bishop (2009)
Northern bobwhite	National Bobwhite Technical Committee (http://bringbackbobwhites.org/)[f]

Notes. This is an abbreviated list since there are many other plans developed by individual states and agencies. Although state employees were involved in all these plans, nongovernmental organizations often take a leadership role (Church et al. 1994).

[a] Leadership role from the Ruffed Grouse Society and Wildlife Management Institute.
[b] Leadership role from the Wildlife Management Institute.
[c] Leadership role from the North American Grouse Partnership.
[d] Leadership role from the National Wild Turkey Federation.
[e] Leadership role from Pheasants Forever.
[f] Leadership role from the National Bobwhite Technical Committee.

and upland game hunters remain supportive of numerous organizations committed to conservation of habitat and maintaining upland game hunting traditions and culture. Pheasants and Quail Forever, the Ruffed Grouse Society, the National Wild Turkey Federation, the Quail and Upland Wildlife Federation, the Wildlife Management Institute, and the North American Grouse Partnership are among NGOs dedicated to these causes. These organizations are national in scope but are generally supported by affiliates or chapters at the local level. Not only do these organizations participate in habitat management activities at specific locations, but they also influence policy for upland game conservation. The Conservation Leaders for Tomorrow organization also contributes to the future of upland/small game conservation by training natural resource university students and agency professionals without a personal background in hunting on the essential role hunting plays in wildlife conservation.

Hunter Recruitment

Upland/small game hunting has historically been the gateway for many first-time hunters. However, as mentioned above, numbers of upland/small game hunters and days spent hunting have declined in recent years (US Department of the Interior et al. 2011). A few overarching factors have likely influenced these trends. First, a rural-to-urban shift in demography has occurred over the past few generations. In addition, competition for youth and parents' time with extramural activities may have significant impact on youth hunter recruitment. People from rural communities have a much higher likelihood of participating in hunting activities (US Department of the Interior et al. 2011). Second, there is currently a baby boomer generational cohort participating in hunting activities at higher rates than previous or subsequent generations (Winkler and Warnke 2012). During the post–World War II era, more leisure time allowed for increased hunting activities from the baby boomers. When hunter age is accounted for in the frequency of upland/small game hunters, the baby boomer generational cohort becomes readily evident (Kansas Department of Wildlife, Parks, and Tourism 2010). This generation is currently retired or near retirement, and their hunting activities have begun to or will soon fade. Lastly, and possibly most significant, is the decline of upland/small game hunting opportunities that has occurred in many parts of the country. For example, many states in the Midwest, such as Indiana,

Illinois, and Ohio, had significant pheasant harvests during the mid-1900s. As agricultural practices became more intensive and urban sprawl occurred, pheasant populations declined precipitously, as did the upland hunters pursuing them. In many of these areas new hunters are being introduced to other types of hunting, such as big game, rather than upland/small game, owing to the recent widespread recovery and increase in deer populations.

It has been estimated that in 2011 there were over 400 recruitment programs for hunting, shooting, and fishing nationwide (Responsive Management and National Wild Turkey Federation 2011). These programs are coordinated by state agencies, as well as conservation organizations and sportsmen's clubs. Many recruitment programs focus the hunting experience on upland/small game. They range in scope and complexity from single-day events, such as youth dove hunts or rabbit hunts, to mentoring programs that span all aspects of hunting from scouting and hunt preparation to game cleaning and cooking. Goals of these programs are often expressed as hunter recruitment, although retention of young hunters and reactivation of lapsed hunters have been increasingly emphasized in recent years. Hunting is largely a contagious activity learned from mentors and shared with peers. The difficulty of attracting new hunters from nonhunting families has been noted many times and remains an obstacle to youth recruitment programs. It is not yet clear whether these recruitment activities truly attract new hunters or simply provide opportunities for youth who would have become hunters regardless. Youth hunts for upland/small game and migratory birds are still popular activities within recruitment programs. Advantages to starting with upland/small game hunting include low initial investment of gear and clothing; abundant game populations in some areas, resulting in higher likelihood of success (i.e., sighting or harvesting game); and more harvest opportunities per field time, engaging young hunters. Further, new and young hunters often make connections with hunting dogs in pursuit of upland/small game like grouse, pheasants, rabbits, and quail.

Summary

Upland/small game programs have made significant contributions to the conservation, management, and current population status of many species across our nation. In the case of some species, the efforts have been prominent and substantial successes have occurred. In other instances, management for species has been passive and state agencies have merely monitored harvest where possible. We anticipate that the role of conservation for upland/small game species will continue to grow for state agencies, especially as landscape-level habitat degradation and the need for mitigating measures increase. Harvest management will likely continue to be a dominant theme for state agencies even as the wildlife conservation funding "tent" is enlarged and more nonconsumptive partners engage. Furthermore, we expect that harvest management strategies for upland/small game will continue to evolve and the need for more science-based adaptive harvest management regulations will increase. Inherent with adaptive harvest management is the need for population-level monitoring data, and finding the resources for this more in-depth management approach will be a challenge to state agencies and their upland/small game programs. Recruitment of and sustaining upland/small game hunters will also provide a significant test for state wildlife agencies. However, if populations of upland/small game are at risk or few in number, there will be little reason for upland and small game hunter recruitment, and thus population and habitat conservation should remain the primary objective for state wildlife agency upland/small game programs.

LITERATURE CITED

Alpizar-Jara, R., E. N. Brooks, K. H. Pollock, D. E. Steffen, J. C. Pack, and G. W. Norman. 2001. An eastern wild turkey population dynamics model for Virginia and West Virginia. Journal of Wildlife Management 65:415–424.

Barker, R. J. 1991. Nonresponse bias in New Zealand waterfowl harvest surveys. Journal of Wildlife Management 55:126–131.

Barrowclough, G. F., J. G. Groth, L. A. Mertz, and R. J. Gutierrez. 2004. Phylogeographic structure, gene flow and species status in blue grouse (Dendragapus obscurus). Molecular Ecology 13:1911–1922.

Bouzat, J. L., J. A. Johnson, J. E. Toepfer, S. A. Simpson, T. L. Esker, and R. L. Westemeier. 2009. Beyond the beneficial effects of translocations as an effective tool for the genetic restoration of isolated populations. Conservation Genetics 10:191–201.

Boyce, M. S., A. R. E. Sinclair, and G. C. White. 1999. Seasonal compensation of predation and harvesting. Oikos 87:419–426.

Braun, C. E., D. A. Budeau, and M. A. Schroeder. 2015. Fall population structure of sage-grouse in Colorado and Oregon. Wildlife Technical Report 005-2015. Oregon Department of Fish and Wildlife. www.dfw.state.or.us/wildlife/research/docs/Fall_Popn_Structure_Sage-grouse_v3182015.pdf.

Braun, C. E., and M. A. Schroeder. 2015. Age and sex identification from wings of sage-grouse. Wildlife Society Bulletin 39:182–187.

Broms, K., J. R. Skalski, J. J. Millspaugh, C. A. Hagen, and J. H. Shulz. 2010. Using statistical population reconstruction to estimate demographic trends in small game populations. Journal of Wildlife Management 74:310–317.

Buckland, S. T., D. R. Anderson, K. P. Burnham, and J. L. Laake. 1993. Distance sampling: Estimating abundance of biological populations. Chapman & Hall, London, UK.

Caughley, G. 1985. Harvesting of wildlife: Past, present and future. Pages 3–14 in S. L. Beasom and S. F. Roberson, editors. Game harvest management. Caesar Kleberg Wildlife Research Institute, Kingsville, Texas, USA.

Church, K. E., J. W. Connelly, and J. W. Enck. 1994. The role of non-governmental organizations in gamebird conservation. Transactions of the North American Wildlife and Natural Resource Conference 59:488–493.

Connelly, J. W., A. D. Apa, R. B. Smith, and K. P. Reese, 2000. Effects of predation and hunting on adult sage-grouse *Centrocercus urophasianus* in Idaho. Wildlife Biology 6:227–232.

Connelly, J. W., J. Gammonley, and J. M. Peek. 2005. Harvest Management. Pages 658–690 in C. E. Braun, editor. Techniques for wildlife investigation and management, 6th edition. The Wildlife Society, Bethesda, Maryland, USA.

Connelly, J. W., K. P. Reese, E. O. Garton, and M. L. Commons-Kemner. 2003a. Response of greater sage-grouse *Centrocercus urophasianus* populations to different levels of exploitation in Idaho, USA. Wildlife Biology 9:335–340.

Connelly, J. W., K. P. Reese, and M. A. Schroeder. 2003b. Monitoring of greater sage-grouse habitats and populations. Station Bulletin 80, College of Natural Resources Experiment Station, Moscow, Idaho, USA.

Crosby, A. D., R. D. Elmore, D. M. Leslie, and R. E. Will. 2015. Looking beyond rare species as umbrella species: Northern bobwhite (*Colinus virginianus*) and conservation of grassland and shrubland birds. Biological Conservation 186:233–240.

Dahlgren, D. K. 2010. Greater sage-grouse population estimation study: Deseret Land and Livestock and Parker Mountain. Preliminary Report, Community-Based Conservation Program, Utah State University Extension, Logan, Utah, USA.

Dahlgren, D. K., M. G. Guttery, T. A. Messmer, D. Caudill, R. D. Elmore, R. Chi, and D. N. Koons. 2016a. Evaluating vital rate contributions to greater sage-grouse population dynamics to inform conservation. Ecosphere 7(3):e01249. 10.1002/ecs2.1249.

Dahlgren, D. K., R. D. Rodgers, R. D. Elmore, and M. R. Bain. 2016b. Grasslands of western Kansas, north of the Arkansas River. Pages 259–279 in D. Haukos and C. Boal, editors, Ecology and conservation of lesser prairie-chickens. Studies in Avian Biology 48. CRC Press, Boca Raton, Florida, USA.

Dessecker, D. R., G. W. Norman, and S. J. Williamson, editors. 2006. Ruffed grouse conservation plan. Association of Fish and Wildlife Agencies. Ruffed Grouse Society, Coraopolis, Pennsylvania, USA.

Dickson, J. G. 1992. The wild turkey: Biology and management. National Wild Turkey Federation, Edgefield, South Carolina.

Drummer, T. D., R. G. Corace, III, and S. J. Sjogren. 2011. Sharp-tailed grouse lek attendance and fidelity in upper Michigan. Journal of Wildlife Management 75:311–318.

Eggebo, S. L., K. F. Higgins, D. E. Naugle, and F. R. Quamen. 2003. Effects of CRP field age and cover type on ring-necked pheasants in eastern South Dakota. Wildlife Society Bulletin 31:779–785.

Ellison, L. N. 1991. Shooting and compensatory mortality in tetraonids. Ornis Scandinavica 22:229–240.

Emmons, S. R., and C. E. Braun. 1984. Lek attendance of male sage-grouse. Journal of Wildlife Management 48:1023–1028.

Errington, P. L. 1945. Some contributions of a fifteen-year study of northern bobwhite to a knowledge of population phenomena. Ecological Monographs 15:1–34.

Errington, P. L., and F. N. Hamerstrom, Jr. 1935. Bob-white winter survival on experimentally shot and unshot areas. Iowa State College Journal of Science 9:625–639.

Fedy, B. C., and K. E. Doherty. 2011. Population cycles are highly correlated over long time series and large spatial scales in two unrelated species: Greater sage-grouse and cottontail rabbits. Oecologia 165:915–924.

Flanders-Wanner, B. L., G. C. White, and L. L. McDaniel. 2004. Validity of prairie grouse harvest-age ratios as production indices. Journal of Wildlife Management 68:1088–1094.

Gunnison Sage-Grouse Rangewide Steering Committee. 2005. Gunnison sage-grouse rangewide conservation plan. Colorado Division of Wildlife, Denver, Colorado, USA.

Guthery, F. S., M. J. Peterson, and R. R. George. 2000. Viability of northern bobwhite populations. Journal of Wildlife Management 64:646–662.

Hagen, C. A., and T. M. Loughin. 2008. Productivity estimates from upland bird harvests: Estimating variance and necessary sample sizes. Journal of Wildlife Management 72:1369–1375.

Hansen, C. P., J. J. Millspaugh, and M. A. Rumble. 2011. Occupancy modeling of ruffed grouse in the Black Hills National Forest. Journal of Wildlife Management 75:71–77.

Hefley, T. J. 2014. New statistical method for analysis of historical data from wildlife populations. PhD diss., University of Nebraska, Lincoln, Nebraska, USA.

Hefley, T. J., A. J. Tyre, and E. E. Blankenship. 2013. Statistical indicators and state-space population models predict extinction in a population of bobwhite quail. Theoretical Ecology 6:319–331.

Hudson, P. J., D. Newborn, and P. A. Robertson. 1997. Geographical and seasonal patterns of mortality in red grouse *Lagopus lagopus scoticus* populations. Wildlife Biology 3: 79–87.

Humberg, L. A., T. L. Devault, and O. E. Rhodes, Jr. 2009. Survival and cause specific mortality of wild turkeys in northern Indiana. American Midland Naturalist 161:313–322.

IAFWA (International Association of Fish and Wildlife Agencies). 2002. Economic importance of hunting in America. International Association of Fish and Wildlife Agencies, Washington, DC, USA.

Jenkins, D., A. Watson, and G. R. Miller. 1964. Predation and red grouse populations. Journal of Applied Ecology 1:183–195.

Kansas Department of Wildlife, Parks, and Tourism. 2010. Small game hunter activity surveys 2006–2010. Pratt, Kansas, USA. http://ksoutdoors.com/Services/Research-Publications/Wildlife-Research-Surveys/Small-Game-Surveys.

Kennamer, J. E., M. Kennamer, and R. Brenneman. 1992. History. Pages 6–17 *in* J. G. Dickson, editor. The wild turkey: Biology and management. Stackpole Books, Harrisburg, Pennsylvania.

Knick, S. T., and J. W. Connelly. 2011. Greater sage-grouse: Ecology and conservation of a landscape species and its habitats. Studies in Avian Biology 38. University of California Press, Berkeley, California, USA.

Leif, A. P. 1994. Survival and reproduction of wild and pen-reared ring-necked pheasant hens. Journal of Wildlife Management 58:501–506.

Leopold, A. 1933. Game management. Charles Scribner's Sons, New York, New York, USA.

———. 1938. Chukaremia. Outdoor America 3:3.

Manly, B. F. 1998. Randomization, bootstrap and Monte Carlo methods in biology. Chapman & Hall, London, UK.

Mehls, C. L., K. C. Jensen, M. A. Rumble, and M. C. Wimberly. 2014. Multi-scale habitat us of male ruffed grouse in the Black Hills National Forest. Prairie Naturalist 46:21–33.

Midwest Pheasant Study Group. 2012. National wild pheasant conservation plan. N. B. Veverka, editor. Association of Fish and Wildlife Agencies.

Morgan, J. P., K. Duren, and T. V. Dailey. 2014. NBCI Coordinated Implementation Program. Addendum, The National Bobwhite Conservation Initiative: A range-wide plan for recovering bobwhites. National Bobwhite Technical Committee Technical Publication, Version 2.0. Knoxville, Tennessee, USA.

Nielson, R. M., L. L. Mcdonald, J. P. Sullivan, C. Burgess, D. S. Johnson, D. H. Johnson, S. Bucholtz, S. Hyberg, and S. Howlin. 2008. Estimating the response of ring-necked pheasants (*Phasianus colchicus*) to the Conservation Reserve Program. Auk 125:434–444.

Nusser, S. M., W. R. Clark, J. Wang, and T. R. Bogenschutz. 2004. Combining data from state and national monitoring surveys to assess large-scale impacts of agricultural policy. Journal of Agricultural, Biological, and Environmental Statistics 9:381–397.

Pack, J. C., G. W. Norman, C. I. Taylor, D. E. Steffen, D. A. Swanson, K. H. Pollock, and R. Alpizar-Jara. 1999. Effects of fall hunting on wild turkey populations in Virginia and West Virginia. Journal of Wildlife Management 63:964–975.

Pendleton, G. W. 1992. Nonresponse patterns in the federal waterfowl hunter questionnaire survey. Journal of Wildlife Management 56:344–348.

Reese, K. P., and J. W. Connelly. 1997. Translocations of sage grouse in North America. Wildlife Biology 3:87–93.

Responsive Management and National Wild Turkey Federation. 2011. Effectiveness of hunting, shooting, and fishing recruitment and retention programs. Final report. Harrisonburg, Virginia, USA.

Rice, C. G. 2003. Utility of pheasant call counts and brood counts for monitoring population density and predicting harvest. Western North American Naturalist 63:178–188.

Riffell, S. D., and L. W. Burger. 2008. Effects of the Conservation Reserve Program on northern bobwhite and grassland birds. Environmental Monitoring and Assessment 146:309–323.

Rodgers, R. D., and R. W. Hoffman. 2005. Prairie grouse response to Conservation Reserve Grasslands: An overview. Pages 122–130 *in* A. W. Allen and M. W. Vandever, editors. The Conservation Reserve Program: Planting for the future: Proceedings of a national conference. US Geological Survey, Biological Resources Division, Scientific Investigation Report 2005-5145.

Roseberry, J. L., and W. D. Klimstra. 1992. Further evidence of differential harvest rates among bobwhite sex-age groups. Wildlife Society Bulletin 20:91–94.

Sandercock, B. K., E. B. Nilsen, H. Brøseth, and H. C. Pedersen. 2011. Is hunting mortality additive or compensatory to natural mortality? Effects of experimental harvest on the survival and cause-specific mortality of willow ptarmigan. Journal of Animal Ecology 80:244–258.

Schroeder, M. A., and C. E. Braun. 1992. Greater prairie-chicken attendance at leks and stability of leks in Colorado. Wilson Bulletin 104:273–284.

Schroeder, M. A., and M. Vander Haegen. 2011. Response of greater sage-grouse to the Conservation Reserve Program in Washington State. Chapter 22 *in* S. T. Knick and J. W. Connelly, editors. Greater sage-grouse: Ecology and conservation of a landscape species and its habitats. Studies in

Avian Biology 38. University of California Press, Berkeley, California, USA.

Sedinger, J. S, G. C. White, S. Espinosa, E. T. Partee, and C. E. Braun. 2010. Assessing compensatory versus additive harvest mortality: An example using greater sage-grouse. Journal of Wildlife Management 74:326–332.

Seidel, S. A., C. E. Comer, W. C. Conway, R. W. Deyoung, J. B. Hardin, and G. E. Calkins. 2013. Influence of translocations on eastern wild turkey population genetics in Texas. Journal of Wildlife Management 77:1221–1231.

Small, R. J., J. C. Holzwart, and D. H. Rusch. 1991. Predation and hunting mortality of ruffed grouse in central Wisconsin. Journal of Wildlife Management 55:512–520.

Smith, A., and T. Willebrand. 1999. Mortality causes and survival rates of hunted and unhunted willow grouse. Journal of Wildlife Management 63:722–730.

Snyder, J. W., E. C. Pelren, and J. A. Crawford. 1999. Translocation histories of prairie grouse in the United States. Wildlife Society Bulletin 27:428–432.

Stiver, S. J., A. D. Apa, J. R. Bohne, S. D. Bunnell, P. A. Deibert, S. C. Gardner, M. A. Hilliard, C. W. McCarthy, and M. A. Schroeder. 2006. Greater sage-grouse comprehensive conservation strategy. Western Association of Fish and Wildlife Agencies. Cheyenne, Wyoming, USA.

Tapley, J. L., R. K. Abernethy, and J. E. Kennamer. 2001. Status and distribution of the wild turkey in 1999. Proceedings of the National Wild Turkey Symposium 8:15–22.

Trefethen, J. B. 1975. An American crusade for wildlife. Winchester Press, New York, New York, USA.

US Department of the Interior, US Fish and Wildlife Service, and US Department of Commerce, US Census Bureau. 2011. National survey of fishing, hunting, and wildlife-associated recreation.

US Fish and Wildlife Service. 2010. Attwater's prairie-chicken recovery plan, 2nd revision. Albuquerque, New Mexico, USA.

Van Pelt, W. E., S. Kyle, J. Pitman, D. Klute, G. Beauprez, D. Schoeling, A. Janus, and J. Haufler. 2013. The lesser prairie-chicken range-wide conservation plan. Western Association of Fish and Wildlife Agencies, Cheyenne, Wyoming, USA.

Vodehnal, W. L., and J. B. Haufler, editors. 2008. A grassland conservation plan for prairie grouse. North American Grouse Partnership, Fruita, Colorado, USA.

Walsh, D. P., G. C. White, T. E. Remington, and D. C. Bowden. 2004. Evaluation of the lek-count index for greater sage-grouse. Wildlife Society Bulletin 32:56–68.

Wellendorf, S. D., W. E. Palmer, and P. T. Bromley. 2004. Estimating call rates of northern bobwhite coveys and censusing population. Journal of Wildlife Management 68:672–682.

Williams, C. K., F. S. Guthery, R. D. Applegate, and M. J. Peterson. 2004. The northern bobwhite decline: Scaling our management for the twenty-first century. Wildlife Society Bulletin 32:861–869.

Williams, C. K., A. R. Ives, and R. D. Applegate. 2003. Population dynamics across geographical ranges: Time-series analyses of three small game species. Ecology 84:2654–2667.

Williamson, S. J., D. Keppie, R. Davison, D. Budeau, S. Carrière, D. Rabe, and M. A. Schroeder. 2008. Spruce grouse continental conservation plan. Association of Fish and Wildlife Agencies, Washington, DC, USA.

Winkler, R., and K. Warnke. 2012. The future of hunting: An age-period-cohort analysis of deer hunter decline. Population and Environment 34:460–480.

World Pheasant Association and IUCN/SSC Re-introduction Specialist Group, editors. 2009. Guidelines for the re-introduction of galliformes for conservation purposes. IUCN, Gland, Switzerland, and World Pheasant Association, Newcastle-upon-Tyne, UK.

Zornes, M, and R. A. Bishop. 2009. Western quail conservation plan. S. J. Williamson, editor. Wildlife Management Institute, Cabot, Vermont, USA.

8

State Management of Furbearing Animals

Tim L. Hiller,
H. Bryant White,
and John D. Erb

Furbearing animals have been consumptively used by humans for cultural, religious, subsistence, economic, and other purposes for thousands of years (Hiller and Vantassel 2015). During centuries of exploration and colonization by Europeans in North America, predator control, extensive land-use changes and habitat loss, and unregulated commercial harvest led to substantial population declines (e.g., North American beaver [*Castor canadensis*], gray wolf [*Canis lupus*]) or extirpation (e.g., sea mink [*Mustela macrodon*]) of many furbearing species. Important precursors to contemporary furbearer management essentially did not exist until protective laws became increasingly implemented after 1900 under state authority following the public trust doctrine and the North American Model of Wildlife Conservation (Sanderson 1982; chaps. 1 and 2 of this volume).

Furbearer management and research have a rich and important history. Fritzell and Johnson (1982) suggested that furbearer management may be the oldest form of wildlife management in North America, with early examples related to efforts by Hudson's Bay Company to incorporate a form of sustained-yield harvest into furbearer management. Furthermore, some wildlife management principles that remain important today have early roots in furbearer research. For example, the concept of compensatory mortality and its application in wildlife management was proposed by Errington (1946, 1956) based in part on his observations of muskrats (*Ondatra zibethicus*).

In the first half of the twentieth century, furbearer management involved a mix of protective or restorative measures and control or eradication, depending on species. Early protective efforts through state furbearer management focused on rare species (e.g., American marten [*Martes americana*], fisher [*Pekania pennanti*], beaver) and closely regulating (or prohibiting) harvest of certain furbearer populations (Sanderson 1982; Batcheller et al. 2000). Concurrently, bounties and intense control programs were in place for many predatory furbearing species. Although the need or public support for this type of management has somewhat subsided in modern times, the large and diverse array of species and conditions under the umbrella of furbearer management suggests that a mix of management objectives will continue to characterize these programs into the foreseeable future. Federal involvement in furbearer management includes, but is not limited to, species listed under the Convention on International Trade of Endangered Species (CITES; CITES 2013), species listed under the Endangered Species Act of 1973 (US Fish and Wildlife Service 2013), wildlife damage management and human–wildlife conflict resolution (US Department of Agriculture 2015), general prohibition of harvest on certain federally managed lands (e.g., National Parks), habitat management, population monitoring, and research. These activities often occur in partnership or consultation with state wildlife agencies.

Although furbearing animals may be defined as the

group of mammalian species either currently or historically harvested primarily for their pelts, a definition related more to management than to ecology, this does not imply that current management objectives of state wildlife agencies are limited only to sustainable harvest. Furbearer management continues to include restoration and conservation of species of concern (e.g., reintroduction of fishers and North American river otters [*Lontra canadensis*]; Lewis 2014; Mowry et al. 2015), but it has also morphed to include issues related to damage management of abundant (or over-abundant) species, invasive species management (e.g., nutria [*Myocaster coypus*]; Kendrot 2013), management to reduce negative impacts on endangered species, and other components while concurrently addressing increasing social and political pressures from diverse stakeholder groups (Batcheller et al. 2000).

Furbearing animals encompass diverse taxa (≥27 species in North America in ≥10 families within the orders Carnivora, Didelphimorphia, and Rodentia) in equally diverse ecosystems and across trophic levels (Deems and Pursley 1978; Fritzell and Johnson 1982; White et al. 2015). Although the advancement of knowledge of furbearer ecology and management has been somewhat episodic in the past half century and has varied widely across the array of furbearing species, substantial attention has been placed on this important and diverse group of species (e.g., Chapman and Pursley 1981; Novak et al. 1987a; Clark and Fritzell 1992).

To achieve regulatory efficacy, management of furbearing animals includes unique approaches but also some common with other taxa. Whereas the economic value, if any, related to harvest of other taxa (e.g., guiding fees for big game species, sale of antlers) typically is not directly related to species population management, the economic value of furbearers must be directly considered when setting harvest management objectives for many species (Fritzell and Johnson 1982). High market demand for furs can create a complex management scenario of balancing sustainable harvest with high interest and effort from trappers and hunters. Annual market demand has fluctuated widely, directly affecting harvest of furbearers, including at the species level. Retail sales of wild and ranched furs in the United States alone total $1.4 billion per year as part of a $35.8 billion global industry (Fur Information

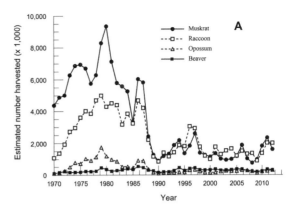

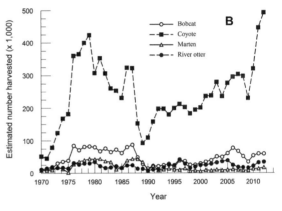

Furbearer harvest from 1970 to 2012 for eight species.
Source: Association of Fish and Wildlife Agencies (2015c).

Council of America 2010), with sales not limited to the fashion industry, but also of pragmatic importance in societies with colder climates. The global fur industry is generally composed of 15–20 percent wild furs (Fur Institute of Canada, n.d.). Secondary markets also exist for sale of furbearer parts (e.g., animal glands, beaver castor), as well as localized markets for meat (e.g., raccoon [*Procyon lotor*], muskrat).

Furbearer harvest includes both hunting (e.g., predator calling, pursuit with dogs) and trapping, where the former is largely limited to canids, felids, and raccoons. Regulated trapping currently occurs in 49 of 50 US states (Hawaii has no furbearers; Armstrong and Rossi 2000). Although regulations can vary widely among states, each state requires some form of license to participate. Licenses for hunting furbearers (or predators) vary by state and may or may not be furbearer specific or required for species classified as predators, thereby complicating inferences drawn from furbearer-hunting license sales across the United States. States also often require licenses for fur buyers, private wildlife control operators, and taxidermists to ensure regulatory compliance and collect additional data for making informed furbearer management decisions.

Avocational trappers and hunters have numerous reasons for going afield to harvest furbearers, including tradition and lifestyle, nature appreciation, recreation, and, for some, the opportunity to profit economically (Todd and Boggess 1987; Daigle 1997). However, for most trappers the annual income derived from trapping is under $1,000, and approximately 80 percent of trappers indicate that trapping is not an important source of income (AFWA 2005). Fluctuations in annual license revenues and furbearer harvest levels may be affected by anti-trapping activities (e.g., ballot initiatives that restrict furbearer hunting or trapping), long-term recruitment of trappers, fluctuations in pelt values, and other factors (Payne 1980; Armstrong and Rossi 2000). Arguably, no other group of state-managed wildlife is as directly linked to economics for individual consumptive users throughout the United States.

It is important for furbearer managers to understand and stay current on factors influencing market values, many of which are associated with global economies, global weather, and local conditions (e.g., gasoline prices). While it may be difficult to predict these conditions in advance of annual regulatory decisions, managers can consult with those best suited for making such predictions (e.g., international fur auction companies). Importantly, responses of trappers to changing markets may exhibit time lags that need to be considered (e.g., high pelt prices last year may result in high effort next year, even if early projections are for low pelt prices). The aforementioned economic factors affect annual participation through the number of licensed trappers, which has shown a long-term decline in the United States. In 1979, there were an estimated 500,000 trappers (Todd and Boggess 1987), steadily declining to 300,000 in 1987 (International Association of Fish and Wildlife Agencies 1992) and to about 142,000 in 2003 (AFWA 2005). Currently there are an estimated 180,000 licensed trappers in the United States (AFWA 2015e).

As of 2007, some form of voluntary or mandatory education requirement is in place for trappers and fur-bearer hunters in 48 states, through either the state wildlife agency or a state trappers association (AFWA 2007). There are also numerous trapper (and hunter) education materials, online courses, and hands-on workshops available through state trappers associations and wildlife agencies (e.g., Wisconsin Cooperative Trapper Education Program), national trappers organizations (e.g., Fur Takers of America [2015] Trapper's College), and AFWA (2015d; e.g., Trapper Education Manual). In addition, AFWA (2015c) offers the online North American Trapper Education Course, which is being used by some states to fulfill their trapper education requirements. The general purpose of trapper education is to help individuals acquire the basic knowledge, skills, and judgment to selectively and responsibly trap furbearers. Trapper education emphasizes trapper ethics and animal welfare and helps new trappers understand state trapping regulations. Hunter education covers topics such as safe gun handling, hunter ethics, wildlife identification, and hunting techniques; options vary by state and may include an online course and exam, workbook, shooting range participation, and written exam.

Harvest regulations (e.g., season dates and lengths, harvest limits) may be species specific or encompass groups of furbearers, depending on the objectives of a given state wildlife agency. In situations where species-specific regulations are in place, state agencies need to assess how regulatory changes may affect other furbearing animals. Regulatory adjustments designed to change harvest of one species may affect harvest of another species at some level. To avoid incidental harvest of furbearing species that are not allowed to be harvested owing to conservation concerns or because of different season timing or length, methods have been developed and distributed to trappers (e.g., Golden and Krause 2003; Hiller and White 2013). Wildlife managers should also carefully consider ways to structure harvest seasons that minimize incidental harvests. For example, noncontiguous seasons for two species commonly caught in the same trap sets should be avoided when consistent with the conservation goals for each species, and combined harvest limits can be considered for two or more species if limited-quota harvests (i.e., individual trapper quotas or maximum total harvest allowed for a given area) are necessary for those species.

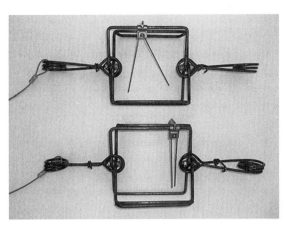

Body-gripping trap with trigger set in common "V" configuration (*top*) and offset to one side (*bottom*) to avoid smaller bodies of nontarget furbearing species. *All photos this page courtesy of Tim L. Hiller, Wildlife Ecology Institute.*

Capture selectivity is normally included in trapper education materials and focuses on trapping methods to reduce incidental captures. Techniques to improve selectivity of traps include simple modifications of trigger configuration or placement (e.g., moving triggering devices to one side on body-gripping traps may allow capture of larger-bodied beavers but reduce captures of river otters), adjustment of pan tension on foothold traps (to reduce the capture of furbearers of lesser body weights), avoidance of baits and lures that may be more attractive to nontarget species, and placement of traps (e.g., trap sets in trees to avoid nonclimbing furbearers), to name but a few methods. Snares and cable-restraint devices (AFWA 2009) may also be used selectively by adjusting the size of the capture loop, adjusting the height of the capture loop from the ground, incorporating a break-away device (which allows the cable lock to release when greater forces than the target species can generate are exerted; see AFWA 2009), and more generally being selective with set locations based on animal sign (e.g., tracks, scat). Further, foot-encapsulating traps (often called "dog-proof" traps) are specifically designed to capture only animals that have enough manual dexterity (e.g., raccoons) to pull or push the enclosed trigger within a small-diameter opening. The use of these techniques and devices can help trappers and furbearer managers improve capture selectivity, resulting in more precise management of some species.

Foothold trap in (*A*) set position and (*B*) sprung position. Tension on the circular pan can be adjusted using the nut and bolt at its pivot point, providing increased pan tension to avoid nontarget species of lesser body weights.

Season timing and lengths are generally chosen to correspond with high pelt quality (i.e., pelt primeness) during colder fall and winter months, although some states implement less restrictive year-round seasons for very common species to meet other management objectives, such as wildlife damage management. Other considerations include timing seasons to minimize conflicts with other outdoor enthusiasts, effects of season timing and length on sex/age ratios in harvest, population resilience or the sensitivity of a species to overharvest, and ultimately the population objective (abundance or trend) for each species. Finally, managers must consider trade-offs in changing season timing, season length, or harvest quotas. Ultimately, these decisions are driven by the goal of maintaining populations at desirable levels through sustainable harvest. However, routinely changing season attributes can hamper the ability to use harvest data as a population index, especially if data about hunter or trapper effort are not collected. Factors such as conservation status, population resilience, and potential for overabundance to cause damage must be balanced with the opportunity to gain knowledge through regulatory stability. Furbearer (and wildlife) management and policy decisions increasingly involve strong reliance on science, but there are also social, political, economic, and other factors that increasingly affect these decisions (Batcheller et al. 2000).

Data Collection Methods

Wolfe and Chapman (1987:101) asked "whether wildlife management, after 50 years, has evolved beyond the stage of an art and become a science," in reference to Leopold's (1933) description of wildlife management as an art. Now, many decades later, we describe how science is implemented in modern furbearer conservation and management decisions. Admittedly, the application of appropriate scientific and statistical methods for furbearer management has at times lagged behind that of large mammalian game species, largely owing to limited state resources and the relative difficulty of collecting and analyzing data across the multitude of furbearing species within a jurisdiction. However, substantial advancements have been made in recent years and for many species, particularly through research projects funded by state hunting and trapping license revenues; federal excise taxes on firearms, ammunition, and archery equipment (i.e., Wildlife Restoration Act of 1937; US Fish and Wildlife Service 2015); and other sources.

Harvest Data

Harvest data are critical for making informed management decisions for many furbearing species. These data may be collected through voluntary or mandatory harvest reporting, hunter-trapper surveys, or more coarse information from fur buyers and fur auctions. Manda-

tory harvest reports typically yield the most useful data in terms of quality and can include number of each species harvested, effort (e.g., to estimate the number of captures per trap night for each species or the number harvested per day hunted), location (e.g., county) of harvest, and other information. Although harvest data may contain biases that must be understood, such data can be useful as long-term indices for assessing change in distribution, abundance, or other demographic characteristics. When estimating the number of each species harvested, particularly for species with lower reproductive potential, managers may view the results conservatively, including an assumption that harvest values are minimums or by using the upper bound of a confidence interval. Each state typically maintains a harvest database, and an electronic database was constructed in 2010 as an open data source (AFWA 2015b), providing a wealth of basic harvest information. Historic data on furbearer harvests in North America can also be found in Novak et al. (1987b).

The CITES international agreement has the goal of ensuring that international trade does not threaten the viability of listed species (see Wijnstekers 2011; CITES 2013). In the United States, the US Fish and Wildlife Service, with cooperation from state wildlife agencies, administers responsibilities related to the agreement. Procedures are in place to ensure that harvest of CITES-listed species is not detrimental to their survival at the national level and that legally acquired specimens are appropriately identified prior to export. CITES has implications for furbearer management as a result of the international trade in pelts of several CITES-listed furbearers (e.g., bobcats [*Lynx rufus*], gray wolves, river otters). Some furbearing species are listed under CITES Appendix II simply because they are similar in appearance to Appendix I species (i.e., those species threatened with extinction) and therefore difficult for trade officials to differentiate upon export. As a tool to identify species and legal acquisition thereof, the US Fish and Wildlife Service provides authorized state and tribal authorities with plastic locking tags that must be applied to pelts of CITES-listed furbearers before they can be exported from the United States. Tags are not federally required if the pelts are not exported, but some states may require these (or other) tags for state management purposes. For a jurisdiction to obtain blanket CITES export authority for a species, the US CITES program requires state and tribal authorities to closely monitor the harvest of Appendix II species in their jurisdictions to ensure that harvests are not negatively impacting populations. However, if a jurisdiction does not have blanket CITES authority for a species, individual trappers and hunters can also apply directly for an export permit through the US Fish and Wildlife Service if harvest of the species is legal.

Some state wildlife agencies also implement a non-CITES species tagging program or permit system to help manage harvest of certain furbearing species. Tags can assist in assuring that harvest is limited to desired levels, is reliably enumerated, and can facilitate acquisition of more accurate biological data from each specimen. Tagging may be implemented for species of naturally low population densities, high vulnerability of harvest, and low reproductive rates (e.g., fisher and marten in many states). Although tags can be distributed to trappers and hunters in various ways, many states have a mandatory check-in process where pelts are tagged and data are collected from harvested furbearers.

Whether mandatory or voluntary, collection of biological samples and data from harvested furbearers can often be a cost-effective way to advance our ecological knowledge and reduce uncertainty about effects of various harvest regulations on furbearer populations. A tooth, jawbone, or skull from a harvested furbearer may be collected to age individuals through cementum annuli analysis or other approaches (e.g., Jenks et al. 1984; Erb et al. 1999). The sex of harvested individuals is also typically recorded, although errors in sex identification may occur and should be considered during management decisions (e.g., Hiller et al. 2014). Sex and age data may be used to describe the composition of harvest, as potential population indices or inputs to population models, or to directly estimate population size through methods such as statistical population reconstruction (SPR) when suitable data are collected. Advantages of using SPR include utilization of harvest data that may already be routinely collected for certain furbearing species over large spatial and temporal extents, ability to estimate abundance and trend over large geographic areas, updating estimates

through integration of additional data, and estimation of natural and harvest mortality and recruitment (Gove et al. 2002; Skalski et al. 2011). Originally developed using fisheries data and called virtual population analysis (see Fournier and Archibald 1982), SPR has only recently been applied to furbearer management. For example, Skalski et al. (2011) used SPR to suggest that a downward trend in American marten populations in Michigan's Upper Peninsula warranted continued monitoring and consideration of regulatory adjustments to ensure sustainable harvests.

Across taxa, juveniles are often considered the most vulnerable age class to harvest, and adult females are typically the most closely monitored sex/age class because many populations, especially those of longer-lived (K-selected) species, are most sensitive to changes in adult female survival (e.g., Strickland 1994; Gorman et al. 2008; for general discussions, see Dixon and Swift 1981; Clark and Fritzell 1992). Although differential vulnerability may not affect analysis of trends in harvest ratios (assuming that vulnerability does not exhibit a trend), it must be considered before any attempt to use harvest data to infer population ratios. Comparatively inexpensive yet reliable data to index reproductive output can also be obtained from analysis of reproductive organs collected from harvested females (Gilbert 1987). Diet can also be assessed through collection of biological samples from pelts or carcasses of harvested furbearers (e.g., hair, muscle, or claw samples; stomach contents) using methods such as stable isotope analysis (e.g., Ben-David et al. 1997; Urton and Hobson 2005) or visual or microscopic examination. Sample type ultimately determines the temporal scope of the diet analysis.

Noninvasive Surveys

Noninvasive surveys encompass a variety of field data collection methods such as track and scat surveys, camera-based surveys, house or den surveys, spotlight surveys, and collection of genetic samples using noninvasive techniques (e.g., hair-snagging devices). These data can be used to assess species distribution, occupancy, abundance, and other metrics depending on the quality and quantity of data, individual species' characteristics, and other factors (see Long et al. 2008). Survey data can be collected by state and federal agency biologists, trained or experienced citizens (e.g., furbearer observations by bowhunters), or a broad array of volunteer citizen scientists (e.g., camera surveys). Because it is not always possible to design a survey that provides sufficiently reliable data for all species of interest, furbearer managers must often prioritize needs and design a survey that maximizes the number of species with sufficiently reliable data, or produces reliable data for the highest-priority species.

Track surveys (or other sign surveys) have been commonly used for furbearing animals in part because multiple species can often be detected from the same survey (Long et al. 2008), an important consideration for furbearer managers. Sign typically refers to the visual detection and identification of tracks or scats (Heinemeyer et al. 2008), but it can also involve auditory detection for some species (e.g., coyotes; Dunbar and Giordano 2003). Survey design will be determined by species of interest, suitable times or methods for adequately detecting those species, and other logistical and statistical considerations (e.g., Zielinski and Kucera 1995; Gompper et al. 2006). Although most sign surveys are based on ground searches, aerial sign surveys (e.g., snow tracks or lodge-house counts) are possible for some furbearers in sufficiently open landscapes (e.g., Magoun et al. 2007; Johnston and Windels 2015). Although scent-station surveys can utilize a wide variety of olfactory attractants, tracking mediums, and sampling designs (Long et al. 2008), the term historically referred to a survey that included attractants placed in circular (about 1 meter in diameter) stations of sifted soil and distributed along transects (Roughton 1982; Roughton and Sweeny 1982). As with any survey, proper sampling design and power to detect trends of interest should be considered for all sign surveys (Sargeant et al. 1998, 2003).

Cameras are not new to wildlife monitoring (Kays and Slauson 2008; Kucera and Barrett 2011), but the simultaneous development of improved remotely triggered cameras and rigorous analytical methods has bolstered their applied value (O'Connell et al. 2011). Cameras allow the remote collection of temporally replicated detection data necessary for occupancy analysis, with survey costs generally independent of the number of temporal replicates or number of species detected.

Furthermore, cameras appear well suited to detection of most carnivore or furbearing species, as evidenced by their use in monitoring a wide range of species in different landscapes (e.g., Kays and Slauson 2008). Nevertheless, careful thought is required in the design of camera-based surveys (Burton et al. 2015).

Track and camera-based surveys share similarities in design and implementation. Each survey type often incorporates olfactory attractants, but survey sites may be deployed randomly or on wildlife trails or other landscape features to increase detectability without use of attractants (Schlexer 2008; Cusack et al. 2015). Although there are exceptions (e.g., bobcats, Heilbrun et al. 2006; wolverines [Gulo gulo], Magoun et al. 2011), most furbearing animals cannot be individually identified through track surveys or digital images from camera surveys. However, depending on survey design, spatial and temporal extent of study, and quantity and quality of data collected, objectives related to distribution, detectability, abundance, and occupancy can be met, even without individually identifiable animals (e.g., Sargeant et al. 2005; Chandler and Royle 2013; Lesmeister et al. 2015).

Regardless of the method of detection utilized, new and practical methods now exist (MacKenzie et al. 2002, 2006; MacKenzie and Royle 2005) for addressing the concern of imperfect and temporally varying detection likely present in many traditional furbearer surveys. Although this commonly involves increased cost and effort to conduct repeat visits to survey sites within a season, the improved rigor of the detection-corrected estimates offers furbearer managers greater confidence in using the results for annual management decisions. Although our experience has been that many traditional large-sample, single-visit furbearer surveys can often detect large annual changes or long-term trends for more common furbearers, they are less desirable in management applications involving rare species or species that disproportionally respond to regulatory changes, or for use as the sole metric in making annual harvest management decisions. Regardless of whether single- or multiple-visit surveys are employed, collection of both site (e.g., vegetation structure, topography) and survey (e.g., weather) covariates will help improve inference from survey results.

Another consideration in data collection protocols,

and one rarely given critical thought, is the frequency with which to collect survey or other biological data (Hauser et al. 2006). Ideally, reliable annual data are desirable for all furbearing species. However, practical limitations may pose challenges in accomplishing this, or management needs may not require it. In the former case, managers must carefully consider whether periodic but more rigorous surveys are a better choice than annual but potentially less rigorous surveys. Here, prioritization of management needs based on limited resources is warranted.

Radio-Marking

As with other groups of wildlife, capturing and radio-marking furbearing animals can yield valuable information on space use (e.g., home-range size), spatial and temporal resource selection patterns (e.g., use-availability), movements (e.g., dispersal, travel corridors and barriers), and demographics (e.g., survival, reproduction). Telemetry studies are valuable, if not necessary, to meet many objectives for furbearer management. Radio-marking involves capturing individuals and affixing a VHF or GPS unit (e.g., neck collar, backpack, or surgically implanted device, depending on species) that ultimately allows researchers to estimate locations of individual animals. Some transmitters also include sensors to detect mortality (or inactivity), or collect temperature, activity, and other data. Transmitters can also play a role in carnivore depredation management by using virtual fences to alert managers when animals are in a certain area (e.g., Jachowski et al. 2014), or through negative conditioning (e.g., Hawley et al. 2013). For more information on radio-marking animals and telemetry equipment, see Millspaugh et al. (2012) and Silvy et al. (2012).

Appropriate live-capture techniques are dependent on species but may include standard or specialized cage or box traps (e.g., beavers, felids, kit [*Vulpes macrotis*] and swift [*V. velox*] foxes, muskrats, mustelids, raccoons, skunks [*Mephitis* spp. and *Spilogale* spp.], Virginia opossums [*Didelphis virginiana*]), cable restraints (e.g., beavers, coyotes [*Canis latrans*], red fox [*V. vulpes*]), or foothold traps (e.g., coyotes, bobcats, gray wolves, red foxes, river otters; Schemnitz et al. 2012). Similarly, handling techniques that are safe for both the trained handler and the captured animal range from chemical immobilization to physical restraint. As with any research project, capturing and radio-marking animals must be done with careful consideration of animal welfare, with advice sought from those highly experienced in both capture and choice of radio-marking methods, and with all necessary state and federal permits approved prior to implementing a project. Protocols should also follow Sikes et al. (2011), include Best Management Practices (BMPs) for Trapping (see below) if applicable, and, particularly if associated with an academic institution, receive project approval from their Institutional Animal Care and Use Committee (IACUC). Adhering to such protocols will help ensure sound treatment of animals and ensure that field methodology will be acceptable to the scientific community. Data collected from animals legally harvested by trappers and hunters, and therefore not from animals specifically captured or killed for research purposes, seem to meet acceptable scientific standards without following Sikes et al. (2011) or IACUC.

Integrating Data Analysis with Management Decisions

Complex analyses are not always required for sound inference; sometimes a simple graph can tell the story as well as a complex analysis. Nevertheless, statistical analyses are often necessary and have become increasingly complex, requiring continual effort by managers and researchers to remain familiar with their use and limitations. However, a newly developed technique must be useful and provide an advantage over previous techniques if it is to be integrated into applied wildlife research and management. A particular statistical technique must also be selected based on the design and objectives of the research, including testing research hypotheses, as opposed to developing or modifying objectives simply to use a currently trendy method. Proven methods that are selected for analyzing data collected annually and in the same format have long-term benefits owing to their consistency and ease of application following initial implementation. Early consultation with a trained statistician, particularly one who is familiar with wildlife studies, will help ensure that the desired objectives can be met and avoid issues that cannot be addressed after project initiation or during data analysis.

Management decisions are also becoming increasingly informed, less subjective, and more defensible through integration of structured decision-making and adaptive management. Structured decision-making is a formal process appropriate in situations where decisions must be made based on complex problems that are high in uncertainty. An iterative form of structured decision-making may be implemented in an adaptive management framework. However, different approaches to adaptive management exist (McFadden et al. 2011), so managers should review each approach to decide which may be best for their situation.

Wolf recovery seems to provide a good example of adaptive management applied to furbearer management and conservation. For example, adaptive management has been used to assess predator–prey relationships during and after recovery of gray wolves in Yellowstone National Park (Varley and Boyce 2006). For more information on structured decision-making and adaptive management for natural resource decisions, see the seminal and detailed work of Holling (1978), Walters (1986), and Williams et al. (2007).

Harvest Assessment

Analysis of furbearer harvest data can take many forms and utilize numerous metrics. As with any assessment, statistical methods should be considered in advance of actual data collection, be driven by clear objectives and prior knowledge of metric utility (or part of an assessment of utility), and be suitable for the manner in which the data were collected.

Magnitude of harvest, perhaps by sex/age class, is sometimes useful as an index of population size, but analyses that control for fluctuations in the number of participants and their harvest effort are likely to be much more reliable. Catch-per-unit-effort data can also be used in certain circumstances to directly estimate population size (Skalski et al. 2005) or may be an important covariate in other methods of population estimation (e.g., SPR). Hence, incorporation of data on hunter and trapper effort should be considered in any harvest analyses. Harvest ratios (e.g., juveniles to adult females, males to females) can also be informative.

For example, juvenile-to-adult-female (greater than or equal to two years) harvest ratios of less than five to six for fishers and martens have been suggested as indicators of potential overharvest for these previously unexploited species in Algonquin Park, Ontario (Douglas and Strickland 1987; Strickland and Douglas 1987). However, Eberhardt (1977) previously demonstrated that interpretation of age ratios is fraught with potential error in the absence of additional information, so caution is warranted.

Increasingly sophisticated approaches for estimating population size are becoming available and utilize age data commonly collected on some harvested furbearing species (for an example, see Skalski et al. 2011). In addition to numeric harvest metrics, changes in spatial distribution of harvest can also be informative. One must be cognizant that harvest locations do not necessarily reflect actual population distribution, but harvest data can be a cost-effective way to monitor range contractions or expansions. Determining appropriate spatial resolution required from hunters and trappers will depend on goals, on whether population or habitat management subunits are used in decision-making, and on the likelihood of getting reliable information at the desired scale from hunters and trappers.

Occupancy Modeling

Historically, occupancy referred simply to the confirmed presence of a species, but it is now more formally used to refer to studies of site occupancy that estimate and control for imperfect detection of the target species (MacKenzie et al. 2002, 2006; MacKenzie and Royle 2005). Occupancy surveys typically require multiple assessments (e.g., multiple site visits or independent observers) of species detections on at least a subset of survey sites within a season. Presence can be assessed in many ways, including identification of tracks and collection of digital images. Although large-scale camera surveys can have a high initial cost (e.g., purchase of cameras and equipment), cameras are well suited for furbearer occupancy surveys owing to their ability to detect many species and because a large number of repeat surveys can be conducted without physically revisiting a site (Lesmeister et al. 2015). Whatever the final choice, furbearer managers are confronted with the task of monitoring many species, so strong consideration should be given to methods that are capable of detecting a wide array of, or a subset of, priority species.

With an appropriate survey design, rigorous estimates of the proportion of a study area that is occupied by a species are possible and may serve as reliable population indices for many species. With additional covariate data and incorporation of spatial correlation functions, spatially explicit maps of species distribution can also be developed (Sargeant et al. 2005; Magoun et al. 2007). Furthermore, multiple approaches have been developed that may allow estimation of actual abundance from site-occupancy data (Royle and Nichols 2003; Rowcliffe et al. 2008; Chandler and Royle 2013); further research is currently needed to evaluate accuracy and precision of the various approaches in empirical situations. Depending on survey design, if habitat covariates are recorded or available for each survey site, occupancy surveys can also provide reliable habitat selection information for surveyed species.

If it is determined that multiple-visit occupancy surveys are not practical, it may be wise to initially conduct a well-designed research project to evaluate which biotic or abiotic factors have the greatest effect on probability of detecting a species. The results can be used to help design the most appropriate single-visit survey and ensure that relevant covariate data are collected at the appropriate temporal scales to help distinguish changes in indices that may be driven simply by detection variability rather than population fluctuations.

Genetic Analyses

When used in conjunction with (or independently of) camera surveys, the collection of genetic samples can identify or assess individuals, sex, origin, population connectivity, effective or actual population size, site occupancy, and other metrics (DeYoung and Honeycutt 2005; Schwartz and Monfort 2008). Genetic samples can be obtained from pelts or carcasses of harvested furbearers, from animals live-captured as part of research efforts, or noninvasively from collection of scat or hair samples (Long et al. 2008). For example, Kierepka et al. (2012) used tissue samples from har-

vested American badgers (*Taxidea taxus*) in Michigan to determine that dispersal was not sex biased. Using microsatellites and mitochondrial DNA collected from jaw samples of harvested bobcats, Reding et al. (2013) provided evidence to support regional management of this species in Oregon, where two regions were defined based on the putative distribution of two subspecies with substantially different pelt values. Lastly, DNA can also be collected from wound sites from preyed-upon animals to help identify the predator (species or individual) involved in the attack (Wengert et al. 2014).

Population Modeling

Population models have been used in furbearer management programs for several decades (Sanderson 1982; Frederick and Cobb 1992; Runge 1999), though the principles of sound and useful modeling (e.g., Starfield 1997; Addison et al. 2013) transcend taxonomy. Models may be either phenomenological or mechanistic in nature and can vary in complexity (Owen-Smith 2007). Although population size is the output most often of interest, other metrics can also be examined. Ultimately, model type, complexity, and output metrics should be based on a clearly stated purpose, availability and quality of data for use as potential model inputs, and availability of data or funds necessary to evaluate accuracy of model projections. Oftentimes the greatest value of a modeling effort comes not from any one specific output, but rather from forcing the modeler to describe exactly how they think the system of interest operates, the consistency of those predictions with existing data, or the sensitivity of the system to changes in model parameters; the latter is useful for identifying and prioritizing knowledge gaps that should be the focus of future research or data collection. This process can also help determine inefficiencies in existing data collection efforts. For example, collection of biological data from carcasses of harvested furbearers is a common practice; models can help evaluate whether such data are worth the effort or whether efforts would be better expended for collection of other information.

Common inputs to mechanistic population models include age- and sex-specific natural mortality rates, age- and sex-specific harvest levels, and age-specific reproductive rates. Models can be deterministic (i.e., no variance in rates) or stochastic (e.g., survival and reproductive inputs have associated measures of variance), and certain annual inputs may be based on previously established relationships with covariates (e.g., an established relationship between survival and winter severity). Inputs are ideally based on data collected from within the jurisdiction of interest, but financial or logistical constraints often result in the incorporation of estimates published from other studies. When substantial uncertainty exists in how the system of interest operates, development of multiple working hypotheses (i.e., models), combined with an adaptive management approach, may provide the most efficient route to improved understanding of the system.

Current Issues

The management of many species of wildlife or their habitat relies on trapping, and this is particularly true of furbearer management. Furbearers are captured in both live-restraining and kill traps by avocational trappers for fur, by agency staff and others for nuisance abatement, and by biologists and researchers for ecological studies. Even though fur trapping is highly regulated and only abundant species are trapped, trapping has often been maligned because it is perceived to be inhumane, unsafe, or nonselective with regard to individual species. Nevertheless, it is acknowledged as an important component of wildlife management (The Wildlife Society 2015). To address both real and perceived concerns, state fish and wildlife agencies, via the AFWA and with federal funding, established a program in 1997 to scientifically evaluate traps and trapping systems through the development of BMPs for Trapping in the United States (AFWA 2015a). The BMP program includes an evaluation of traps based on five criteria: animal welfare, efficiency, selectivity, practicality, and safety.

The protocol used for testing and evaluating traps for the BMP program was developed by the International Organization for Standardization (1999a, 1999b). No other technique for harvesting wild animals (e.g., hunting, fishing) has included an internationally accepted protocol to evaluate animal welfare. In the United States, traps are evaluated against five performance criteria, including numeric thresholds for animal welfare and trap efficiency, and must meet all criteria before being included in a species-specific trapping BMP. Trap testing has been a collaborative effort between the trapping community and US and Canadian researchers and has scientifically demonstrated that furbearers can be trapped humanely and selectively (as well as efficiently, practically, and safely) if the proper traps and trapping systems are used. In the United States, individual BMP documents are currently available for 22 species of furbearers. In addition to the goal of improving animal welfare in US trapping programs, the BMP program serves as the scientifically based mechanism by which the United States is addressing international concerns and thereby maintaining the trade in wild fur between the United States and the European Union (Hamilton et al. 1998), without which some furbearer management activities would be much more costly, if not prohibitive. Nationally, trapper education programs have included information about the BMP process and approved traps to use for various species. In addition to their importance to avocational trappers, traps are critical tools in wildlife management and research. Scientific efforts to evaluate and improve trap performance will continue to play a vital role in all realms of furbearer management.

Current issues in furbearer management are not limited to trapping. One example is the formal organization of coyote-calling contests, where participants compete for prizes based on the number of coyotes called and harvested during a given time period in a given area. Proponents cite population control and reduced livestock damage, whereas opponents question whether these events are ethical. Coursing pens are another controversial issue that involves furbearer management decisions. Live-captured coyotes or foxes are sold and placed in large fenced enclosures that often contain escape cover. Hunters then use dogs to pursue these furbearers under controlled conditions for training or field trials. State management agencies generally regulate live sale of furbearers and use in coursing pens, although several states elected to prohibit one or both of these practices prior to their widespread establishment in those respective states owing to concerns over spread of diseases and inhumane treatment of animals.

As previously mentioned, harvest management is one of many components associated with furbearer management. Addressing ecological questions and management needs through a more holistic, multi-species research approach will likely become more complex. The composition and distribution of furbearers (and wildlife communities in general) have changed dramatically since European colonization, and will continue to do so, through both positive and negative responses to anthropogenic and environmental changes. Landscape-scale habitat fragmentation and loss in the Midwest and the eastern United States have been extensive through agricultural practices, infrastructure, development, and other anthropogenic factors. These land-use changes have resulted in a loss of large carnivores and an increase in food availability, thereby positively influencing populations (and distribution) of some medium- and small-sized furbearing species. For example, raccoons have increased their distribution northward into Canadian prairies (Larivière 2004), and an epizootic of raccoon rabies has been spreading for decades in the eastern United States (Guerra et al. 2003); coyotes have expanded into the northeastern United States and are now commonplace in major metropolitan areas throughout the country (e.g., Boer 1992; Gehrt et al. 2009); and the urban wildlife control industry has expanded to address increased human–wildlife conflicts.

The western United States has experienced (and will also continue to experience) major environmental changes through energy development, forestry practices, climate change, drought, wildfire, and other factors. Climate change predictions generally include increases in both temperature and precipitation throughout the geographic distribution of fishers during the next century, which may result in a southward contraction but northward expansion in their distribution (Lawler et al. 2012). Recently reintroduced and peninsular populations of fishers may be particularly sensitive to habitat loss, including through the increased frequency and severity of stand-replacing wildfires in recent years. Northward contractions have also been predicted for martens and Canada lynx (*Lynx canadensis*) as a result of climate change and habitat alterations (Carroll 2007). Cascades (*Vulpes vulpes cascadensis*) and Sierra Nevada (*V. v. necator*) red foxes, two sub-species each generally specialized to high-elevation montane sky islands, may be negatively impacted by climate change, for example, if coyotes (habitat generalists) increase their elevational distribution (Perrine et al. 2010).

Aquatic furbearers also continue or are likely to be affected by environmental contamination, habitat loss, and climate change. Contaminants (e.g., mercury, polychlorinated biphenyls) can affect aquatic furbearers such as American mink (*Neovison vison*) and river otters through reproductive dysfunction, morphology, and failure (e.g., Wren 1991; Harding et al. 1999; Basu et al. 2007). Historic and continued loss of wetlands (Teal and Peterson 2011) has undoubtedly had substantial negative impacts on muskrats, as has hydrology directly or indirectly through effects on aquatic vegetation dynamics (Clark 2000; Erb and Perry 2003; Ervin 2011). Although not cause and effect, apparent declines in muskrat populations since the mid-1980s (Roberts and Crimmins 2010) have occurred simultaneously with increasing variability or extremity in precipitation events, particularly in the eastern half of the United States (Melillo et al. 2014). Aquatic furbearers, especially muskrats, will undoubtedly be affected by increasing droughts or flooding, though effects will vary depending on the type of aquatic system (e.g., lentic vs. lotic), its baseline hydrology, and the magnitude of changes in both mean and variability.

These are but a few examples of current management challenges and issues faced by today's furbearer manager. Managers are expected to effectively address these emerging issues in a highly complex and technical environment. Development of effective communication, critical thinking, problem solving, and networking skills will be necessary for the successful furbearer manager. Although university faculty are well trained in wildlife ecology, many have not been directly exposed to the reality of state management issues. Therefore, close collaboration between university faculty (and continued close or improved collaborations between US Geological Survey Cooperative Wildlife Research Units) and state agency managers would be beneficial to convey extensive training beyond ecological topics to provide a foundation for our future professionals to become competent leaders.

Case Studies

Case Study 1: Beaver Management, a Balancing Act

Beavers have a long and storied history throughout North America. They were an important source of food and other supplies to many Native Americans, an important impetus for European exploration of North America, and one of the earliest examples of both the negative consequences of unregulated resource consumption and the ability of modern management policies to restore and conserve wildlife populations. Furthermore, beavers are nature's quintessential ecosystem engineers (Rosell et al. 2005; Johnston 2012); their activities not only alter the local environment but also functionally affect landscape and ecosystem dynamics. This capability is at the heart of many ecological and social debates that a furbearer manager must often confront. Depending on location or one's values and perspective, the engineering activities of beavers can be desirable or undesirable; considerable efforts have been and continue to be made to not only increase beaver activity for the many positive ecosystem services they provide but also reduce or mitigate beaver activity in areas where conflicts exist. While still the subject of ongoing research and debate, beaver restoration efforts, now most common in areas of the western United States (Pollock et al. 2015), often seek to achieve the potential benefits of water retention and conservation (Pollock et al. 2003; Rosell et al. 2005; Johnston 2012), incised stream restoration (Pollock et al. 2014), or improved fish or wildlife habitat (Pollock et al. 2003; Rosell et al. 2005; Johnston 2012). Sometimes it may be possible to simultaneously reduce a beaver conflict through programs that trap and translocate problem beavers to areas where their ecosystem services are desired (e.g., Oregon Department of Fish and Wildlife 2012; Pollock et al. 2015), though success is dependent on the survival and movements of translocated beaver (Petro et al. 2015).

Throughout the wetter and more densely populated areas of the eastern United States, beavers have recolonized most of their historic range and are abundant in most areas. In many jurisdictions, beavers are a common furbearing species for which state agencies receive nuisance complaints (Southwick Associates 1993). Hence, management often relies on avocational trappers or nuisance wildlife control operators to help address local conflicts or to maintain populations at levels that minimize nuisance situations while still maintaining the positive effects of beaver activity on the landscape. In some situations, nonlethal abatement options may be effective at minimizing a conflict without substantially reducing any desirable effects of beavers at the local site (Pollock and Lewallen 2015). Whether the goal is to increase, maintain, or reduce beaver abundance (or their effects), sound scientific data are critical in both setting objectives and determining whether those objectives are being met.

Since the earliest days of the fur trade, beavers have been an important provider of goods and services to humans and our natural environment. In more recent times, state agencies have been forced to deal with complex arrays of both positive and negative effects from beavers that vary across space and time. Many human–beaver conflicts past and present can be traced to decisions to situate individual houses or whole towns close to bodies of water, the establishment of networks of drainage ditches for agriculture or flood control, or the ensuing development of transportation networks and associated infrastructure (e.g., culverts) that crisscross many bodies of water. Variability in both positive and negative effects of beavers across time can also be driven by external factors difficult to predict or manage, such as the frequency and severity of rainfall events or global economic conditions that influence fur values and harvest effort. Hence, desired management direction can rapidly shift, and furbearer managers must be prepared to make recommendations that simultaneously maintain the numerous ecosystem services provided by beavers while also providing sustainable harvest opportunities and minimizing conflicts with humans.

Case Study 2: Fishers in North America, Finding Conservation Success

Unregulated harvest, logging, and other factors contributed to substantial losses in distribution of fishers across North America. However, the distribution of fishers across much of northern North America generally has been stable or increasing since the 1950s,

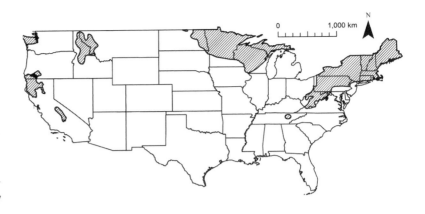

Current estimated distribution of fishers (*hatched areas*) in the contiguous United States, including potential range expansion between two separate populations in Oregon (*black area*) composed of indigenous (CA-OR) and non-native (translocated; southwestern OR) fishers. The map is based on Lewis et al. (2012) and was modified using input from state and federal agency biologists.

with successful population recovery seemingly linked to areas with continuous boreal forests (Powell 1981; Gibilisco 1994), as well as state management and protection and federal agency collaboration. Relatively small and isolated populations currently exist at the edges of their distribution, resulting in population viability concerns (Zielinski et al. 1995; Aubry and Lewis 2003; Lewis et al. 2012).

To address conservation concerns in areas where fisher populations have not recovered, reintroductions have been implemented. Among carnivores, the fisher is one of the most successfully reintroduced species. Based on an assessment of 30 reintroduction efforts in Canada and the United States, 77 percent with known outcomes were successful (Lewis et al. 2012). Such reintroduction efforts date back to the 1940s and often involve cooperation among state wildlife agencies, and occasionally with Canadian provincial agencies, depending on location of source populations. Failed attempts seemed to be linked to too few individuals released, whereas successful reintroductions were associated with more individuals released (especially females) and proximity of the source population and release area (Lewis et al. 2012). The conservation goals of reintroducing fishers include establishing additional viable populations and broadening their distribution within their historical range and over time. In the long term, goals also include increasing both local and large-scale connectivity of different populations.

Current reintroduction efforts are under way in California (California Department of Fish and Wildlife 2017) and Washington (Washington Department of Fish and Wildlife 2015) and may be implemented in the near future in Oregon (Hiller 2015). These reintroductions are based on conservation efforts to address fragmented and peninsular populations in Pacific coastal states, an area where fishers are being considered for listing under the Endangered Species Act of 1973 by the US Fish and Wildlife Service (2014). An immense amount of planning is necessary to define project objectives; identify potential threats to reintroductions; incorporate stakeholder input; develop multi-agency coordination; secure appropriate funding and permits; assess pre-release candidate areas and disease-associated risks; identify, assess, and secure individuals from a source population; and implement post-release population monitoring to ensure conservation success. Although translocations are not a simple endeavor, this level of planning and implementation will help managers yield a measurable conservation benefit to the population, the species, or an ecosystem.

LITERATURE CITED

Addison, P. F. E., L. Rumpff, S. S. Bau, J. M. Carey, Y. E. Chee, F. C. Jarrad, M. F. McBride, and M. A. Burgman. 2013. Practical solutions for making models indispensable in conservation decision-making. Biodiversity Review 19:490–502.

AFWA (Association of Fish and Wildlife Agencies). 2005. Ownership and use of traps by trappers in the United States in 2004. www.fishwildlife.org/files/AFWA-FINAL-TRAPPING-Report.pdf.

———. 2007. Summary of trapping regulations for fur harvesting in the United States. Furbearer Conservation Technical Work Group, Association of Fish and Wildlife Agencies, Washington, DC, USA.

———. 2009. Modern snares for capturing mammals: Definitions, mechanical attributes and use considerations. http://fishwildlife.org/?section=furbearer_management_resources/Modern_Snares_final.pdf.

———. 2015a. Best management practices. http://fishwildlife.org/?section=best_management_practices.

———. 2015b. National fur harvest database. www.fishwildlife.org/index.php?section=furbearer_management&activator=27.

———. 2015c. Online TrapperEd. www.fishwildlife.org/index.php?section=furbearer_management&activator=27.

———. 2015d. Trapper education manual. www.fishwildlife.org/index.php?section=furbearer_management&activator=27.

———. 2015e. Trap use, furbearers trapped, and trapper characteristics in the United States in 2015. www.fishwildlife.org/files/AFWA_Trap_Use_Report_2015_ed_2016_02_29.pdf.

Armstrong, J. B., and A. N. Rossi. 2000. Status of avocational trapping based on the perspectives of state furbearer biologists. Wildlife Society Bulletin 28:825–832.

Aubry, K. B., and J. C. Lewis. 2003. Extirpation and reintroduction of fishers (*Martes pennanti*) in Oregon: Implications for their conservation in the Pacific states. Biological Conservation 114:79–90.

Basu, N., A. M. Scheuhammer, S. J. Bursian, J. Elliot, K. Rouvinen-Watt, and H. M. Chan. 2007. Mink as a sentinel species in environmental health. Environmental Research 103:130–144.

Batcheller, G. R., T. A. Decker, D. A. Hamilton, and J. F. Organ. 2000. A vision for the future of furbearer management in the United States. Wildlife Society Bulletin 28:833–840.

Ben-David, M., R. W. Flynn, and D. M. Schell. 1997. Annual and seasonal changes in diets of martens: Evidence from stable isotope analysis. Oecologia 111:280–291.

Boer, A. H., editor. 1992. Ecology and management of the eastern coyote. Wildlife Research Unit, Fredericton, New Brunswick, Canada.

Burton, A. C., E. Neilson, D. Moreira, A. Ladle, R. Steenweg, J. T. Fisher, E. Bayne, and S. Boutin. 2015. Wildlife camera trapping: A review and recommendations for linking surveys to ecological processes. Journal of Applied Ecology 52:675–685.

California Department of Fish and Wildlife. 2017. Fisher translocation project. www.wildlife.ca.gov/Regions/1/Fisher-Translocation.

Carroll, C. 2007. Interacting effects of climate change, landscape conversion, and harvest on carnivore populations at the range margin: Marten and lynx in the northern Appalachians. Conservation Biology 21:1092–1104.

Chandler, R. B., and J. A. Royle. 2013. Spatially explicit models for inference about density in unmarked or partially marked populations. Annals of Applied Statistics 7:936–954.

Chapman, J. A., and D. Pursley, editors. 1981. Worldwide furbearer conference proceedings. Worldwide Furbearer Conference, Frostburg, Maryland, USA.

CITES (Convention on International Trade in Endangered Species of Wild Fauna and Flora). 2013. How CITES works. www.cites.org/eng/disc/how.php.

Clark, W. R. 2000. Ecology of muskrats in prairie wetlands. Pages 287–313 in H. R. Murkin, A. G. van der Valk, and W. R. Clark, editors. Prairie wetland ecology: The contribution of the Marsh Ecology Research Program. Iowa State University Press, Ames, Iowa, USA.

Clark, W. R., and E. K. Fritzell. 1992. A review of population dynamics of furbearers. Pages 899–910 in D. R. McCullough and R. H. Barrett, editors. Wildlife 2001: Populations. Elsevier Science, London, UK.

Cusack, J. J., A. J. Dickman, J. M. Rowcliffe, C. Carbone, D. W. Macdonald, and T. Coulson. 2015. Random versus game trail-based camera trap placement strategy for monitoring terrestrial mammal communities. PLoS ONE 10(5): e0126373.

Daigle, J. J. 1997. The sociocultural importance of fur trapping in six northeastern states. PhD diss., University of Massachusetts, Amherst, Massachusetts, USA.

Deems, E. F., Jr., and D. Pursley. 1978. North American furbearers: Their management, research and harvest status in 1976. International Association of Fish and Wildlife Agencies. University of Maryland, College Park, Maryland, USA.

DeYoung, R. W., and R. L. Honeycutt. 2005. The molecular toolbox: Genetic techniques in wildlife ecology and management. Journal of Wildlife Management 69:1362–1384.

Dixon, K. R., and M. C. Swift. 1981. The optimal harvesting concept in furbearer management. Pages 1524–1551 in J. A. Chapman and D. Pursley, editors. Worldwide furbearer conference proceedings. Worldwide Furbearer Conference, Frostburg, Maryland, USA.

Douglas, C. W., and M. J. Strickland. 1987. Fisher. Pages 511–529 in M. Novak, J. A. Baker, M. E. Obbard, and B. Malloch, editors. Wild furbearer management and conservation in North America. Ministry of Natural Resources, Ontario, Canada.

Dunbar, M. R., and M. R. Giordano. 2003. Abundance and condition indices of coyotes on Hart Mountain National Antelope Refuge, Oregon. Western North American Naturalist 62:341–347.

Eberhardt, L. L. 1977. Interpretation of age ratios. Journal of Wildlife Management 38:557–562.

Erb, J., R. D. Bluett, E. K. Fritzell, and N. F. Payne. 1999. Aging muskrats using molar indices: A regional comparison. Wildlife Society Bulletin 27:628–635.

Erb, J., and H. R. Perry. 2003. Muskrats. Pages 311–348 in G. A. Feldhamer, B. C. Thompson, and J. A. Chapman, editors. Wild mammals of North America: Biology, management, and conservation. Johns Hopkins University Press, Baltimore, Maryland, USA.

Errington, P. L. 1946. Predation and vertebrate populations. Quarterly Review of Biology 21:144–177, 221–245.

———. 1956. Factors limiting higher vertebrate populations. Science 124:304–307.

Ervin, M. D. 2011. Population characteristics and habitat selection of muskrats (*Ondatra zibethicus*) in response to water level management at the Summerberry Marsh Complex, The Pas, Manitoba, Canada. Thesis, Iowa State University, Ames, Iowa, USA.

Fournier, D., and C. P. Archibald. 1982. A general theory for analysing catch at age data. Canadian Journal of Fisheries and Aquatic Sciences 39:1195–1207.

Frederick, R. B., and D. T. Cobb. 1992. Computer simulation of furbearer population dynamics. Pages 911–921 in D. R. McCullough and R. H. Barrett, editors. Wildlife 2001: Populations. Elsevier, London, UK.

Fritzell, E. K., and N. F. Johnson. 1982. A perspective on furbearer management. Pages 1–9 in G. C. Sanderson, editor. Midwest furbearer management. North Central Section, Central Mountains and Plains Section, and Kansas Chapter of The Wildlife Society.

Fur Information Council of America. 2010. FICA facts. www.fur.org/fica-facts/.

Fur Institute of Canada. n.d. Canada's fur trade: Facts and figures. http://fur.ca/fur-trade/canadas-fur-trade-fact-figures/.

Fur Takers of America. 2015. Fur Takers of America trapper's college. www.furtakersofamerica.com/college.html.

Gehrt, S. D., C. Anchor, and L. A. White. 2009. Home range and landscape use of coyotes in a metropolitan landscape: Conflict or coexistence? Journal of Mammalogy 90:1045–1057.

Gibilisco, C. J. 1994. Distributional dynamics of modern *Martes* in North America. Pages 59–71 in S. W. Buskirk, A. S. Harestad, M. G. Raphael, and R. A. Powell, editors. Martens, sables, and fishers: Biology and conservation. Cornell University Press, Ithaca, New York, USA.

Gilbert, F. F. 1987. Methods for assessing reproductive characteristics of furbearers. Pages 180–190 in M. Novak, J. A. Baker, M. E. Obbard and B. Malloch, editors. Wild furbearer management and conservation in North America. Ministry of Natural Resources, Ontario, Canada.

Golden, H., and T. Krause. 2003. How to avoid incidental tale of lynx while trapping or hunting bobcats and other furbearers. International Association of Fish and Wildlife Agencies, Washington, DC, USA.

Gompper, M. E., R. W. Kays, J. C. Ray, S. D. LaPoint, D. A. Bogan, and J. R. Cryan. 2006. Comparison of noninvasive techniques to survey carnivore communities in northeastern North America. Wildlife Society Bulletin 34:1142–1151.

Gorman, T. A., B. R. McMillan, J. D. Erb, C. S. DePerno, and D. J. Martin. 2008. Survival and cause-specific mortality of a protected population of river otters in Minnesota. American Midland Naturalist 159:98–109.

Gove, N. E., J. R. Skalski, P. Zager, and R. L. Townsend. 2002. Statistical models for population reconstruction using age-at-harvest data. Journal of Wildlife Management 66:310–320.

Guerra, M. A., A. T. Curns, C. E. Rupprecht, C. A. Hanlon, J. W. Krebs, and J. E. Childs. 2003. Skunk and raccoon rabies in the eastern United States: Temporal and spatial analysis. Emerging Infectious Diseases 9:1143–1150.

Hamilton, D. A., B. Roberts, G. Linscombe, N. R. Jotham, H. Noseworthy, and J. L. Stone. 1998. The European Union's wild fur regulation: A battle of politics, cultures, animal rights, international trade and North America's wildlife policy. Transactions of the North American Wildlife and Natural Resources Conference 63:572–588.

Harding, L. E., M. L. Harris, C. R. Stephen, and J. E. Elliot. 1999. Reproductive and morphological condition of wild mink (*Mustela vison*) and river otter (*Lutra canadensis*) in relation to chlorinated hydrocarbon contamination. Environmental Health Perspectives 107:141–147.

Hauser, C. E., A. R. Pople, and H. P. Possingham. 2006. Should managed populations be monitored every year? Ecological Applications 16:807–819.

Hawley, J. E., S. T. Rossler, T. M. Gehring, R. N. Schultz, P. A. Callahan, R. Clark, J. Cade, and A. P. Wydeven. 2013. Developing a new shock-collar design for safe and efficient use on wild wolves. Wildlife Society Bulletin 37:416–422.

Heilbrun, R. D., N. J. Silvy, M. K. Peterson, and M. E. Tewes. 2006. Estimating bobcat abundance using automatically triggered cameras. Wildlife Society Bulletin 34:69–73.

Heinemeyer, K. S., T. J. Ulizio, and R. L. Harrison. 2008. Natural sign: Tracks and scats. Pages 45–74 in R. A. Long, P. MacKay, W. J. Zielinski, and J. C. Ray, editors. Noninvasive survey methods for carnivores. Island Press, Washington, DC, USA.

Hiller, T. L. 2015. Feasibility assessment for the reintroduction of fishers in western Oregon, USA. US Fish and Wildlife Service, Portland, Oregon, USA.

Hiller, T. L., D. M. Reding, W. R. Clark, and R. L. Green. 2014. Misidentification of sex among harvested bobcats. Wildlife Society Bulletin 38:752–756.

Hiller, T. L., and S. M. Vantassel. 2015. The global consumptive use of small carnivores: Social, cultural, religious, economic, and subsistence trends from prehistoric to modern times. In E. Do Linh San, J. Sato, J. L. Belant, and M. Somers, editors. Small carnivores: Evolution, ecology, behaviour, and conservation. Wiley-Blackwell, Hoboken, New Jersey, USA.

Hiller, T. L., and H. B. White. 2013. How to avoid incidental take of wolverine during regulated trapping activities. Association of Fish and Wildlife Agencies, Washington, DC, USA.

Holling, C. S. 1978. Adaptive environmental assessment and management. John Wiley and Sons, New York, New York, USA.

International Association of Fish and Wildlife Agencies. 1992. Ownership and use of traps by trappers in the United States in 1992. The Fur Resources Committee of the International Association of Fish and Wildlife Agencies and the Gallup Organization, Washington, DC, USA.

International Organization for Standardization. 1999a. Animal (mammal) traps—part 4: Methods for testing killing-trap systems on land or underwater (ISO-10990-4:1999E). Geneva, Switzerland.

——. 1999b. Animal (mammal) traps—part 5: Methods for testing restraining traps (ISO 10990-5:1999E). Geneva, Switzerland.

Jachowski, D. S., R. Slotow, and J. J. Millspaugh. 2014. Good virtual fences make good neighbors: Opportunities for conservation. Animal Conservation 17:187–196.

Jenks, J. A., R. T. Bowyer, A. G. Clark. 1984. Sex and age-class determination for fisher using radiographs of canine teeth. Journal of Wildlife Management 48:626–628.

Johnston, C. A. 2012. Beaver wetlands. Pages 161–172 in D. P. Batzer and A. H. Baldwin, editors. Wetland habitats of North America: Ecology and conservation concerns. University of California Press, Berkeley, California, USA.

Johnston, C. A., and S. K. Windels. 2015. Using beaver works to estimate colony activity in boreal landscapes. Journal of Wildlife Management 79:1072–1080.

Kays, R. W., and K. M. Slauson. 2008. Remote cameras. Pages 110–140 in R. Long, P. Mackay, J. Ray, and W. Zielinski, editors. Noninvasive survey methods for carnivores. Island Press, Washington, DC, USA.

Kendrot, S. 2013. An overview of the Chesapeake Bay nutria eradication project. Proceedings of the Wildlife Damage Management Conference 15:70.

Kierepka, E. M., E. K. Latch, and B. J. Swanson. 2012. Influence of sampling scheme on the inference of sex-biased gene flow in the American badger (Taxidea taxus). Canadian Journal of Zoology 90:1231–1242.

Kucera, T., and R. H. Barrett. 2011. A history of camera trapping. Pages 9–26 in A. F. O'Connell, J. D. Nichols, and K. U. Karanth, editors. Camera traps in animal ecology: Methods and analyses. Springer, New York, New York, USA.

Larivière, S. 2004. Range expansion of raccoons in the Canadian prairies: Review of hypotheses. Wildlife Society Bulletin 32:955–963.

Lawler, J. J., H. D. Safford, and E. H. Girvetz. 2012. Martens and fishers in a changing climate. Pages 371–397 in K. B. Aubry, W. J. Zielinski, M. G. Raphael, G. Proulx, and S. W. Buskirk, editors. Biology and conservation of martens, sables, and fishers: A new synthesis. Cornell University Press, Ithaca, New York, USA.

Leopold, A. 1933. Game management. Charles Scribner's Sons, New York, New York, USA.

Lesmeister, D. B., C. K. Nielsen, E. M. Schauber, and E. C. Hellgren. 2015. Spatial and temporal structure of a mesocarnivore guild in midwestern North America. Wildlife Monographs 191:1–61.

Lewis, J. C. 2014. Post-release movements, survival, and resource selection of fishers (Pekania pennanti) translocated to the Olympic Peninsula of Washington. PhD diss., University of Washington, Seattle, Washington, USA.

Lewis, J. C., R. A. Powell, and W. J. Zielinski. 2012. Carnivore translocations and conservation: Insights from population models and field data for fishers (Martes pennanti). PLoS ONE 7(3):e32726.

Long, R. A., P. MacKay, W. J. Zielinski, and J. C. Ray, editors. 2008. Noninvasive survey methods for carnivores. Island Press, Washington, DC, USA.

MacKenzie, D. I., J. D. Nichols, G. B. Lachman, S. Droege, J. A. Royle, and C. A. Langtimm. 2002. Estimating site occupancy rates when detection probabilities are less than one. Ecology 83:2248–2255.

MacKenzie, D. I., J. D. Nichols, J. A. Royle, K. H. Pollock, L. L. Bailey, and J. E. Hines. 2006. Occupancy estimation and modeling: Inferring patterns and dynamics of species occurrence. Elsevier, London, UK.

MacKenzie, D. I., and J. A. Royle. 2005. Designing occupancy studies: General advice and allocating survey effort. Journal of Applied Ecology 42:1105–1114.

Magoun, A. J., C. D. Long, M. K. Schwartz, K. L. Pilgrim, R. E. Lowell, and P. Valkenburg. 2011. Integrating motion-detection cameras and hair snags for wolverine identification. Journal of Wildlife Management 75:731–739.

Magoun, A. J., J. C. Ray, D. S. Johnson, P. Valkenburg, F. N. Dawson, and J. Bowman. 2007. Modeling wolverine occurrence using aerial surveys of tracks in snow. Journal of Wildlife Management 71:2221–2229.

McFadden, J. E., T. L. Hiller, and A. J. Tyre. 2011. Evaluating the efficacy of adaptive management approaches: Is there a formula for success? Journal of Environmental Management 92:13541359.

Melillo, J. M., T. C. Richmond, and G. W. Yohe, editors. 2014. Climate change impacts in the United States: The third national climate assessment. US Global Change Research Program.

Millspaugh, J. J., D. C. Kesler, R. W. Kays, R. A. Gitzen, J. H. Schulz, C. T. Rota, C. M. Bodinof, J. L. Belant, and B. J. Keller. 2012. Wildlife radiotelemetry and remote monitoring. Pages 258–283 in N. J. Silvy, editor. The wildlife techniques manual, 7th edition. Vol. 1. Johns Hopkins University Press, Baltimore, Maryland, USA.

Mowry, R. A., T. M. Schneider, E. K. Latch, M. E. Gompper, J. Beringer, and L. S. Eggert. 2015. Genetics and the successful reintroduction of the Missouri river otter. Animal Conservation 18:196–206.

Novak, M., J. A. Baker, M. E. Obbard, and B. Malloch, editors. 1987a. Wild furbearer management and conservation in North America. Ministry of Natural Resources, Ontario, Canada.

Novak, M., M. E. Obbard, J. G. Jones, R. Newman, A. Booth, A. J. Satterthwaite, and G. Linscombe. 1987b. Furbearer harvests in North America, 1600–1984. Wild furbearer management and conservation in North America. Ministry of Natural Resources, Ontario, Canada.

O'Connell, A. F., J. D. Nichols, and K. U. Karanth. 2011.

Camera traps in animal ecology: Methods and analyses. Springer, New York, New York, USA.

Oregon Department of Fish and Wildlife. 2012. Guidelines for relocation of beaver in Oregon. Oregon Department of Fish and Wildlife, Salem, Oregon, USA. www.dfw.state.or.us/wildlife/living_with/beaver.asp.

Owen-Smith, N. 2007. Introduction to modeling in wildlife and natural resource conservation. Blackwell, Oxford, UK.

Payne, N. F. 1980. Furbearer management and trapping. Wildlife Society Bulletin 8:345–348.

Perrine, J. D., L. A. Campbell, and G. A. Green. 2010. Sierra Nevada red fox (*Vulpes vulpes necator*): A conservation assessment. US Department of Agriculture, Forest Service, R5-FR-010.

Petro, V., J. Taylor, and D. Sanchez. 2015. Evaluating landowner-based beaver relocation as a tool to restore salmon habitat. Global Ecology and Conservation 3:477–486.

Pollock, M., T. J. Beechie, J. M. Wheaton, C. E. Jordan, N. Bouwes, N. Weber, and C. Volk. 2014. Using beaver dams to restore incised stream ecosystems. BioScience 64:279–290.

Pollock, M., M. Heim, and D. Werner. 2003. Hydrologic and geomorphic effects of beaver dams and their influence on fishes. Pages 213–233 in S. V. Gregory, K. Boyer, and A. Gurnell, editors. The ecology and management of wood in world rivers. American Fisheries Society, Bethesda, Maryland, USA.

Pollock, M., and G. Lewallen. 2015. Non-lethal options for mitigating the unwanted effects of beaver. Pages 103–112 in M. Pollock, G. Lewallen, K. Woodruff, C. E. Jordan, and J. M. Castro, editors. Beaver restoration guidebook: Working with beaver to restore streams, wetlands, and floodplains. Version 1.0. US Fish and Wildlife Service, Portland, Oregon, USA.

Pollock, M., G. Lewallen, K. Woodruff, C. E. Jordan, and J. M. Castro, editors. 2015. The beaver restoration guidebook: Working with beaver to restore streams, wetlands, and floodplains. Version 1.0. US Fish and Wildlife Service, Portland, Oregon, USA.

Powell, R. A. 1981. *Martes pennanti*. Mammalian Species No. 156. American Society of Mammalogists, Northampton, Massachusetts, USA.

Reding, D. M., C. E. Carter, T. L. Hiller, and W. R. Clark. 2013. Using population genetics for management of bobcats in Oregon. Wildlife Society Bulletin 37:342–351.

Roberts, N. M., and S. M. Crimmins. 2010. Do trends in muskrat harvest indicate widespread population declines? Northeastern Naturalist 17:229–238.

Rosell, F., O. Bozser, P. Collen, and H. Parker. 2005. Ecological impact of beavers *Castor fiber* and *Castor Canadensis* and their ability to modify ecosystems. Mammal Review 35:248–276.

Roughton, R. D. 1982. A synthetic alternative to fermented egg as a canid attractant. Journal of Wildlife Management 46:230–234.

Roughton, R. D., and M. W. Sweeny. 1982. Refinements in scent-station methodology for assessing trends in carnivore populations. Journal of Wildlife Management 46:217–229.

Rowcliffe, J. M., J. Field, S. T. Turvey, and C. Carbone. 2008. Estimating animal density using camera traps without the need for individual recognition. Journal of Applied Ecology 45:1228–1236.

Royle, J. A., and J. D. Nichols. 2003. Estimating abundance from repeated presence-absence data or point counts. Ecology 84:777–790.

Runge, M. C. 1999. Design and analysis of a population model for beaver (*Castor canadensis*). Cornell Biometrics Unit Technical Series BU-1462. Cornell University, Ithaca, New York, USA.

Sanderson, G. C. 1982. Midwest furbearer management. North Central Section, Central Mountains and Plains Section, and Kansas Chapter of The Wildlife Society.

Sargeant, G. A., D. H. Johnson, and W. E. Berg. 1998. Interpreting carnivore scent station surveys. Journal of Wildlife Management 62:1235–1245.

———. 2003. Sampling designs for carnivore scent-station surveys. Journal of Wildlife Management 67:289–298.

Sargeant, G. A., M. Solvada, C. Slivinski, and D. Johnson. 2005. Markov chain Monte Carlo estimation of species distributions: A case study of the swift fox in western Kansas. Journal of Wildlife Management 69:483–497.

Schemnitz, S. D., G. R. Batcheller, M. J. Lovallo, H. B. White, and M. W. Fall. 2012. Capturing and handling wild animals. Pages 64–117 in N. J. Silvy, editor. The wildlife techniques manual, 7th edition. Vol. 1. Johns Hopkins University Press, Baltimore, Maryland, USA.

Schlexer, F. V. 2008. Attracting animals to detection devices. Pages 263–292 in R. A. Long, P. Mackay, W. J. Zielinski, and J. C. Ray, editors. Noninvasive survey methods for carnivores. Island Press, Washington, DC, USA.

Schwartz, M. K., and S. L. Monfort. 2008. Genetic and endocrine tools for carnivore surveys. Pages 238–262 in R. A. Long, P. MacKay, W. J. Zielinski, and J. C. Ray, editors. Noninvasive survey methods for carnivores. Island Press, Washington, DC, USA.

Sikes, R. S., W. L. Gannon, and the Animal Care and Use Committee of the American Society of Mammalogists. 2011. Guidelines of the American Society of Mammalogists for the use of wild mammals in research. Journal of Mammalogy 92:235–253.

Silvy, N. J., R. R. Lopez, and M. J. Peterson. 2012. Techniques for marking wildlife. Pages 230–247 in N. J. Silvy, editor. The wildlife techniques manual, 7th edition. Vol. 1. Johns Hopkins University Press, Baltimore, Maryland, USA.

Skalski, J. R., J. J. Millspaugh, M. V. Clawson, J. L. Belant, D. R. Etter, B. J. Frawley, and P. D. Friedrich. 2011. Abundance trends of American martens in Michigan based on statistical population reconstruction. Journal of Wildlife Management 75:1767–1773.

Skalski, J. R., K. E. Ryding, and J. J. Millspaugh. 2005. Wildlife demography: Analysis of sex, age, and count data. Elsevier, London, UK.

Southwick Associates, Inc. 1993. 1993 state-by-state survey of furbearers with emphasis on nuisance animals. Furbearer Resources Committee, International Association of Fish and Wildlife Agencies, Washington, DC, USA.

Starfield, A. M. 1997. A pragmatic approach to modeling for wildlife management. Journal of Wildlife Management 61:261–270.

Strickland, M. A. 1994. Harvest management of fishers and American martens. Pages 149–164 in S. W. Buskirk, A. S. Harestad, M. G. Raphael, and R. A. Powell, editors. Martens, sables, and fishers: Biology and conservation. Cornell University Press, Ithaca, New York, USA.

Strickland, M. A., and C. W. Douglas. 1987. Marten. Pages 531–546 in M. Novak, J. A. Baker, M. E. Obbard, and B. Malloch, editors. Wild furbearer management and conservation in North America. Ministry of Natural Resources, Ontario, Canada.

Teal, J. M., and S. Peterson. 2011. U.S. wetland protection and restoration: Have we made a difference? Ecological Restoration 29:22–24.

Todd, A. W. and E. K. Boggess. 1987. Characteristics, activities, lifestyles and attitudes of trappers in North America. Pages 59–76 in M. Novak, J. A. Baker, M. E. Obbard, and B. Malloch, editors. Wild furbearer management and conservation in North America. Ministry of Natural Resources, Ontario, Canada.

Urton, E. J. M., and K. A. Hobson. 2005. Intrapopulation variation in gray wolf isotope (δ^{15} N and δ^{13} C) profiles: Implications for the ecology of individuals. Oecologia 145:316–325.

US Department of Agriculture. 2015. Wildlife damage. US Department of Agriculture, Animal and Plant Health Inspection Service, Wildlife Services. www.aphis.usda.gov/wps/portal/aphis/ourfocus/wildlifedamage.

US Fish and Wildlife Service. 2013. Endangered Species Act, overview. www.fws.gov/endangered/laws-policies/.

———. 2014. Endangered and threatened wildlife and plants; threatened species status for West Coast Distinct Population Segment of fisher. Docket No. FWS-R8-ES-2014-0041; 4500030113. Federal Register 79(194):60419–60443.

———. 2015. Wildlife and sport fish restoration program, Wildlife Restoration Act. http://wsfrprograms.fws.gov/Subpages/GrantPrograms/WR/WR_Act.htm.

Varley, N., and M. S. Boyce. 2006. Adaptive management

for reintroductions: Updating a wolf recovery model for Yellowstone National Park. Ecological Modeling 193:315–339.

Walters, C. J. 1986. Adaptive management of renewable resources. Blackburn, Caldwell, New Jersey, USA.

Washington Department of Fish and Wildlife. 2015. Fishers in Washington. http://wdfw.wa.gov/conservation/fisher/.

Wengert, G. M., M. W. Gabriel, S. M. Matthews, J. M. Higley, R. A. Sweitzer, C. M. Thompson, K. L. Purcell, R. H. Barrett, L. W. Woods, R. E. Green, S. M. Keller, P. M. Gaffney, M. Jones, and B. N. Sacks. 2014. Using DNA to describe and quantify interspecific killing of fishers in California. Journal of Wildlife Management 78:603–611.

White, H. B., T. Decker, M. J. O'Brien, J. F. Organ, and N. M. Roberts. 2015. Trapping and furbearer management in North American wildlife conservation. International Journal of Environmental Studies. doi:10.1080/00207233.2015.1019297.

Wijnstekers, W. 2011. The Evolution of CITES, 9th edition. International Council for Game and Wildlife Conservation, Budapest. http://www.cites.org/eng/resources/publications.php.

The Wildlife Society. 2015. Standing position statement: Traps, trapping, and furbearer management. http://wildlife.org/position-statements/.

Williams, B. K., R. C. Szaro, and C. D. Shapiro. 2007. Adaptive management: The US Department of the Interior Technical Guide. Adaptive Management Working Group, US Department of the Interior, Washington, DC, USA.

Wolfe, M. L., and J. A. Chapman. 1987. Principles of furbearer management. Pages 101–112 in M. Novak, J. A. Baker, M. E. Obbard and B. Malloch, editors. Wild furbearer management and conservation in North America. Ministry of Natural Resources, Ontario, Canada.

Wren, C. D. 1991. Cause-effect linkages between chemicals and populations of mink (Mustelavison) and otter (Lutra canadensis) in the Great Lakes basin. Journal of Toxicology Environmental Health 33:549–585.

Zielinski, W. J., and T. E. Kucera. 1995. American marten, fisher, lynx, and wolverine: Survey methods for their detection. General Technical Report PSW-GTR-157. US Department of Agriculture, Forest Service, Pacific Southwest Research Station, Albany, California, USA.

Zielinski, W. J., T. E. Kucera, and R. H. Barrett. 1995. Current distribution of the fisher, Martes pennanti, in California. California Fish and Game 81:104–112.

9 — State Management of Migratory Game Birds

Mark P. Vrtiska and
Shaun L. Oldenburger

The conservation and management of migratory game birds are unique in wildlife management because state, provincial, and federal agencies have statutory responsibility for these species. The overarching authority and responsibility for ducks, geese, swans, coots, cranes, doves, pigeons, snipe, woodcock, and rails falls under the Migratory Bird Treaty Act, and thus under the federal governments of the United States, Canada, and Mexico. While other wildlife may also migrate across state or international boundaries, the degree of migration seldom exceeds that of migratory birds, and levels of cooperation mostly exist between entities whose boundaries the species cross. This shared responsibility has developed into a state and federal cooperative management and decision-making process (i.e., the flyway system), also unique to migratory game birds.

Migration of game birds across multiple international and state or provincial borders necessitates cooperative conservation, management, and monitoring programs at those same scales. For example, management of migratory goose populations dictates that managers in Canada and the United States not only monitor harvest across all borders to ensure healthy goose populations but also have some agreement about sustaining and providing equitable harvest opportunity for all. Concerned parties may also provide resources to monitor goose population (e.g., aerial surveys) and harvest (e.g., banding programs) levels.

However, mutual cooperation and management extend beyond harvest. Managers (and hopefully hunters) in southern states need to be cognizant of possible changes in habitat conditions in Canada and the United States, given that that is where the majority of ducks they harvest in the fall are produced. Conversely, because a substantial number of ducks winter in coastal Louisiana, managers in Canada and the United States need to be concerned with wintering habitat conditions in the southern United States and Mexico. These birds' cross-seasonal dependency on habitats across vast regions of North America again makes cooperative management necessary, but it also envelops another group of individuals who primarily focus on the habitat components of migratory game birds. Therefore, the resources for management and conservation programs and the efforts of state wildlife agencies are interwoven with these linkages and partnerships for populations, habitat, and harvest.

Of note, integration and partnership of conservation and management by local, state, and federal governments, nongovernmental organizations, and other entities tend to obscure contributions of all partners, including state wildlife agencies. A tendency toward obscurity regarding individual contributions may actually be an indication of the strength of the partnership. Nonetheless, it must be acknowledged that state wildlife agencies have played, and continue to play, lead roles in conservation and management of migratory game birds.

The final, implicit point underlying much of this

chapter is that migratory game birds are hunted based on tenets of the North American Model of Wildlife Conservation (Geist et al. 2001). Debate about hunting or the model itself is not germane here, but it does need to be recognized that hunting migratory game birds has generated and continues to generate hundreds of millions of dollars toward habitat and management programs. Suffice it to say that there would be serious loss of conservation and management programs as a result of potential decreases in economic, political, and social capital currently derived from hunting these species if hunting were ended (Vrtiska et al. 2013). Management of some populations, such as temperate-nesting Canada geese (*Branta canadensis*), also would become more problematic. Lastly, beyond economics and management, various traditions of hunting migratory game birds, from marshes or a dove field, are simply integral, instinctual parts of the fabric of life in North America for participants.

Flyway Councils and Technical Committees

Formation of the flyway councils (Atlantic, Mississippi, Central, and Pacific) and associated technical committees arose from the various legal and legislative mandates, recognition of regional migration patterns, and shared management responsibilities necessitated by the linkages of species or populations across multiple boundaries previously mentioned. Delineations of the flyways were largely based on Fredrick Lincoln's assessment of waterfowl banding data that indicated four major avian migration routes (Lincoln 1935, 1939). Slight alterations to adapt the four flyways to other migratory game bird management units (see below) also used banding assessments (Kiel 1959; Krohn and Clark 1977). Thus, although the flyway councils are formed along administrative boundaries to accommodate regulation setting, they are based on biological foundations.

Although flyway councils and their respective technical committees have been previously reviewed (e.g., Hawkins 1980; Hawkins et al. 1984; Reeves 1993; Wagner 1995; Baldassarre and Bolen 2006), a basic understanding of the structures (i.e., flyways) within which state agencies operate is warranted. These structures also determine the various monitoring and

survey or other programs that state agencies conduct or participate in. The flyway structure provides an effective means for interaction among representatives from state and provincial wildlife agencies. Additionally, interaction with federal representatives from the United States and Canada, as well as other pertinent entities such as nongovernmental agencies (e.g., Ducks Unlimited), allows for coordination on waterfowl management activities, even across flyway boundaries. Finally, the regular meeting schedule of flyways allows coordination of special programs or issues or dissemination of information for those not typically associated with flyways. For example, the large surveillance effort conducted for highly pathogenic avian influenza was largely facilitated through the flyway system.

The primary duty of flyway councils and associated technical groups is to establish migratory game bird hunting regulations. While this is an indispensable role for state and provincial wildlife agencies and is directly connected to various programs they conduct, we refer the reader to previous reviews of the process (Hawkins 1980; Hawkins et al. 1984; Reeves 1993; Baldassarre and Bolen 2006) and focus here on the roles, responsibilities, and programs of the state and provincial agencies. However, we note that timing of the annual regulatory process has changed from previous descriptions (US Fish and Wildlife Service 2013).

Flyway Councils

Flyway councils are composed of at least one representative from higher administration levels within each participating agency. The councils consider and vote on recommendations from the various technical groups regarding hunting regulations or other relevant matters that are forwarded to the US Fish and Wildlife Service (USFWS) or other entities. Councils provide the official stance or "voice" of the flyways and are the formal linkage to the USFWS with regard to regulations. For example, a recommendation by a technical group may be altered by the council, which then becomes the official recommendation that goes to the USFWS for its consideration. Councils also can direct their technical committees to examine or consider specific issues for their consideration. Thus, an important role of the councils is to direct flyway policy.

Not mentioned in other reviews, however, is the National Flyway Council (NFC), in which one council member from each flyway serves as a representative. This group discusses and resolves broad-scale issues involving most or all flyways. Typically, councils meet twice per year, and the NFC may meet three to four times per year.

Waterfowl Technical Groups

Most activities occurring in flyways deal with management and regulations concerning waterfowl. State and provincial waterfowl biologists may meet two to three times per year with federal counterparts and other biologists or managers to assimilate or disseminate biological and population demographic information; discuss various aspects of regulations, harvest objectives, strategies, or management; and develop or update various management plans. Preliminary discussion and initial formulation of hunting regulation recommendations occur and are forwarded to the council for their consideration and approval. However, many other waterfowl-related issues, such as monitoring programs, research projects, and habitat, may also be addressed. Correspondence may be directed from the flyways to appropriate agencies or individuals regarding these issues. Essentially, the flyway technical groups synthesize biological and technical information to formulate regulations, management or program actions, or position stances for their respective councils to consider.

Webless Groups

Responsibility for migratory shore and upland game birds known as "webless" migratory game birds, including Wilson's snipe (*Gallinago delicata*), American woodcock (*Scolopax minor*), rails (*Porzana carolina* and *Rallus* spp.), and sandhill cranes (*Grus canadensis*), may be handled by a separate webless committee. Currently, the eastern three flyways have webless groups, but webless species in the Pacific Flyway are managed by a subcommittee of their waterfowl technical group. Webless groups typically meet once per year to discuss regulations/issues, and they have similar responsibilities to and follow the same processes as waterfowl technical groups.

Dove Management Units

Harvest management for doves is accomplished through three established management units (Eastern, Central, and Western; Reeves 1993), which do not align with the typical structure of other flyway groups. Although the structure is slightly different, roles and responsibilities are analogous to waterfowl and webless groups, meeting once a year.

Nongame Migratory Bird Technical Groups

Nongame migratory bird technical committees were formed with a primary function of facilitating discussion and coordination among states on regulatory issues that affect nongame migratory bird species within and among flyways. Although they do not deal directly with migratory game birds and hunting regulations, discussion of this technical group within the flyway context is important as they may have ramifications across technical groups. For example, in the Central Flyway, issues with whooping cranes (*Grus americana*) may dictate communication and discussion among the nongame, waterfowl, and webless groups. Nongame technical groups are relatively new within flyways, beginning around 2006.

Other Groups

Less recognized than the flyway councils and technical committees are other technical groups or committees that have been created to help address management problems or technical issues and also serve to improve communications between the USFWS and the flyways. State representatives not only serve as liaisons between these technical groups and their respective flyways but also may conduct analysis or other work for these groups. The first and longest-running technical group is the [Adaptive] Harvest Management Working Group (HMWG), formed in the early 1990s to implement adaptive harvest management as the formal process of setting duck seasons. Two state representatives from each flyway are official members of this group.

A parallel group to the HMWG, the National Dove Task Force, was formed to address harvest strategies for mourning doves (*Zenaida macroura*) and is represented by state representatives from each management unit. Currently, the group focuses on more technical aspects of harvest strategies.

The most recent technical group to be formed is the Human Dimensions Working Group (HDWG). The HDWG was formed to include not only flyway representation but also other state personnel such as those involved in human dimensions aspects. The HDWG was formed to examine issues related to hunters and regulations, as well as to help address the "people" goal in the revised North American Waterfowl Management Plan (NAWMP; North American Waterfowl Management Plan, Plan Committee 2012).

Ad hoc groups also may be formed when specific topics or issues arise. For example, a group of state and federal agency personnel was formed to assess the harvest potential of teal, primarily associated with possible changes to regulations for September teal seasons (Teal Harvest Potential Working Group 2013).

Survey and Monitoring Programs
Surveys
DUCKS

Many monitoring projects enumerate different waterfowl species during the same survey (ducks, geese, swans, and coots) owing to their shared habitats across the United States. For duck species, the primary method to monitor breeding duck populations has been the Waterfowl Breeding Population and Habitat Survey, which is a cooperative aerial and ground survey conducted by the USFWS, the Canadian Wildlife Service (CWS), and state, provincial, and tribal biologists, pilots, and technicians (US Fish and Wildlife Service and Canadian Wildlife Service 1987). This survey was developed in the late 1940s by USFWS personnel (then the US Biological Survey) and became operational in 1955. Although all surveys are flown by USFWS pilot-biologists, state and provincial wildlife biologists have assisted as aerial and ground observers. However, prior to this survey being implemented, several state wildlife agencies had separate breeding waterfowl surveys, including North Dakota (starting in 1948) and California (starting in 1949). In addition to breeding waterfowl surveys, North Dakota has conducted July production

surveys since 1958 and with the consistent current routes since 1965.

Many state wildlife agencies reproduced methodology similar to the cooperative waterfowl survey, including Minnesota, Michigan, and Wisconsin in the Mississippi Flyway, where no federal surveys are conducted. These surveys are now included in the Midcontinent mallard population estimate (US Fish and Wildlife Service 2015). In the Pacific Flyway, California has conducted a state breeding waterfowl survey since 1949; it changed methodology to reflect USFWS protocols in 1992, but it incorporated visibility-correction factor surveys using helicopters. Oregon conducted a combination fixed-wing and helicopter transect survey starting in 1994, but then it continued with a helicopter-only transect survey in 1996. Nevada, Utah, and Washington have all developed or conducted breeding waterfowl surveys in their respective states in the past 20 years. Nebraska conducted an aerial survey of the Sandhills region beginning in 1966. Beginning in 1999, USFWS survey protocols were adopted, and double-observer methodology for visibility correction was incorporated after that (Vrtiska and Powell 2011). The survey was discontinued in 2006 owing to lack of pilots and aircraft.

Separate surveys also have been developed for other duck populations not in the continental breeding population. Since 2008, the Gulf Coast population of mottled ducks (A. fulvigula) has been monitored by a cooperative visibility-corrected transect survey between the Louisiana Department of Wildlife and Fisheries, the Texas Parks and Wildlife Department (TPWD), and the USFWS.

GEESE

Similar to what was found in the previously mentioned monitoring projects on Canada geese (e.g., Mississippi Valley population), some goose populations cannot be surveyed on their breeding grounds as a result of logistical constraints. Furthermore, some goose populations may be staging during migration. Thus, most of the population is concentrated, and there are fewer logistical constraints. Aleutian cackling geese (B. hutchinsii leucopareia) are monitored by mark-resight survey by the California Department and Fish and Wildlife, the Oregon Department of Fish and Wildlife, and the

USFWS (Sanders and Trost 2013). Cackling geese (*B. hutchinsii minima*), breeding on the Yukon–Kuskokwim Delta in Alaska, have been monitored by a variety of methods, including roost counts, breeding transect surveys, midwinter counts, and mark-resight monitoring programs by both state and federal agency personnel on both the breeding and wintering grounds.

For both the Pacific and Midcontinent greater white-fronted goose populations, separate staging surveys are used for their respective management plans (Pacific Flyway Council 2003; Central Flyway Council et al. 2005). A fall (October) survey is conducted for the Pacific population, when nearly the entire population is located in either Oregon or California. This population is enumerated via fixed-wing aircraft by USFWS and state wildlife agency personnel with a cruise survey. For the Midcontinent population, US-FWS pilot-biologists and state wildlife agency personnel conduct cruise surveys in Saskatchewan during late September, when this population is staging along the South Saskatchewan River and other important staging locations.

Separate white goose (e.g., lesser snow, Ross's geese) surveys are conducted in several flyways by state and federal wildlife agency personnel to inventory the number of wintering geese separate from the Midwinter Waterfowl Survey (MWS). Most of these surveys are conducted either by fixed-wing aircraft or from the ground. In some instances, additional operational surveys conduct scanned samples of white geese to differentiate snow and Ross's geese.

Currently, state wildlife agency personnel, along with USFWS partners, participate in the MWS. The MWS was initiated in 1935 but had a hiatus during World War II; surveying resumed after the war and has been operational on an annual basis ever since. As the longest-standing waterfowl survey in North America, it was originally used to provide population indices to ducks, geese, and swan populations. From 1935 to 1954, the MWS was used to guide the regulatory process in setting waterfowl hunting regulations, since surveying the breeding grounds was considered logistically infeasible. Survey designs differ significantly across many states, but most surveys are cruise surveys in fixed-wing aircraft, where abundant waterfowl are enumerated by trained observers at important wintering locations along rivers, reservoirs, wildlife management areas (WMAs), national wildlife refuges, private duck clubs, bays, estuaries, and other important wintering locations. The MWS has been used to assist in conservation planning with Joint Ventures, as well as addressing state issues (Sharp et al. 2002).

MOURNING, WHITE-TIPPED, AND WHITE-WINGED DOVES

The Southeastern Association of Game and Fish Commissioners (1957), a 10-state initiative to obtain life history information on mourning doves and to produce a standardized survey methodology, developed a monitoring program for mourning dove populations from 1950 to 1952. This methodology was developed from studies conducted on mourning dove life history in Iowa (McClure 1939). This methodology requires each observer to drive a 20-mile (32.2-kilometer) route, recording the number of doves heard and seen, including the number at 20 three-minute stops located at 1-mile intervals. Beginning in 1959, all 48 conterminous states conducted survey routes across the nation. Currently, mostly state wildlife agency personnel conduct the surveys. This survey was conducted annually by state and federal agencies until 2013, when the primary monitoring of mourning dove populations was done using banding and harvest data (Otis 2006; Seamans and Sanders 2014). Approximately 1,000 surveys were completed on an annual basis until 2013.

Although no national, coordinated surveys are conducted for white-winged doves, specific state wildlife agencies have conducted separate surveys to monitor both the western and eastern populations of white-winged doves. Both the Arizona Fish and Game Department (AFGD) and the TPWD have done considerable work monitoring these populations. Since 1962, the AFGD has conducted 25–30 white-winged dove call-count survey routes along rural secondary roads as an index to abundance (Rabe and Sanders 2010; A. Anoude, AFGD, personal communication). Currently, the TPWD conducts point-count surveys with distance sampling in both urban (~3,500 points) and rural locations (~2,500) across the state to estimate abundance of white-winged doves (Oldenburger et al. 2014).

Owing to its limited range in southern Texas, the TPWD conducted a species-specific call-count survey for white-tipped doves from 1983 to 2000. However, this survey was discontinued because of possible randomization and lack of effectiveness.

BAND-TAILED PIGEONS

Band-tailed pigeons are managed within the Pacific and Four Corners populations in western North America. The Pacific population was historically monitored by several state wildlife agencies using different methodologies. Since 1950, Oregon has employed mineral-site surveys (MSSs). In the 1960s, California conducted September surveys at select locations across the state. From 1975 to 2003, Washington used call-count surveys on 50 selected routes.

Since 2003, the Pacific population has been monitored by these three states using an MSS during the month of July, and British Columbia now utilizes a similar survey (Casazza et al. 2005; Sanders 2014). Band-tailed pigeons are counted at select mineral-site locations across their range. These numbers provide an index; mostly state wildlife agency personnel conduct these surveys, along with the assistance of US Forest Service and USFWS staff in California, Oregon, and Washington.

Currently, no band-tailed pigeon surveys are conducted for the Four Corners population located in Arizona, Colorado, New Mexico, and Utah. However, in the late 1940s and 1950s, biologists, technicians, and volunteers from these states conducted annual counts from May to October on the Four Corners population. These counts were later truncated to a five-day period in September (Pacific Flyway Study Committee and Central Flyway Webless Migratory Game Bird Technical Committee 2001). The Colorado Division of Wildlife conducted separate surveys, but these were later discontinued as well.

AMERICAN WOODCOCK

The North American Woodcock Singing-Ground Survey has been conducted in the eastern United States and Canada since 1968 (Cooper and Rau 2015). The survey has approximately 1,500 3.6-mile (5.4-kilometer) routes randomly selected along secondary roads in the species breeding range. Observers record the number of woodcock heard during each of the 10 stops along the route shortly after sunset. The number of singing males heard is used as an index to breeding population density, and thus changes in this index are assumed to be equal to changes in the breeding population. Annually, approximately 800–900 surveys are conducted during the survey period, and state wildlife agency personnel represent approximately two-thirds of the surveyors. This is the primary method of inventorying the population, and currently it is the major monitoring program in the harvest strategy (Cooper and Rau 2015).

RAILS AND GALLINULES

Many different survey methodologies were used by both state and federal wildlife agency personnel from the 1950s to the 1990s to enumerate both rails and gallinules across the United States (Tacha and Braun 1994). Currently, no nationally coordinated survey is conducted to estimate or index abundance of rails and gallinules. Breeding Bird Surveys (Sauer et al. 2014; completed by federal, state, and volunteer personnel) are used to index these populations for harvest management. In the 1990s, only four states had consistent surveys for rails or gallinule species (Virginia, New Jersey, Ohio, and California; Conway and Eddleman 1994). Surveys for rails and gallinules have been difficult to perform as a result of their cryptic behavior and use of coastal and freshwater wetlands with dense emergent vegetation. Owing to the varying degree of surveys conducted in the past, federal and state wildlife managers desired a standard protocol for monitoring these populations. Playback calls during the breeding season were considered a highly promising methodology. Conway (2009) developed these protocols to conduct consistent surveys for secretive marsh birds, which includes rails and gallinules. Because of their likely declining populations due to habitat loss and wetland degradation across their range, a renewed effort to obtain trends and population estimates of these species has occurred in the past decade. Although no nationwide coordinated survey occurs at this time, a large number of state wildlife agencies conduct or are actively engaged in secretive marsh bird surveys at the WMA or regional level, including Arizona, Florida, Kansas, Nebraska, Ohio, Louisiana, Michigan, Missouri, New Jersey, and Wisconsin.

SANDHILL CRANES

All sandhill crane surveys are conducted cooperatively among state and federal wildlife agencies, as stated in respective flyway management plans. There are seven recognized sandhill crane populations in the United States and Canada. Two of these populations are non-migratory and not hunted (i.e., those in Florida and Mississippi) and will not be mentioned further. Additionally, the Central Valley population breeding in California, Oregon, and Washington is not hunted, so minimal annual monitoring has occurred. Of the remaining four populations, the Lower Colorado River population comprises a small assemblage of sandhill cranes that breed in the Intermountain West (Nevada, Oregon, and Idaho) and winter in southern California and Arizona near the Lower Colorado River. This population has been monitored by both state and federal personnel, who conduct counts on the primary wintering grounds annually (Pacific Flyway Council 1995).

The largest population, the Midcontinent population, breeds from Siberia in the west to Ontario in the east and winters primarily in Texas, New Mexico, and Mexico. This population stages in large concentrations (~600,000) along the Platte River in Nebraska annually in late February–March. This population is monitored by a photo-corrected, aerial transect survey conducted by the USFWS in late March. Additional state and federal wildlife agency personnel conduct ground counts across the population's range in an attempt to index the timing of the survey period (Central Flyway Webless Migratory Game Bird Technical Committee 2006). Additional surveys were conducted by the TPWD from the 1960s to the 1990s to enumerate sandhill cranes on their wintering grounds and perform age ratios, but none remain operational.

State wildlife agencies are more engaged in monitoring of other sandhill crane populations, primarily as a result of their widespread breeding, staging, or wintering range. The Rocky Mountain population, which breeds in the Intermountain West in both the Central and Pacific Flyways, is monitored by both aerial and ground counts. During a survey period of three to five days each September, the USFWS monitors the main fall staging areas by fixed-wing aircraft. However, as a result of the widespread breeding range, federal and state wildlife agency personnel conduct ground sur-

veys of other smaller populations and concentrations outside the survey area. During this time, federal and state wildlife agency personnel also conduct age ratios on those flocks (Subcommittee on Rocky Mountain Greater Sandhill Cranes 2007). Another population that has been monitored during fall staging is the Eastern sandhill crane population. This population has grown substantially in both range and number in the past two decades. Across this population's range, many state wildlife agencies have historically conducted separate sandhill crane surveys. Currently, however, a cooperative federal and state fall survey is conducted in late October to enumerate the number of eastern sandhill cranes on important fall staging areas. In Kentucky and Tennessee, additional surveys are conducted by state wildlife agency personnel on important habitats because sandhill cranes are hunted in these states. These surveys are used as the primary monitoring index to this population (Ad Hoc Eastern Population Sandhill Crane Committee 2010).

OTHER SPECIES

State and federal wildlife agency staff conducted a midwinter Wilson's snipe survey during a period of one to five days in late January from 1952 to 1964 throughout the southern states in all four flyways. This survey was discontinued owing to inconsistency in methodology and limited population status information being obtained. Currently, no nationwide coordinated survey occurs for Wilson's snipe.

Banding

Besides abundance surveys, migratory game bird banding has contributed more to the basic understanding of population dynamics and general life history than any other monitoring program in North America. Migratory game bird banding data have allowed managers, biologists, and researchers to examine migratory pathways, annual survival rates, fidelity, population size, reporting rates, recovery rates, and harvest rates (Williams et al. 2002). Lincoln (1935) used migratory pathways from banding recoveries to develop the flyway concept in the early 1930s. A few state wildlife agencies began banding ducks and geese in the 1930s (e.g., Michigan Department of Natural Resources, New

York Department of Environmental Conservation), but with the creation of the flyway system in the late 1940s and early 1950s, duck and goose banding became more widespread, with nearly every state wildlife agency contributing to continental banding programs. Starting in 1959, the USFWS's Office of Migratory Bird Management began to coordinate banding activities among flyways, the CWS, state wildlife agencies, private individuals and organizations, and nongovernmental organizations (US Fish and Wildlife Service 1986). Also beginning that year, duck banding goals were reviewed and adjusted by the USFWS, the CWS, and the flyways on a five-year basis, with the last review occurring in 1989. Banding stations in breeding locations, winter banding, species-specific banding goals, and many other details of banding operations were established through this process. Additionally, through these reviews and duck banding goals, species-specific banding reference areas have been developed by federal and state wildlife biologists to gain insight into population dynamics and movements of particular duck species (Anderson 1975; Szymanski and Dubovsky 2013). Duck banding also lets researchers and managers examine various hypotheses regarding population dynamics, including the impacts of harvest (e.g., Burnham and Anderson 1984). Finally, harvest rates obtained from banding are an integral part of harvest management strategies such as adaptive harvest management used for Midcontinent mallards (US Fish and Wildlife Service 2015).

DUCKS

Between 1950 and 2014, state wildlife agencies banded nearly 5.5 million ducks in all 50 states, including 2 million mallards. Because mallards are the most harvested duck in North America, they are the basis of most historical and current hunting regulation frameworks. Thus, mallards have been a primary target for many banding projects throughout Canada and the continental United States. Northern pintail (A. acuta) banding has occurred in large numbers on both the breeding and wintering grounds to gain insight into their population dynamics (Rice et al. 2010). Other recent investigations into population dynamics of teal depended greatly on many of the state wildlife agency–banded blue- (A. discors) and green-winged teal (A. crecca) and cinnamon teal (A. cyanoptera; Teal Harvest Potential

Working Group 2013). Duck banding plays a critical role in the harvest strategies for mallards, American black ducks (A. rubripes), and wood ducks (Aix sponsa; US Fish and Wildlife Service 2015).

Many coordinated banding projects have occurred through flyway efforts in recent time. For instance, from 1996 to 2001, the Central Flyway Council initiated a large-scale project to band birds outside of previous banding locations in order to obtain more thorough information on survival and movements of mallards and other ducks from the northern US prairies. This project has banded approximately 139,000 ducks and nearly 31,000 mallards and is still active. In the Atlantic and Mississippi Flyways, there is an annual coordinated effort to meet banding needs for wood ducks, and since 1970, state agencies in these two flyways have banded nearly 791,000 wood ducks. State or state-sponsored banding, along with banding conducted on USFWS National Wildlife Refuges in the Pacific Flyway, has provided the primary data used to estimate harvest rates in the western mallard model used to manage this population (US Fish and Wildlife Service 2015). State wildlife agencies have played a critical role in banding waterfowl through state duck stamp and Wildlife Restoration Program projects.

CANADA GEESE

Many state wildlife agencies have conducted independent or cooperative projects on banding other goose species in the United States, such as black brant (Branta bernicla), emperor (Chen canagica), greater white-fronted (Anser albifrons), Ross (C. rossii), and lesser snow geese (C. caerulescens). Canada goose restoration and efforts to monitor both resident and migratory populations have occurred in all four flyways; thus, we concentrate on these efforts. The Michigan Department of Natural Resources was the first state wildlife agency to band Canada geese in 1933. Some state wildlife agencies began banding of Canada geese in the 1940s, but Canada goose banding did not become widespread until the 1950s, after establishment of the flyway system. Starting in the 1950s, state agencies invested large amounts of resources into banding and translocating Canada geese to increase resident populations. State wildlife agencies banded about 2.5 million Canada geese in the United States from 1950 to 2014, which is approximately 69 percent of all Canada geese banded in North America during this time period. Mississippi Flyway states banded approximately 1.4 million Canada geese, and Iowa, Illinois, Ohio, and Michigan banded over 100,000 birds, with many of these birds being involved in the ongoing restoration of giant Canada geese in the Mississippi Flyway (Giant Canada Goose Subcommittee 1996).

Many state agencies have monitored population dynamics of resident Canada geese by banding in various states among all flyways (e.g., Rexstad 1992; Powell et al. 2004; Heller 2010; Beston et al. 2014). Additionally, the flyways have investigated the population dynamics of a number of subarctic migratory populations from Canada and Alaska to the conterminous United States. In these investigations, not only have state wildlife agencies used trapping and banding, but many migratory populations have also been monitored by mark-resight of neck collars on summer- or winter-banded Canada geese by both state and federal wildlife agency personnel. In the Atlantic Flyway, state and federal wildlife agencies have invested significant time and resources to monitoring the decreasing population of wintering Atlantic Canada geese in the Chesapeake region starting in the 1950s, primarily with winter banding (Hestbeck 1994). In the Mississippi Flyway, neck collars were monitored on Mississippi Valley and Eastern Prairie populations of Canada geese in the late 1970s and 1980s (Samuel et al. 1991). Aleutian cackling geese have been monitored for their population status using neck collars, primarily in California and Oregon since 1996 (Sanders and Trost 2013). Also in the Pacific Flyway, neck-collared cackling geese have recently been monitored to estimate abundance in Oregon and Washington. In both projects, state wildlife agency personnel play critical roles in trapping, marking, and resighting neck-collared geese.

MOURNING AND WHITE-WINGED DOVES

Although waterfowl banding started in the 1930s with state wildlife agencies, dove banding also has a long tradition. The Alabama Department of Conservation was the first state wildlife agency to band mourning doves in 1936. However, it was not until the 1950s that dove banding became more widespread within state wildlife agencies. The banding program in the

1950s assisted in the delineation of the Eastern, Central, and Western Management Units (Kiel 1959). A cooperative nationwide banding program occurred from 1967 to 1975 (Dunks et al. 1982; Tomlinson et al. 1988). Critical population demographic information was obtained from this long-term banding operation, including annual survival rates, recovery rates, recovery distributions (i.e., migratory pathways), and derivation of harvest. Many state agencies continued to band mourning doves after the completion of this nationwide program, although fewer in the 1980s and 1990s.

Beginning in 2003, nationwide mourning dove banding became operational again based on its ability to gain more insight into the population dynamics of mourning doves (Otis 2009). Given a new mourning dove harvest strategy (US Fish and Wildlife Service 2014), mourning dove banding continues to be a cooperative endeavor, with state and federal participation. From 1950 to 2014, state wildlife agencies in Eastern, Central, and Western Management Units banded approximately 518,000, 377,000, and 187,000 mourning doves, respectively.

Based on original work by Lincoln (1930) for estimating abundance from bird bands and harvest information, Otis (2006) developed a framework for determining abundance of mourning doves from banding and harvest data. Starting with the 2014–2015 hunting season, this framework has been used as the primary monitoring program for mourning dove populations in the United States. Currently, respective flyways, technical committees, and the National Dove Task Force, composed of both state and federal wildlife agency personnel, continue to refine monitoring methods and management of mourning doves through banding data and other pertinent information.

The AFGD and the TPWD banded approximately 55,000 and 130,000 white-winged doves (Z. asiatica), respectively, in June–August from 1950 to 2014. These banding data have been critical in understanding migration pathways, annual survival, harvest rates, and other population dynamics of this species (George et al. 2000; Collier et al. 2012a, 2012b). White-tipped doves (Leptotila verreauxi) have been banded by the TPWD, but they have not been banded in sufficient numbers to provide satisfactory information on life history.

State wildlife agencies in the western United States trapped and banded band-tailed pigeons (*Columba fasciata*) on a consistent basis starting in the 1950s. Oregon and Washington had long-term banding projects starting in 1952 and 1958, respectively. Pacific and Four Corners population states banded band-tailed pigeons consistently until the 1970s. Braun et al. (1975) and Kautz and Braun (1981) analyzed these data for distribution, recovery, and annual survival rates.

AMERICAN WOODCOCK

American woodcock banding has occurred throughout the eastern United States on both breeding and wintering grounds. Currently, banding by both state and federal agencies is used to establish current boundaries of the Central and Eastern Management Units (Krohn and Clark 1977). To gain insight into migration pathways and population dynamics of American woodcock, the Michigan Department of Natural Resources initiated a long-term woodcock banding project in 1953. Since that time, they have banded 39,064 woodcock, approximately 29 percent of all woodcock banded in the United States and Canada, providing valuable insight into American woodcock population dynamics (Krementz et al. 2003). The Wisconsin Department of Natural Resources has also conducted long-term banding programs on woodcock, with 11,380 birds (approximately 9% of all banded birds) banded from 1967 to 2014. Approximately 6,500 woodcock have been banded in Louisiana since 1952, mostly at Sherburne Wildlife Management Area in southern Louisiana. Many other state wildlife agencies have banded woodcock in substantial numbers (>500), including Alabama, Indiana, Maine, Massachusetts, New York, North Carolina, Pennsylvania, and West Virginia.

OTHER SPECIES

State agencies, with federal partners, banded Wilson's snipe in the southern states during the 1950s and 1960s to investigate recovery rates, since minimal information had been gathered about this species and its respective harvest. This project revealed extremely low recovery rates on snipe, providing an inference to the level of harvest on this population; the project was later discontinued.

Parts Collection Surveys
WATERFOWL

As evidenced above, state wildlife agencies have played a critical role in the monitoring of migratory game birds. Harvest surveys provide another essential element of migratory game bird population management. Although many state wildlife agencies have initiated harvest surveys since the 1940s, the USFWS's Harvest Survey Branch has had the lead role in estimating harvest of migratory game birds.

The Harvest Inventory Program (HIP), first implemented in 1999, required every state wildlife agency to begin collecting migratory game bird hunter information. State wildlife agencies must provide the USFWS with this information on a regularly scheduled basis prior to and during the hunting season for possible inclusion in various harvest surveys (Ver Steeg and Elden 2002).

Furthermore, annual flyway wing bees have been largely attended by state wildlife agency personnel. Randomly selected hunters send the USFWS their harvested waterfowl parts to a respective location in each flyway. When possible, the species, gender, and age are determined for each waterfowl part, which allows the USFWS to estimate the age and sex ratio for each species in the harvest in each state and flyway. Wing bees have been held in every flyway since the 1950s. Age ratio information for mallards derived from wing bees has taken on a larger role in harvest regulations, including incorporation into the adaptive harvest management protocol (US Fish and Wildlife Service 2015).

MOURNING DOVE

Prior to the beginning of the HIP in 1999, webless migratory game bird harvest was estimated solely from federal duck stamps purchased by waterfowl hunters. Because there are more dove than waterfowl hunters, the sampling frame was unknown to be representative of the dove hunters. Similar to waterfowl wing bees, an annual mourning dove wing bee is held to determine age ratios in the harvest. The Missouri Department of

Conservation hosts the annual wing bee at James A. Reed Memorial Wildlife Area in Lee's Summit, Missouri.

AMERICAN WOODCOCK

Similar to the other wing bees, age and gender can be determined from wings in American woodcock. The USFWS's Parts Collection Survey for woodcock uses wings collected from hunters on an annual basis. State and federal agency personnel then meet every March or April at a state-sponsored location to determine the age and gender of each hunter-provided woodcock wing. These wings provide a good indicator of the total recruitment that occurred during the previous breeding season. State wildlife agency personnel from both management units participate in this survey. In total, trained participants determine age and sex of woodcock from approximately 9,000–14,000 wings annually (Cooper and Rau 2015).

Habitat Surveys

One of the biggest, and perhaps most overlooked, contributions that states have made to migratory game birds is that of conservation, restoration, and management of habitats. Habitat conservation programs may occur at various scales, from restoration or technical assistance on private lands, to owning and or managing public WMAs, to participation in NAWMP Joint Ventures (North American Waterfowl Management Plan, Plan Committee 2012) or other flyway or national efforts. Combined, all of these efforts have contributed to millions of hectares of habitat used by migratory game birds and numerous other species. Indeed, state programs and efforts effecting habitat conservation, restoration, and management may exceed those of some federal agencies.

PRIVATE LANDS

Restoration, conservation, and management of migratory game bird habitat on private lands led by states and provinces may number into the hundreds of thousands if not millions of hectares. Unfortunately, an accurate accounting or assessment of state agency impacts on private lands is unavailable, primarily because states

use programs (e.g., Wetland Reserve Program) or assistance from a variety of partners, including, but not limited to, the USFWS, the Natural Resources Conservation Service, Joint Ventures, tribal governments, nongovernmental organizations (e.g., Ducks Unlimited), and local natural resource organizations and agencies. Reporting of hectares of habitat restored or conserved may be conducted outside the state agency itself, or perhaps not at all, and is definitely not accounted for across the United States. As an example of the level of potential impact, the USFWS's Partners for Fish and Wildlife Program, which targets habitat work on cooperative private lands and typically partners with state agencies, has restored or enhanced over 304,000 hectares of wetlands and 755,000 hectares of prairie, shrub, and forest upland habitat since the program's inception in 1987 (US Fish and Wildlife Service 2010). Thus, state and provincial agencies have provided and continue to provide significant contributions to habitat for migratory game birds on private lands in the United States and Canada.

STATE WILDLIFE MANAGEMENT AREAS AND REFUGES

Similar to the national wildlife refuge system, an important aspect of habitat contributions to migratory game birds is the creation of state WMAs and refuges that provide not only habitat but also hunting opportunities. Sauvie Island in Oregon, Cheyenne Bottoms in Kansas, Bayou Meto in Arkansas, and Pymatuning in Pennsylvania are examples of important WMAs that are well known to hunters. State agencies also may lease or manage areas owned or controlled by other agencies, such as the US Army Corps of Engineers. Bellrose (1980) compiled records of state WMAs and refuges primarily used by waterfowl and estimated a total of over 2,076,099 hectares of habitat. More current estimates of state-owned habitat are not available. However, it should be noted that Bellrose's (1980) original estimate was not exhaustive and only contained those areas of relatively large size (>400 hectares). It understandably also does not contain recent acquisitions of new or existing WMAs and refuges since 1980. Thus, more area exists than Bellrose's estimate of 2.1 million hectares. Additionally, his totals also did not consider

WMAs or refuges that may be used by doves, cranes, snipe, or woodcock, thus adding more hectares to the total.

Another frequent positive impact of state WMAs arises when they lie adjacent to or near national wildlife refuges or other federal areas that create contiguous or a complex of habitats. Duck Creek Conservation Area in southeast Missouri, for example, lies adjacent to Mingo National Wildlife Refuge and provides a contiguous block of wetland and bottomland hardwood habitat of about 11,300 hectares under public ownership. Numerous state-owned lands in other states provide similar benefits.

REGIONAL OR NATIONAL

Given the international coverage of habitats, all states and provinces are involved in some fashion in NAWMP Joint Ventures that are more eco-regionally based and typically extend beyond their jurisdictional boundaries. Similar to private lands work, states may provide resources that are intertwined with issues or goals identified in Joint Ventures programs. States are represented on management boards and technical committees that may provide some level of coordination and influence on habitats at larger scales, and this may also extend to evaluation and monitoring programs. Finally, states also may send representatives to assist in efforts of the NAWMP's National Science Support Team (NSST), which provides technical advice to the NAWMP Plan Committee. The NSST provides biological technical support to the NAWMP to facilitate improvement of NAWMP-related conservation efforts.

A relatively unknown program is the state grants program that is coordinated through the Association of Fish and Wildlife Agencies (AFWA). This program was initiated as a result of recognizing the importance of Canadian prairies to duck production and that most of the duck harvest occurs in the United States. State agencies allocate funding, typically through the sale of their state duck stamps, to Ducks Unlimited, Canada, which then uses the funding to conserve, restore, and manage upland and wetland habitats in Canada. Those funds can be used to match other funding, such as North American Wetlands Conservation Act grants,

which in effect doubles the investment. Since 1986 and the establishment of the NAWMP, state agencies have contributed $65 million to Canadian projects that have helped conserve and manage 4 million hectares of habitat (MacCallum and Melinchuk 2011).

Other Roles of State Wildlife Agencies
Restoration of Populations

Similar to efforts with other wildlife species, state and provincial agencies have attempted to augment or restore migratory bird populations, most notably Canada geese and trumpeter swans (*Cygnus buccinator*). Indeed, restoration of Canada geese began with the rediscovery of giant Canada geese (*B. c. maxima*) by Hanson (1965), and state efforts began soon after. Most programs entailed maintaining captive flocks, translocating birds, erecting nesting structures, and altering hunting regulations to allow populations to expand (e.g., Aldrich et al. 1998; Zenner and LaGrange 1998). Establishment programs were vastly successful, with current populations in the eastern three flyways of over 3 million birds (USFWS 2005). Most efforts to restore Canada geese were ended by the 1990s.

Similar efforts toward restoring populations of trumpeter swans have occurred in several midwestern states and provinces (Matteson et al. 2007). Most efforts began later than Canada goose programs, but success has been similar. The 2010 continental estimate of trumpeter swan abundance was over 46,000 swans (Groves 2012).

Nuisance and Depredation Management

While the aforementioned restoration of Canada goose populations is considered a wildlife management success story, it conversely also created substantial nuisance and depredation problems (US Fish and Wildlife Service 2005). States and provinces are typically the initial or primary point of contact regarding nuisance and depredation issues, running the gamut of problems from feces on sidewalks to human health and safety concerns (e.g., geese at airports). Although perhaps not viewed as a positive contribution, dealing with nuisance and depredation problems is nonetheless important because they can have serious effects, both economically and emotionally. Resolution of nuisance problems can also be quite controversial, as it may involve dispatching offending animals (Conroy 2002). Even when dispatching nuisance birds is not considered, confrontations still may occur, possibly making nuisance or depredation issue management an unpleasant experience. Finally, depending on how states deal with these issues, considerable resources—both personnel and financial—may be expended. For example, the South Dakota Game, Fish and Parks Department spent over $446,000 in 2014 on handling depredation complaints regarding Canada geese (R. Murano, South Dakota Game Fish, and Parks Department, unpublished report).

The response and degree of involvement in resolving nuisance and depredation issues vary among state and provincial agencies. In some states, Wildlife Services, a division of the US Department of Agriculture Animal and Plant Health Inspection Service, may be the lead agency in dealing with nuisance problems, in which case state agencies no longer deal with nuisance calls. A recent Environmental Impact Statement by the USFWS for management of resident Canada geese has allowed states various permit activities that facilitate resolution of problem geese (US Fish and Wildlife Service 2005).

Disease Management

Similar to nuisance and depredation issues, states and provinces are typically the initial or primary point of contact to handle disease issues or outbreaks. Disease issues may involve investigating die-offs, submitting samples or conducting surveillance programs, and collecting dead or dying birds. An extensive discussion regarding state agency involvement is given in chapter 12. However, it should be noted here that coordination of one of the largest national disease surveillance efforts for detection of avian influenza viruses (AIVs) was facilitated through the flyway system. Additionally, a large number of the 155,535 migratory waterfowl samples collected from 2006 to 2009 for AIVs in the United States were taken by state agency individuals (Groepper et al. 2014).

Hunter Outreach

Although not exclusive to migratory game bird hunters, a large investment in hunter education and recruitment and retention programs has been made by, or in cooperation with, state wildlife agencies (Byrne 2009). More specific efforts have been made for some migratory game birds, such as waterfowl, in terms of recruitment and retention efforts (Case 2004). Given the additional economic and habitat conservation ramifications of declining waterfowl hunter numbers (Vrtiska et al. 2013), recruitment and retention programs aimed at migratory game bird hunters are likely to be important in the future.

Related to hunter recruitment and retention efforts are programs that allow access to private land. Again, the emphasis is not solely on migratory game bird hunters, but opportunities do exist. While federal funding may assist with these programs (US Department of Agriculture, Natural Resource Conservation Service 2014), state wildlife agencies are the primary administrators.

Probably an overlooked aspect of state and provincial agencies in hunter outreach is the dissemination of information. States and provinces are obligated to publish guides or brochures regarding hunting regulations, but inclusion of information on local, state, regional, and national issues in these brochures is also common. In addition, television shows, magazines, web pages, and other social media outlets can provide more in-depth and possibly more state-specific information to hunters and the general public.

Additional Roles

States regularly participate in the migratory bird subcommittees of the AFWA and its associated regional groups. These subcommittees and sometimes ad hoc groups examine more specific issues such as nontoxic shot, examination of the Federal Migratory Bird Hunting Conservation Stamp (i.e., duck stamp) program, and others. The work and coordinated action of subcommittees have driven policy-level changes in migratory game bird management.

LITERATURE CITED

Ad Hoc Eastern Population Sandhill Crane Committee. 2010. Management plan for the Eastern sandhill cranes. Atlantic and Mississippi Flyway Councils, Minneapolis, Minnesota, USA.

Aldrich, J. W., C. M. Potter, J. L. Dorr, and A. D. Stacey. 1998. Homesteading giant Canada geese in the Sooner state: Oklahoma's establishment program. Pages 311–317 in D. H. Rusch, M. D. Samuel, D. D. Humburg, and B. D. Sullivan, editors. Biology and management of Canada geese. Proceedings of the International Canada Goose Symposium, Milwaukee, Wisconsin, USA.

Anderson, D. R. 1975. Population ecology of the mallard. V: Temporal and geographic estimates of survival, recovery, and harvest rates. US Fish and Wildlife Service Resource Publication 125, Washington, DC, USA.

Baldassarre, G. A., and E. G. Bolen. 2006. Waterfowl ecology and management, 2nd edition. Krieger, Malabar, Florida, USA.

Bellrose, F. C. 1980. Ducks, geese, and swans of North America. Stackpole Books, Harrisburg, Pennsylvania, USA.

Beston, J. A., T. C. Nichols, P. M. Castelli, and C. K. Williams. 2014. Survival of Atlantic Flyway resident population Canada geese in New Jersey. Journal of Wildlife Management 78:612–619.

Braun, C. E., D. E. Brown, J. C. Pederson, and T. P. Zapatka. 1975. Results of the Four Corners cooperative band-tailed pigeon investigation. US Fish and Wildlife Service Resource Publication 126, Washington, DC, USA.

Burnham, K. P., and D. R. Anderson. 1984. Tests of compensatory vs. additive hypotheses of mortality in mallards. Ecology 65:105–112.

Byrne, R. L., project manager. 2009. Hunting heritage action plan, recruitment and retention assessment survey report. Wildlife Management Institute and Association of Fish and Wildlife Agencies. D. J. Case and Associates, Mishawaka, Indiana, USA.

Casazza, M. L., J. L. Yee, M. R. Miller, D. L. Orthmeyer, D. R. Yparraguirre, R. L. Jarvis, and C. T. Overton. 2005. Evaluation of current population indices for band-tailed pigeons. Wildlife Society Bulletin 33:606–615.

Case, D. J. 2004. Waterfowl hunter satisfaction think tank: Understanding the relationship between waterfowl hunting regulations and hunter satisfaction/participation, with recommendations for improvements to agency management and conservation programs. Wildlife Management Institute, Washington, DC, USA.

Central Flyway Council, Mississippi Flyway Council, Pacific Flyway Council, Canadian Wildlife Service, and US Fish and Wildlife Service. 2005. Management plan for Midcontinent greater white-fronted geese.

Central Flyway Webless Migratory Game Bird Technical Committee. 2006. Management guidelines of the Central, Mississippi, and Pacific Flyways for the mid-continent populations of sandhill cranes. Central Flyway Council, Mississippi Flyway Council, Pacific Flyway Council, Canadian Wildlife Service, and US Fish and Wildlife Service, Denver, Colorado, USA.

Collier, B. A., S. R. Kremer, C. D. Mason, M. J. Peterson, and K. W. Calhoun. 2012a. Survival, fidelity, and recovery rates of white-winged doves in Texas. Journal of Wildlife Management 76:1129–1134.

Collier, B. A., K. L. Skow, S. R. Kremer, C. D. Mason, R. T. Snelgrove, and K. W. Calhoun. 2012b. Distribution and derivation of white-winged dove harvests in Texas. Wildlife Society Bulletin 36:304–312.

Conroy, M. R. 2002. Resolving human–wildlife conflicts: The science of wildlife damage management. CRC Press, Boca Raton, Florida, USA.

Conway, C. J. 2009. Standardized North American marsh bird monitoring protocols, Version 2009-2. Wildlife Research Report #2009-02. US Geological Survey, Arizona Cooperative Fish and Wildlife Research Unit, Tucson, Arizona, USA.

Conway, C. J., and W. R. Eddleman. 1994. Virginia Rail. Pages 193–206 in T. C. Tacha and C. E. Braun, editors. Management of migratory shore and upland game birds in North America. International Association of Fish and Wildlife Agencies, Washington, DC, USA.

Cooper, T. R., and R. D. Rau. 2015. American woodcock population status, 2015. US Fish and Wildlife Service, Laurel, Maryland, USA.

Dunks, J. H., R. E. Tomlinson, H. M. Reeves, D. D. Dolton, C. E. Braun, and T. P. Zapatka. 1982. Migration, harvest, and population dynamics of mourning doves banded in the Central Management Unit, 1967–1977. US Fish and Wildlife Service Special Science Report 249, Washington, DC, USA.

Geist, V., S. P. Mahoney, and J. F. Organ. 2001. Why hunting has defined the North American model of wildlife conservation. Transactions of the North American Wildlife and Natural Resources Conference 66:175–185.

George, R. R., G. L. Waggerman, D. M. McCarty, R. E. Tomlinson, D. Blankinship, and J. H. Dunks. 2000. Migration, harvest, and population dynamics of white-winged doves banded in Texas and northeastern Mexico, 1950–1978. Texas Parks and Wildlife Department, Austin, Texas, USA.

Giant Canada Goose Subcommittee. 1996. Mississippi Flyway giant Canada goose management plan. Mississippi Flyway Council, Minneapolis, Minnesota, USA.

Groepper, S. R., T. J. DeLiberto, M. P. Vrtiska, K. Pedersen, S. R. Swafford, and S. E. Hygnstrom. 2014. Avian influenza virus prevalence in migratory waterfowl in the United States, 2007–2009. Avian Diseases 58:531–540.

Groves, D. J., compiler. 2012. The 2010 North American trumpeter swan survey. US Fish and Wildlife Service, Juneau, Alaska, USA.

Hanson, H. C. 1965. The giant Canada goose. Southern Illinois University Press, Carbondale, Illinois, USA.

Hawkins, A. S. 1980. The role of hunting regulations. Pages 59–63 *in* F. C. Bellrose. Ducks, geese, and swans of North America. Stackpole Books, Harrisburg, Pennsylvania, USA.

Hawkins, A. S., R. C. Hansen, H. K. Nelson, and H. M. Reeves, editors. 1984. Flyways. US Department of the Interior, US Fish and Wildlife Service, Washington, DC, USA.

Heller, B. J. 2010. Analysis of giant Canada goose band recovery data in Iowa and the Mississippi Flyway. Thesis, Iowa State University, Ames, Iowa, USA.

Hestbeck, J. B. 1994. Survival of Canada geese banded in winter in the Atlantic Flyway. Journal of Wildlife Management 58:748–756.

Kautz, J. E., and C. E. Braun. 1981. Survival and recovery rates of band-tailed pigeons in Colorado. Journal of Wildlife Management 45:214–218.

Kiel, W. H. 1959. Mourning dove management units, a progress report. US Fish and Wildlife Service Special Scientific Report 42, Washington, DC, USA.

Krementz, D. G., J. E. Hines, and D. R. Luukkonen. 2003. Survival and recovery rates of American woodcock banded in Michigan. Journal of Wildlife Management 67:398–407.

Krohn, W. B., and E. R. Clark. 1977. Band-recovery distribution of eastern Maine woodcock. Wildlife Society Bulletin 5:118–122.

Lincoln, F. C. 1930. Calculating waterfowl abundance on the basis of band returns. US Department of Agriculture Circular 118, Washington, DC, USA.

———. 1935. The waterfowl flyways of North American. US Department of Agriculture Circular 342, Washington, DC, USA.

———. 1939. The migration of American birds. Doubleday, Doran, New York, New York, USA.

MacCallum, W., and R. Melinchuk. 2011. Report of the task force on state contributions to NAWMP/NAWCA projects in Canada. Association of Fish and Wildlife Agencies, Washington, DC, USA.

Matteson, S. W., P. F. Manthey, M. J. Mossman, and L. M. Hartman. 2007. Wisconsin trumpeter swan recovery program: Progress toward restoration, 1987–2005. Pages 11–18 *in* M. H. Linck and R. E. Shea, editors. Selected papers of the 20th trumpeter swan society conference, Council Bluffs, Iowa, USA.

McClure, H. E. 1939. Cooing activity and censusing of the mourning dove. Journal of Wildlife Management 3:323–328.

North American Waterfowl Management Plan, Plan Committee. 2012. North American waterfowl management plan 2012: People conserving waterfowl and wetlands. Canadian Wildlife Service, US Fish and Wildlife Service, Secretaria de Medio Ambiente y Recursos Naturales. http://nawmprevision.org/.

Oldenburger, S. L., M. Frisbie, J. Purvis, N. A. Heger, J. Roberson, and D. Morrison. 2014. Texas Population Status and Harvest: Eurasian-collared, mourning, and white-winged doves. Texas Parks and Wildlife Department. Austin, Texas, USA.

Otis, D. L. 2006. A mourning dove hunting regulation strategy based on annual harvest statistics and banding data. Journal of Wildlife Management 70:1302–1307.

———. 2009. Mourning dove banding needs assessment. US Fish and Wildlife Service. http://www.fws.gov/migratory-birds/NewsPublicationsReports.html.

Pacific Flyway Council. 1995. Pacific Flyway management plan for the greater sandhill crane population wintering along the lower Colorado River Valley. Portland, Oregon, USA.

———. 2003. Pacific Flyway management plan for the greater white-fronted goose. Greater White-fronted Goose Subcommittee, Pacific Flyway Study Committee, Portland, Oregon, USA.

Pacific Flyway Study Committee and Central Flyway Webless Migratory Game Bird Technical Committee. 2001. Pacific and Central Flyways management plan for the Four Corners population of band-tailed pigeons. Pacific Flyway Council, US Fish and Wildlife Service, Portland, Oregon, USA.

Powell, L. A., M. P. Vrtiska, and N. Lyman. 2004. Survival rates and recovery distributions of Canada geese banded in Nebraska. Pages 60–65 *in* T. J. Moser, R. D. Lien, K. C. VerCauteren, K. F. Abraham, D. E. Andersen, J. G. Bruggink, J. M. Coluccy, D. A. Graber, J. O. Leafloor, D. R. Luukkonen, and R. E. Trost, editors. Proceedings of the 2003 International Canada Goose Symposium, Madison, Wisconsin, USA.

Rabe, M. J., and T. A. Sanders. 2010. White-winged dove population status, 2010. US Department of the Interior, Fish and Wildlife Service, Division of Migratory Bird Management, Washington, DC, USA.

Reeves, H. M. 1993. Mourning dove hunting regulations. Pages 429–448 *in* T. S. Baskett, M. W. Sayre, R. E. Tomlinson, and R. E. Mirarchi, editors. Ecology and management of the mourning dove. Stackpole Books, Harrisburg, Pennsylvania, USA.

Rexstad, E. 1992. Effect of hunting on the annual survival rate of Canada geese banded in Utah. Journal of Wildlife Management 56:295–303.

Rice, M. B., D. A. Haukos, J. A. Dubovsky, and M. C. Runge. 2010. Continental survival and recovery rates of northern pintails using band-recovery data. Journal of Wildlife Management 74:778–787.

Samuel, M. D., D. H. Rusch, K. F. Abraham, M. M. Gillespie, J. P. Prevett, and G. W. Swenson. 1991. Fall and winter distribution of Canada geese in the Mississippi Flyway. Journal of Wildlife Management 55:449–456.

Sanders, T. A. 2014. Band-tailed pigeon population status, 2014. US Department of the Interior, Fish and Wildlife Service, Division of Migratory Bird Management, Washington, DC, USA.

Sanders, T. A., and R. E. Trost. 2013. Use of capture-recapture models with mark-resight data to estimate abundance of

Aleutian cackling geese. Journal of Wildlife Management 77:1459–1471.

Sauer, J. R., J. E. Hines, J. E. Fallon, K. L. Pardieck, D. J. Ziolkowski, Jr., and W. A. Link. 2014. *The North American Breeding Bird Survey, Results and Analysis 1966–2013. Version 01.30.2015. USGS Patuxent Wildlife Research Center, Laurel, Maryland, USA.*

Seamans, M. E., and T. A. Sanders. 2014. Mourning dove population status, 2014. US Department of the Interior, Fish and Wildlife Service, Division of Migratory Bird Management, Washington, DC, USA.

Sharp, D. E., K. L. Kruse, and P. P. Thorpe. 2002. The midwinter waterfowl survey in the Central Flyway. US Department of the Interior, Fish and Wildlife Service, Division of Migratory Bird Management, Denver, Colorado, USA.

Southeastern Association of Game and Fish Commissioners. 1957. Mourning dove investigations: 1948–1956. Technical Bulletin 1. Columbia, South Carolina, USA.

Subcommittee on Rocky Mountain Greater Sandhill Cranes. 2007. Management plan of the Pacific and Central Flyways for the Rocky Mountain population of greater sandhill cranes. Pacific Flyway Study Committee Rocky Mountain Population Greater Sandhill Cranes Subcommittee and Central Flyway Webless Migratory Game Bird Technical Committee, US Fish and Wildlife Service, Portland, Oregon, USA.

Szymanski, M. L., and J. A. Dubovsky. 2013. Distribution and derivation of the blue-winged teal (*Anas discors*) harvest, 1970–2003. US Fish and Wildlife Service Biological Technical Publication FWS/BTP-R6017-2013. Washington, DC, USA.

Tacha, T. C., and C. E. Braun, editors. 1994. Migratory shore and upland game bird management in North America. International Association of Fish and Wildlife Agencies, Washington, DC, USA.

Teal Harvest Potential Working Group. 2013. An assessment of the harvest potential of North American teal. www.fws.gov /migratorybirds/pdf/surveys-and-data/FinalTealAssessmentReport.pdf.

Tomlinson, T. E., D. D. Dolton, H. M. Reeves, J. D. Nichols, and L. A. McKibben. 1988. Migration, harvest, and population characteristics of mourning doves banded in the Western Management Unit: 1954–1977. Fish and Wildlife Technical Report 13. US Fish and Wildlife Service, Washington, DC, USA.

US Department of Agriculture, Natural Resource Conservation Service. 2014. The Voluntary Public Access and Habitat Incentive Program. www.nrcs.usda.gov/wps/portal/nrcs/detail /national/programs/farmbill/?cid=stelprdb1242739.

US Fish and Wildlife Service. 1986. The North American duck banding program—a revised approach. US Fish and Wildlife Service and Canadian Wildlife Service, Laurel, Maryland, USA.

———. 2005. Final environmental impact statement: Resident Canada goose management. US Fish and Wildlife Service, Washington, DC, USA.

———. 2010. Strategic plan: The Partners for Fish and Wildlife Program, stewardship of fish and wildlife through voluntary conservation. US Fish and Wildlife Service, Washington, DC, USA.

———. 2013. Final supplement environmental impact statement: Issuance of annual regulations permitting the hunting of migratory birds. US Fish and Wildlife Service, Washington, DC, USA.

———. 2014. Mourning dove harvest strategy. US Department of the Interior, Fish and Wildlife Service, Division of Migratory Bird Management, Washington, DC, USA.

———. 2015. Adaptive harvest management: 2015 hunting season. US Department of the Interior, Washington, DC, USA. www.fws.gov/birds/management/adaptive-harvest -management/publications-and-reports.php.

US Fish and Wildlife Service and Canadian Wildlife Service. 1987. Standard operation procedures (SOP) for aerial waterfowl breeding ground population and habitat surveys in North America. US Fish and Wildlife Service, Washington, DC, USA.

Ver Steeg, J. M., and R. C. Elden, compilers. 2002. Harvest Information Program: Evaluation and recommendations. International Association of Fish and Wildlife Agencies, Migratory Shore and Upland Game Bird Working Group, Ad Hoc Committee on HIP, Washington, DC, USA.

Vrtiska, M. P., J. H. Gammonley, L. W. Naylor, and A. H. Raedeke. 2013. Economic and conservation ramifications from the decline of waterfowl hunters. Wildlife Society Bulletin 37:380–388.

Vrtiska, M. P., and L. A. Powell. 2011. Estimates of duck breeding populations in the Nebraska Sandhills using double observer methodology. Waterbirds 34:96–101.

Wagner, W. C. 1995. Flyway council system review, final project report. Project Steering Committee, International Association of Fish and Wildlife Agencies, Washington, DC, USA.

Williams, B. K., J. D. Nichols, and M. J. Conroy. 2002. Analysis and management of animal populations. Academic Press, San Diego, California, USA.

Zenner, G. G., and T. G. LaGrange. 1998. Giant Canada geese in Iowa: Restoration, management, and distribution. Pages 303–309 *in* D. H. Rusch, M. D. Samuel, D. D. Humburg, and B. D. Sullivan, editors. Biology and management of Canada geese. Proceedings of the International Canada Goose Symposium, Milwaukee, Wisconsin, USA.

— 10 — State Management of Nongame Wildlife

Karie L. Decker,
Mark Humpert,
and J. Scott Taylor

As sport hunting of game species spread across North America during the 1700s and 1800s, early Euro-American conservationists realized that protection and management should be applied to all species of wildlife, whether hunted or not. Thus, an emerging North American Model of Wildlife Conservation, while rooted in game management, helped fuel a similar desire to conserve species never thought of as game, particularly as the model's tenet of no wasteful killing developed. The Migratory Bird Treaty Act (MBTA) of 1918 was the first major federal law requiring protection of nongame species. The act protected all migratory birds from unregulated take and helped codify this tenet of the model. The original MBTA and subsequent modifications, along with a plethora of similar state and federal laws, still remain the cornerstone of nongame bird conservation today.

Early on, conservationists recognized that considerable funding would be necessary to accomplish the goal of conserving wildlife, and they worked diligently to establish mechanisms to provide stable and secure funding. As a result, a "user-pay" concept was born, and legislative actions beginning in the 1930s set an unprecedented stage for fiscal assurance. While the Pittman–Robertson Act of 1937 (officially the Federal Aid in Wildlife Restoration Act; 16 U.S.C. 669–669i; 50 Stat. 917) and similar legislation that followed were grounded in game species sustainability, nongame species were entrenched in the North American Model of Wildlife Conservation and conserved alongside their hunted counterparts, particularly as habitat management became central to wildlife conservation. Through the years, however, various amendments and provisions to the Pittman–Robertson Act clarified eligible activities allowed under this funding source, and many state agencies directed efforts derived from these funds toward managing solely game species.

Formal consideration of funding for nongame species conservation was not initiated until passage of the Endangered Species Preservation Act of 1966 (Public Law 89-669), which provided limited protection for wildlife species at risk of extinction by listing them as endangered. This law also authorized acquisition of land as habitat for endangered species. Then, in 1969, an amendment provided additional protection to species at risk of worldwide extinction, leading to international adoption of endangered species conservation (Public Law 91-135). The Convention on International Trade in Endangered Species of Wild Fauna and Flora convened 80 nations in agreement to monitor and restrict international commerce of sensitive plant and animal species, culminating in passage of the Endangered Species Act (ESA) of 1973 (16 U.S.C. § 1531–1544) and multiple complementary state laws.

Defining state and federal agency roles in managing threatened and endangered species was only a part of the ESA; it also provided a funding mechanism. The ESA required federal agencies to protect listed species and prohibited authorization, funding, or any action that would jeopardize a listed species or its habitat. Section 6 of the ESA (16 U.S.C. § 1535) laid groundwork for the collaborative nature of the act and, through development of cooperative agreements, allowed for an annual allocation of funding from the federal government to state agencies for protection and conservation of listed and at-risk species.

State agencies were still responsible for management of nonmigratory, nongame species not listed as threatened or endangered, but they lacked funding directed specifically toward that cause. The Fish and Wildlife Conservation Act (FWCA) of 1980 (16 U.S.C. § 2901–2911; 94 Stat. 1322 Public Law 96-366) initiated discussion of planning for and funding of nongame (but not endangered) species conservation. The FWCA authorized, but never appropriated, funding to states to develop and implement plans focused on nongame species conservation.

To fill the vacuum, states adopted a number of approaches to fund such work. These included funding from state income tax check-offs, sales of conservation license plates, lottery funds and sales, and taxes. While funds derived from these sources were considerable, especially for the small handful of states that devote specific sales tax revenues to conservation, most did not produce a revenue stream approaching the magnitude of hunting license sales for game species, nor the magnitude of what was needed to manage the hundreds of nongame species in each state.

Since available funding continued to lag behind the need to effectively manage nongame species across the country, states collectively recognized that a more reliable and sustainable funding source was required. To address this issue, the Association of Fish and Wildlife Agencies launched a "Teaming with Wildlife" initiative with the purpose of advocating for dedicated and sustained funding to states to address the unmet needs of nongame fish and wildlife conservation. The initiative included formation of a "Teaming with Wildlife" Coalition that today includes more than 6,400 organizations and businesses representing millions of birdwatchers, hunters, anglers, hikers, campers, and other conservationists who support restoration and conservation of our nation's fish and wildlife resources. The original purpose of the coalition was to gain support for a 0.25–5 percent excise tax on outdoor recreation equipment (e.g., binoculars, wildlife field guides). For more than three decades, the coalition championed a dedicated funding source for endangered species prevention.

It was not until 2000 that Congress nearly passed the Conservation and Reinvestment Act (CARA, H.R. 701) to help finance nongame species conservation. The CARA passed the House of Representatives by a 3–1 margin and had majority support in the Senate but was not allowed to go to the Senate floor for a vote. In Fiscal Year 2001, the Wildlife Conservation and Restoration Program (WCRP) and State and Tribal Wildlife Grants (SWG) program (see below) were created under the banner of "CARA-lite" (2008 Public Law 110-161). Like the Pittman–Robertson Act and the ESA, the WCRP and the SWG program have continued a long history of collaborative efforts between state and federal governments and nongovernmental organizations for conserving wildlife.

Today, states remain responsible for nonmigratory species and those not listed under the ESA. While state and federal funding is relatively limited to manage the vast diversity of nongame species, many programs and partnerships exist to help make efforts successful.

State and Tribal Wildlife Grants Program

The SWG program allocates federal funds for use in developing and implementing programs of benefit to wildlife and their habitats, particularly species not hunted or fished. Emphasizing collaboration among state and federal governments and nongovernmental partners, Congress required each state to develop and periodically revise a State Wildlife Action Plan (SWAP) as a condition of receiving funding through the SWG program. Individual SWAPs must identify and focus on species of greatest conservation need and their habitats. Since its inception, the SWG program has provided states with nearly $1 billion for conservation of at-risk wildlife.

Required elements of each SWAP include

1. information on the distribution and abundance of species of wildlife, including low and declining populations as the state fish and wildlife agency deems appropriate, that is indicative of the diversity and health of the state's wildlife;

2. descriptions of extent and condition of habitats and community types essential to conservation of species identified in item 1;

3. descriptions of problems that may adversely affect species identified in item 1 or their habitats, and priority research and survey efforts needed to identify factors that may assist in restoration and improved conservation of these species and habitats;

4. descriptions of conservation actions proposed to conserve the identified species and habitats and priorities for implementing such actions;

5. proposed plans for monitoring species identified in item 1 and their habitats, for monitoring the effectiveness of the conservation actions proposed in item 4, and for adapting these conservation actions to respond appropriately to new information or changing conditions;

6. descriptions of procedures to review the plan at intervals not to exceed 10 years;

7. plans for coordinating development, implementation, review, and revision of the plan with federal, state, and local agencies and Indian tribes that manage significant land and water areas within the state or administer programs that significantly affect the conservation of identified species and habitats; and

8. broad public participation in developing and implementing these plans, the projects that are carried out while these plans are developed, and the species in greatest need of conservation.

By 2005, all 50 states, the District of Columbia, and five US territories had developed SWAPs to conserve wildlife and habitat through partnerships before they become too rare or costly to restore. As a whole, SWAPs represent the national action agenda for endangered species prevention. The SWG program is considered the core program for keeping species healthy and off the federal threatened and endangered species list, a goal shared by a broad constituency of conservationists, managers, businesses, farmers, ranchers, and land developers.

Prior to SWG funding, most states lacked the resources needed to undertake comprehensive planning for at-risk species. The planning process was an un-

paralleled success that brought together thousands of public and private participants. The result is that for the first time in conservation history, the United States has coast-to-coast conservation planning coverage for fish and wildlife. A key strength of SWAPs is that they can be designed to address specific needs of individual states. Although Congress laid out eight required elements as listed above, states were given discretion to address these elements in a manner that met individual state needs. For example, states were allowed to develop their own criteria for designation of species of greatest conservation need and key habitats. SWAPs represent a groundbreaking effort to bring together the best science available to conserve priority fish and wildlife and their habitats through innovative public–private partnerships. Collectively, initial SWAPs identified 12,000 species at risk of becoming endangered and offered a set of conservation actions to address key threats, providing a voluntary, nonregulatory alternative to the federal listing process.

Eastern Massasauga Rattlesnake Status: An SWG Program Case Study

The eastern massasauga rattlesnake (*Sistrurus catenatus*) is a federal candidate species under the ESA. It is also listed as a state endangered species in each state in which it is thought to occur except Michigan, where it is listed as a species of special concern. Each of the nine applicable SWAPs identifies threats and concerns for this species' persistence. In nearly every state, population numbers are low, hampering the ability of managers to comprehensively evaluate habitat and conservation needs. However, populations of eastern massasauga rattlesnakes in Michigan are suspected to be relatively abundant (Eagle et al. 2005) and create an ideal system for better understanding the needs of the species and population status across its range.

Habitat degradation and human persecution have contributed to range-wide population declines, and survival estimates for eastern massasauga rattlesnake were lacking at the time of development of SWAPs. A series of studies funded through the SWG program and in partnership with the Upper Midwest and Great Lakes LCC, the Michigan Department of Natural Resources, and other agency, nongovernmental, and university partners led to the development of survival probabilities on state-managed lands (Bailey et al. 2011). This study and subsequent research led to a range-wide analysis of eastern massasauga survivorship (Jones et al. 2012) and has provided critical information to assess state and federal listing status across the species' range.

With additional SWG funds, the Michigan Department of Natural Resources and partners are currently developing several Candidate Conservation Agreements with Assurances (CCAAs)—a formal agreement between the US Fish and Wildlife Service (USFWS) and one or more parties to address the conservation needs proposed or candidate species, before they become listed as endangered or threatened. Landowners voluntarily commit to conservation actions that will help stabilize or restore the species, with the goal that listing will become unnecessary. CCAAs benefit landowners in several ways. First, if the actions preclude listing, the landowner is not regulated by the ESA. Second, if the conservation actions are not sufficient and the species is listed, the agreement automatically becomes a permit authorizing the landowner incidental take of the species. Thus, the agreements provide landowners with assurances that their conservation efforts will not result in future regulatory obligations in excess of those they agree to at the time in which they enter into the agreement. Third, for landowners who want to conserve the species or want to manage habitat on their land, the agreement provides an avenue to potential federal or state cost-share programs. In the case of the massasauga rattlesnake, CCAAs cover most of the remaining population, which helps address threats to the species within individual states and potentially preclude the need to list as a federal endangered species.

More recently, eastern massasauga species distribution models for northeastern Ohio and Michigan (McCluskey 2016), along with range-wide climate change vulnerability models (Pomara et al. 2014), provided key pieces of information for consideration by the USFWS. On September 30, 2015, the USFWS published a proposed rule in the *Federal Register* and opened a 60-day public comment period. After analyzing comments and considering new information, the USFWS published a final rule listing the eastern massasauga as a threatened species, effective October 31, 2016.

State wildlife action plans have been largely successful in garnering partnerships, as is exemplified in the eastern massasauga case study. With the help of federal, nonfederal, and nongovernmental organizations, as well as funding leveraged across multiple sources, imperative data were collected to aid the USFWS in the listing decision. State wildlife action plans act as a common thread among states and provide a clear pipeline to federal programs.

Swift Fox Conservation Team: A Western Association of Fish and Wildlife Agencies Initiative Case Study

In 1992, the USFWS received a petition to list the swift fox under the authority of the ESA in the northern portion of the species' range (Montana, North Dakota, South Dakota, and Nebraska), if not the entire range. In 1994, the USFWS concluded that listing was warranted in the entire range. Also in 1994, the 10 affected state wildlife management agencies and interested cooperators formed the Swift Fox Conservation Team (SFCT). The SFCT completed the Conservation Assessment and Conservation Strategy (CACS) for swift fox in the United States in 1997 to guide management and conservation activities. The USFWS's 12-month finding in 1995 designated the swift fox as a federal candidate species, with listing warranted but precluded by higher listing priorities. The CACS has guided activities of state, federal, tribal, and private entities to provide defensible data on swift fox abundance and distribution with a more coordinated approach to range-wide conservation and management. As a result of new information and improved coordination among partners, the USFWS removed the swift fox from the candidate species list in 2001.

The commitment of state and federal agencies, tribes, private organizations, and landowners in the United States and Canada to swift fox conservation continued following removal of the species from the candidate species list. In 2011, the SFCT completed an update to the 1997 CACS, including a conservation assessment of the five listing factors to long-term swift fox sustainability, and concluded that none of the listing factors have risen to the level of a threat. This update also includes a revised strategy section with associated objectives, strategies, and activities to provide a guidance framework for continued monitoring, research, and recovery.

The continuing efforts of the SFCT indicate that management activities for this species will be carefully considered in the future. Recent, successful reintroduction and research efforts have also contributed to understanding the behavior, ecology, and habitat requirements of swift fox.

Migratory Bird Joint Ventures

In 1986, the US secretary of the interior and Canadian minister of environment signed the first North American Waterfowl Management Plan (Canadian Wildlife Service and US Fish and Wildlife Service 1986). This groundbreaking plan recognized that the continuing loss of habitat and declines in waterfowl populations required a unified continental effort to restore this valuable resource to population levels that existed in the 1970s. Engrained in the plan was the call for the establishment of Migratory Bird Joint Ventures to guide collaborative efforts in maintaining habitats for healthy migratory bird populations. Following signing of the 1986 North American Waterfowl Management Plan, the Arctic Goose Joint Venture was created to further the scientific understanding and management of North America's geese. By 1990, eight additional Joint Ventures had been established.

Today, 18 habitat-based Joint Ventures address bird habitat conservation issues found within their geographic area. Additionally, three species-based Joint Ventures, all with an international scope, work to further the scientific understanding needed to effectively manage these individual bird species.

All Migratory Bird Joint Ventures are supported by multiple partners. The USFWS provides a coordinator and basic program infrastructure, while other federal, state, and nongovernmental partners support projects relating to biological planning and prioritization; habitat project development, implementation, monitoring, and evaluation; and applied research activities.

Cerulean Warbler Forest Management Project: A Migratory Bird Joint Venture Case Study

Long-term declines in cerulean warbler (*Setophaga cerulea*) populations (Buehler et al. 2008; Sauer et al. 2011) led to a focus on intense habitat management throughout the eastern United States, particularly in the Appalachian forests. In 2005, the Appalachian Mountains Joint Venture (AMJV), with over 20 agency, nongovernmental, and university partners, was established through the Cerulean Warbler Forest Management Project. The project was initiated to allow scientific and management entities to test forestry methods and use experimental silvicultural approaches to enhance habitat. By studying the response of cerulean warblers and the overall bird community to various harvesting treatments, the AMJV and its partners hoped to provide land managers and biologists with recommendations for enhancing habitat across the species' range. This project led to development of Cerulean Warbler Management Guidelines (Wood et al. 2013). The cerulean warbler guidelines outline land management actions intended to retain and enhance habitat for federal, state, and private foresters, biologists, and other land managers. Guidelines are based on several relevant studies conducted by the Cooperative Cerulean Warbler Forest Management Project and others.

Through development of the guidelines, multiple partnership projects arose. For example, Virginia began an initiative for Important Bird Areas (IBAs)—sites that provide essential habitat for breeding, migrating, or wintering birds. Within the Commonwealth of Virginia alone, there are now six IBAs contained within the AMJV. The Virginia IBA program seeks to conserve and protect these IBAs by supporting on-the-ground conservation efforts and monitoring priority species. The program teams with the National Park Service to promote cerulean warbler surveys in the Upper Blue Ridge Mountains IBA.

In 2009, Ohio announced its intent to purchase the 15,849-acre Vinton Furnace Experimental Forest (supported through private, industry, state, and federal funds). This experimental forest supports one the highest recorded densities of cerulean warblers within their breeding range. The forest was integral to investigating silvicultural effects on cerulean warblers and other forest songbirds in order to ultimately create cerulean warbler best management practices. Several other acquisitions in North Carolina, West Virginia, and Virginia aided the habitat protection of thousands of acres for cerulean warbler and other declining bird populations.

Ultimately, in 2015, the Appalachian Mountains Joint Venture Partnership received $8 million for the Cerulean Warbler Appalachian Forestland Enhancement Project. The grant, awarded by the Regional Conservation Partnership Program through the US Department of Agriculture, will allow partners to work with private landowners to enhance 12,500 acres of forest habitat on private lands for cerulean warblers and other wildlife. The five-year project will be modeled after the Natural Resource Conservation Service Working Lands for Wildlife Program for golden-winged warblers (*Vermivora chrysoptera*) using the recently released Cerulean Warbler Habitat Management Guidelines to target conservation practices in delineated focal areas. Conservation work will take place in West Virginia, Pennsylvania, Kentucky, Ohio, and Maryland.

Trilateral Committee

To more effectively address priorities of continental significance and boost the concerted efforts of the three countries of North America, the Canada/Mexico/US Trilateral Committee for Wildlife and Ecosystem Conservation and Management was established in 1995. The Trilateral Committee is headed by the directors of the Canadian Wildlife Service, the USFWS, and the Ministry of Environment and Natural Resources of Mexico. The goals of the Trilateral Committee are to foster an integrated continental perspective for cooperative conservation and sustainable use of biological resources, contribute to the maintenance of the ecological integrity of North American eco-regions, and promote biodiversity conservation capacity building and cooperative cross-sectoral activities in the three countries that will contribute to the reduction and mitigation of threats to North American shared species and ecosystems. This is achieved through coordination, cooperation, and development of partnerships among wildlife agencies of the three countries and other interested parties. Thirty organizations currently partic-

ipate in various activities of the Trilateral Committee, including federal and state government agencies, research and academic institutions, nongovernmental organizations, and private industry. The Trilateral Committee provides an effective and efficient mechanism to address conservation and management of natural resources and covers concerns from the state level to a continental scale. Delegations from each country meet annually for discussions on a wide range of topics such as joint, on-the-ground projects, climate change adaptation and mitigation, and issues of law enforcement.

The Seabird Restoration on the Baja California Pacific Islands: A Trilateral Committee Case Study

In August of 2011, the Montrose Settlements and S. S. Jacob Luckenbach Trustee Councils selected a partnership of Mexican and US-based organizations to implement a five-year seabird restoration program on the Baja California Pacific Islands, Mexico. This partnership consists of Grupo de Ecología y Conservación de Islas, the National Audubon Society, Cornell Lab of Ornithology, the Friends of the Mexican Fund for the Conservation of Nature, and the National Fish and Wildlife Foundation. Restoration projects included installation of nest boxes, habitat improvement, human disturbance reduction and education, and monitoring of seabird populations on northern and southern islands. Another example in British Columbia, Canada, aims to restore historic seabird nesting colonies on the forested islands of Haida Gwaii. The introduction of rats to the islands in the 1700s has led to devastation of bird, mammal, and invertebrate species. Eradication and restoration (with follow-up monitoring) activities will help maintain ecological integrity in the ecosystems.

Partners in Flight

Partners in Flight (PIF) was launched in 1990 in response to growing concerns about declining populations of landbird species (Robbins et al. 1986; Askins et al. 1990), with an initial focus on neotropical migrants (species that breed in the North American Nearctic and winter in the Neotropics of Central and South America). Recognizing the decline of many other bird species, PIF quickly expanded its focus to include all landbirds. The central premise of PIF has been that the resources of public and private organizations in the Western Hemisphere must be combined, coordinated, and increased in order to achieve success in conserving bird populations. Much like other successful programs, PIF is a cooperative effort involving partnerships among federal, state, and local government; philanthropic foundations; professional organizations; conservation groups; industry; the academic community; and private individuals.

The mission of PIF is expressed through three concepts:

1. *Helping species at risk.*—Species must be conserved before they become imperiled. Allowing species to become threatened or endangered results in long-term, costly recovery efforts whose success is far from guaranteed. Endangered species must be not only protected from extinction but also recovered to once again play their roles in ensuring the future of healthy ecosystems.
2. *Keeping common birds common.*—Common native birds, both resident and migratory, must remain common throughout their natural ranges. These species compose the core of our avian diversity and are integral to the integrity of the ecosystems of which they are a part.
3. *Voluntary partnerships for birds, habitats, and people.*—Conservation of landbirds and their habitats is not a task that can be undertaken alone. Partnerships must be formed with others who are working for conservation on the same landscapes, as well as those who depend on those landscapes for their economic and social well-being. The conservation of natural systems is fundamentally necessary for life on earth, including that of humans.

In 1995, PIF began a comprehensive planning effort to conserve nongame landbirds and their habitats throughout the United States. A critical first step in the planning process was to establish clear and consistent priorities among the several hundred landbird species based on their vulnerability and need for conservation action. To this end, PIF developed a species prioritization process for the southeastern United States and later expanded the effort to include all of North America north of Mexico. The seven parameters in the prioritization process are based on global and local information. Breeding distribution, nonbreeding distribution, and relative abundance represent global parameters; a single value is assigned for the entire range of the species. Values assigned to threats to breeding, threats to nonbreeding, and population trend also represent global values, but they may be superseded by values assigned specifically to a physiographic area when appropriate and possible. Finally, area importance is assigned locally for a specific physiographic area. Scores for each of these seven variables are determined independently. The intended application of priority scores is in developing Bird Conservation Plans for the physiographic areas and/or states of the continental United States.

As such, the PIF North American Landbird Conservation Plan (Rich et al. 2004) provided a continental synthesis of priorities and objectives that guide landbird conservation actions at state, national, and international scales. Nearly 450 bird species that breed regularly in the United States and Canada are encompassed in the plan, and 100 of these species warrant inclusion on the PIF watch list owing to some combination of threats to habitats, declining populations, small population sizes, or limited distributions. Of these, 28 species require immediate action to protect small remaining populations, and 44 more are in need of management to reverse long-term declines.

The PIF North American Landbird Conservation Plan also highlighted the need for stewardship of the species and landscapes characteristic of each portion of the continent, identifying 158 species (including 66 on the watch list) that are particularly representative of large avifaunal biomes, and whose needs should be considered in conservation planning. Taken together, the pool of watch list and stewardship species represents the landbirds of greatest continental importance for conservation action.

In 2004, PIF spearheaded an initiative to bring together the Western Hemisphere through Saving Our Shared Birds (Berlanga et al. 2010). Canada, Mexico, and the continental United States are home to 882 native landbird species, more than one-third of which depend substantially on habitats in more than one country. More than 200 species constituting 83% of individual landbirds rely on habitats in all three countries (Berlanga et al. 2010). Saving Our Shared Birds was the first tri-national comprehensive conservation assessment of landbirds in Canada, Mexico, and the continental United States, encompassing the complete range of many migratory species. It identified a set of continent-scale actions necessary to maintain landbird diversity and abundance. This collaborative effort of PIF was the integral step linking the countries of the Western Hemisphere to help species at risk and to keep common birds common through voluntary partnership. Saving Our Shared Birds builds on PIF's 2004 North American Landbird Conservation Plan (Rich et al. 2004).

In 2010 at the annual meeting of the Trilateral Committee for Wildlife and Ecosystem Conservation and Management, all three nations committed to cooperative conservation through Saving Our Shared Birds Tri-National Assessment. PIF's first tri-national assessment identified 148 bird species in need of immediate conservation attention because of their highly threatened and declining populations. The most imperiled species include 44 species with very limited distributions, mostly in Mexico, that are at greatest risk of extinction; 80 tropical residents dependent on deciduous, highland, and evergreen forests in Mexico; and 24 species that breed in temperate-zone forests, grasslands, and arid land habitats.

Landscape Conservation Cooperatives

In 2009, Secretarial Order 3285 made production and transmission of renewable energy on public lands a priority for the US Department of the Interior (DOI). This order established a department-wide approach for applying scientific tools to increase understanding of climate change and to coordinate an effective response to its impacts on tribes and on the land, water, ocean, fish and wildlife, and cultural heritage resources that DOI manages. To fulfill the nation's vision for a clean energy economy, DOI began managing US public lands and oceans not just for balanced oil, natural gas, and coal development but also to promote environmentally responsible renewable energy development. As part of this plan, DOI launched the Landscape Conservation Cooperatives (LCCs) in 2010 to better integrate science and management to address climate change and other landscape-scale issues (Secretarial Order No. 3289).

LCCs are applied conservation science partnerships with two main functions. The first is to provide the science and technical expertise needed to support conservation planning at landscape scales—beyond the reach or resources of any one organization. Through the efforts of in-house staff and science-oriented partners, LCCs strive to generate the tools, methods, and data managers need to design and deliver conservation using the Strategic Habitat Conservation approach (US Fish and Wildlife Service 2008). Briefly, this approach is guided by four main foci: (1) addressing conservation challenges at ecologically meaningful scales, (2) working in partnerships to maximize effectiveness and efficiency, (3) conserving through an adaptive management framework, and (4) encouraging use of the best science and tools available.

The second function of LCCs is to promote collaboration among members in defining shared conservation goals. With these goals in mind, partners identify where and how they will take action, within their own authorities and organizational priorities, to best contribute to the larger conservation effort. The 22 LCCs collectively form a network of resource managers and scientists who share a common need for scientific information and interest in conservation. Each LCC brings together federal, state, and local governments, along with Tribes and First Nations, nongovernmental organizations, universities, and interested public and private organizations. Successful network cooperatives depend on LCCs to

1. develop and provide integrated science-based information about the implications of climate change and other stressors for the sustainability of natural and cultural resources;
2. develop shared, landscape-level, conservation objectives and inform conservation strategies that are based on a shared scientific understanding about the landscape, including the implications of current and future environmental stressors;
3. facilitate the exchange of applied science in the implementation of conservation strategies and

products developed by the cooperative or their partners;

4. monitor and evaluate the effectiveness of LCC conservation strategies in meeting shared objectives; and

5. develop appropriate linkages that connect LCCs to ensure an effective network.

Distribution Model for Fishers in the Northern US Rocky Mountains: An LCC Case Study

Fishers (*Pekania pennanti*) are common in the Northeast and Midwest, but in the Northern Rockies and Northwest, they are one of the rarest carnivores. Historically this species ranged the northern forests of Canada and the United States, as well as forests in the Appalachian, Rocky, and Pacific Coast Mountains. Today, fishers are found only in parts of their historic range. In 2014, the West Coast Distinct Population Segment of fishers was proposed for listing as a threatened species under the ESA (79 FR 60419). The populations in the Rocky Mountains remain rare but are not under consideration for listing. This may be in part because fisher habitat is likely more contiguous than in the Pacific Coast states. However, the presence/absence of fishers in the Rocky Mountains and their level of connectivity to other fisher populations were largely unknown. Filling this information gap was soon recognized as a priority in Rocky Mountain states; improving knowledge about fishers in the Rocky Mountains could shed light on management of fishers in Pacific Coast states.

Through a grant provided by the Great Northern LCC, multiple agencies and organizations began a collaboration to provide the first comprehensive fisher distribution model in the Northern Rocky Mountains. Using a variety of techniques, the project aimed to serve as a baseline for identifying population trends and changes in distribution and ultimately, through downscale global climate models, develop environmental predictors of the species habitat distribution over time. Since the grant initiation in 2010, multiple products have been developed. For example, researchers established noninvasive genetic sampling techniques to model the distribution of fishers across western Montana and northern Idaho, including future distribution modeling under a global climate model and two

climate change scenarios (Olson et al. 2014). Another study identified multiple scales at which fishers are selecting habitat in the Northern Rocky Mountains, utilizing animals equipped with radio telemetry (Schwartz et al. 2013). Researchers highlighted the importance of late-successional forests, consistent with a recent conservation strategy for fishers, and the importance of both stand- and landscape-level factors when directing forest management of fisher habitat in the Rocky Mountains.

These, along with several other studies, expanded knowledge of fisher populations and distribution in the Rocky Mountains and provided states with forest management recommendations for sustaining populations in this region.

Partnerships Key to Success

Throughout their establishment, state wildlife agencies have relied on the contributions of stakeholders. State and federal agencies, while charged with conservation and protection of nongame species, would not be successful without partnerships with landowners, nongovernmental organizations, industry, and others. The initiatives and programs described above represent a small fraction of those resources available to states for nongame species needs. However, financial support continues to be one of the greatest challenges facing states and their partners. While the funding mechanisms currently in place tend to focus individually on game or nongame species management, states and partners have begun to blur the lines in practice. Emerging threats such as invasive species, climate change, and disease continue to impact species whether they are hunted or not; partnering across taxa (as well as hunting status) will provide novel approaches to century-long natural resource conservation concerns.

LITERATURE CITED

Askins, R. A., J. F. Lynch, and R. Greenberg. 1990. Population declines of migratory birds of eastern North America. Current Ornithology 7:1057.

Bailey, R. L., H. Campa, III, T. M. Harrison, and K. Bissell. 2011. Survival of eastern massasauga rattlesnakes (*Sistrurus catenatus catenatus*) in Michigan. Herpetologica 67:167–173.

Berlanga, H., J. A. Kennedy, T. D. Rich, M. C. Arizmendi, C. J. Beardmore, P. J. Blancher, G. S. Butcher, A. R. Couturier, A. A. Dayer, D. W. Demarest, W. E. Easton, M. Gustafson, E. Iñigo-Elias, E. A. Krebs, A. O. Panjabi, V. Rodriguez Contreras, K. V. Rosenberg, J. M. Ruth, E. Santana Castellón, R. Ma. Vidal, and T. Will. 2010. Saving our shared birds: Partners in Flight tri-national vision for landbird conservation. Cornell Lab of Ornithology, Ithaca, New York, USA.

Buehler, D. A., J. J. Giocomo, J. Jones, P. B. Hamel, C. M. Rogers, T. A. Beachy, D. W. Varble, C. P. Nicholson, K. L. Roth, J. Barg, R. J. Robertson, J. R. Robb, and K. Islam. 2008. Cerulean warbler reproduction, survival, and models of population decline. Journal of Wildlife Management 72:646–653.

Canadian Wildlife Service and US Fish and Wildlife Service. 1986. North American waterfowl management plan. Canadian Wildlife Service, Ottawa, Ontario, Canada, and US Fish and Wildlife Service, Washington, DC, USA.

Conservation and Reinvestment Act; H.R. 701 (106th): Conservation and Reinvestment Act. www.congress.gov/bill/106th-congress/house-bill/701.

Eagle, A. C., E. M. Hay-Chmielewski, K. T. Cleveland, A. L. Derosier, M. E. Herbert, and R. A. Rustem, editors. 2005. Michigan's wildlife action plan. Michigan Department of Natural Resources. Lansing, Michigan, USA.

Endangered Species Act; 1966 Public Law 89-669, 1969 Public Law 91-135, 1973 ESA; 16 U.S.C. § 1531 et seq. www.fws.gov/endangered/laws-policies/esa.html.

Federal Aid in Wildlife Restoration Act (16 U.S.C. 669–669i; 50 Stat. 917) of September 2, 1937. www.fws.gov/laws/lawsdigest/FAWILD.HTML.

Fish and Wildlife Conservation Act of 1980 (16 U.S.C. 2901–2911; 94 Stat. 1322 Public Law 96-366).

Jones, P. C., R. B. King, R. L. Bailey, N. D. Bieser, K. Bissell, H. Campa, T. Crabill, M. D. Cross, B. A. Degregorio, M. J. Dreslik, F. E. Durbian, D. S. Harvey, S. E. Hecht, B. C. Jellen, G. Johnson, B. A. Kingsbury, M. J. Kowalski, J. Lee, J. V. Manning, J. A. Moore, J. Oakes, C. A. Phillips, K. A. Prior, J. M. Refsnider, J. D. Rouse, J. R. Sage, R. A. Seigel, D. B. Shepard, C. S. Smith, T. J. Vandewalle, P. J. Weatherhead, and A. Yagi. 2012. Range-wide analysis of eastern massasauga survivorship. Journal of Wildlife Management 76:1576–1586.

McCluskey, E. 2016. Landscape ecology approaches to Eastern Massasauga Rattlesnake conservation. Electronic Dissertation. Retrieved from https://etd.ohiolink.edu/.

Olson, L. E., J. D. Sauder, N. M. Albrecht, R. S. Vinkey, S. A. Cushman, and M. K. Schwartz. 2014. Modeling the effects of dispersal and patch size on predicted fisher (*Pekania martes pennanti*) distribution in the U.S. Rocky Mountains. Biological Conservation 169:89–98.

Pomara, L. Y., O. E. LeDee, K. J. Martin, and B. Zuckerberg. 2014. Demographic consequences of climate change and land cover help explain a history of extirpations and range contraction in a declining snake species. Global Change Biology 20:2087–2099.

Rich, T. D., C. J. Beardmore, H. Berlanga, P. J. Blancher, M. S. W. Bradstreet, G. S. Butcher, D. W. Demarest, E. H. Dunn, W. C. Hunter, E. E. Iñigo-Elias, J. A. Kennedy, A. M. Martell, A. O. Panjabi, D. N. Pashley, K. V. Rosenberg, C. M. Rustay, J. S. Wendt, and T. C. Will. 2004. Partners in Flight North American landbird conservation plan. Cornell Lab of Ornithology, Ithaca, New York, USA.

Robbins, C. S., D. Bystrak, and P. H. Geissler. 1986. The breeding bird survey: Its first fifteen years, 1965–1979. US Fish and Wildlife Service Resource Publication 157. Washington, DC, USA.

Sauer, J. R., J. E. Hines, J. E. Fallon, K. L. Pardieck, D. J. Ziolkowski, Jr., and W. A. Link. 2011. The North American breeding bird survey, results and analysis 1966–2009. Version 3.23.2011. US Geological Survey. Patuxent Wildlife Research Center, Laurel, Maryland, USA.

Schwartz, M. K., N. J. DeCesare, B. S. Jimenez, J. P. Copeland, and W. E. Melquist. 2013. Sand and landscape scale selection of large trees by fishers in the Rocky Mountains of Montana and Idaho. Forest Ecology and Management 305:103–111.

State and Tribal Wildlife Grants Program. 2008 Public Law 110-161. www.gpo.gov/fdsys/pkg/PLAW-110publ161/pdf/PLAW-110publ161.pdf.

US Fish and Wildlife Service. 2008. Strategic habitat conservation handbook: A guide to implementing the technical elements of strategic habitat conservation. Version 1.0. www.fws.gov/landscape-conservation/pdf/SHCHandbook.pdf.

Wood, P. B., J. Sheehan, P. Keyser, D. Buehler, J. Larkin, A. Rodewald, S. Stoleson, T. B. Wigley, J. Mizel, T. Boves, G. George, M. Bakermans, T. Beachy, A. Evans, M. McDermott, F. Newell, K. Perkins, and M. White. 2013. Management guidelines for enhancing cerulean warbler breeding habitat in Appalachian hardwood forests. American Bird Conservancy. The Plains, Virginia, USA.

11

Kurt VerCauteren,
Daniel Hirchert,
and Scott Hygnstrom

State Management of Human–Wildlife Conflicts

Many positive experiences are associated with wildlife, from passively watching animals in our backyards to actively hunting in publicly owned forests. Unfortunately, wildlife can be a double-edged sword. Human–wildlife conflicts are pervasive in society, and nearly all segments—wealthy and in need, urban and rural, east and west—can experience problems with wildlife. Agricultural producers lose an estimated $45 billion each year as a result of crop and livestock damage caused by big game, predators, waterfowl, and other wildlife species (Conover 2002). Row crops, forages, rangeland, fruits, vegetables, ornamentals, turf, and livestock are susceptible to damage by wildlife at various stages of production. Inhabitants of urban/suburban areas endure significant damage and nuisance problems caused by bears, deer, raccoons, squirrels, pigeons, rabbits, skunks, snakes, and others. In addition, over 75,000 people are injured annually or become ill as a result of wildlife-related incidents, at costs well exceeding $10 billion annually (Conover 2002).

Coexistence with wildlife is a balancing act of dealing with their positive and negative impacts. Many state wildlife agencies have taken on the responsibility of reducing these negative impacts for the betterment of society. Wildlife damage management (WDM) is an increasingly important part of the wildlife profession because of expanding human populations and intensified land-use practices. Concurrent with this growing need to reduce human–wildlife conflicts, public attitudes and environmental regulations are restricting use of some traditional control tools, such as toxicants and traps. Agencies and individuals carrying out control programs are being scrutinized more carefully to ensure that their actions are justified, environmentally safe, humane, and in the public interest. Thus, WDM activities must be based on sound economic, scientific, and sociological principles and carried out as positive, necessary components of overall wildlife management programs (VerCauteren et al. 2012a).

Definitions
Wildlife Damage Management

The term "wildlife damage management" can be specifically defined as the process of dealing with free-ranging vertebrate species that (1) cause economic damage to food, fiber, personal property, and natural resources; (2) threaten human health and safety through attacks, collisions, and zoonotic diseases; and (3) create a nuisance that is less than economically significant.

Integrated Wildlife Damage Management

Agencies have adopted an "integrated wildlife (damage) management" approach that incorporates the timely use of a variety of cost-effective, environmentally safe, and socially acceptable methods that reduce human–wildlife conflicts to tolerable levels. For most wildlife problems, no silver bullets exist for resolving issues. To enhance effectiveness and efficiency, procedures should be applied when problem animals are particularly susceptible, before they establish a pattern of conflict, or before populations become overabundant. Seldom will a single technique effectively reduce problems, and multiple techniques tend to work synergistically to enhance effectiveness. Efficiency is critical and benefits must exceed costs if WDM practices are to be sustainable. Care should be taken to use practices that have the least potential impact on the environment and nontarget animals. State wildlife agencies and the US Environmental Protection Agency (EPA) closely monitor and regulate the materials and practices that are used in WDM. The measure of success in WDM should be reduction of damage, threats, or impacts to tolerable levels, rather than the total elimination of damage or a problem population or species.

Overabundance

Wildlife damage often is caused by the offending behavior of individual animals, which can be dealt with by removing them or modifying their habitat. Equally important are the density-dependent impacts that are caused when populations of wildlife become overabundant and their numbers exceed biological and cultural (social) carrying capacity. Overabundance is caused by high fecundity and survival of a species over time, lead-ing to high rates of population growth (McShea et al. 1997) and associated human–wildlife conflicts in areas where high population levels compete with other land uses or human activities.

Biological Carrying Capacity

Biological carrying capacity is the number of animals in a population that an environment can sustain without long-term detrimental impacts to that environment (Ehrlich and Holdren 1971). For example, when white-tailed deer become overabundant, a browse line appears on shrubs, trees, and ornamentals. The plants have few live branches below 6 feet, undergrowth is dramatically limited, and plant diversity is reduced owing to overbrowsing. Eventually, the population of deer will decline as a result of starvation, disease, and competition. Long-term environmental damage will occur long before the deer population declines.

Cultural (Social) Carrying Capacity

Cultural carrying capacity is defined as the number of animals in a population that people are willing to tolerate based on a balance of environmental and social benefits and costs (Seidl and Tisdell 1999). For example, the public's tolerance of deer–vehicle collisions

Overabundance of wildlife, like these wild turkeys, can result in conflict within urban settings. *Photo courtesy of the USDA APHIS Wildlife Services.*

and agricultural damage is influenced by the benefits they experience from viewing and hunting deer.

Responsibility

State wildlife agencies have been charged with the responsibility of managing our publicly owned wildlife resources through the public trust doctrine (PTD; Batcheller et al. 2010). The PTD entrusts state wildlife agencies to manage wildlife resources for the benefit of the public, who owns these resources. Throughout the twentieth century, the primary focus of state wildlife agencies was on protecting wildlife, managing habitats and consumptive uses, and bringing some species back from the brink of extinction. It also stands to reason that these public agencies should be responsible for managing damage caused by wildlife.

Responsibility can be seen in two contradicting forms: (1) individuals can be responsible for protecting personal property from wildlife damage, and (2) society can be responsible for protecting wildlife by restricting what individuals can do to protect personal property. For example, a farmer cannot simply shoot deer to protect crops. State wildlife agencies carefully control the take of deer with hunting seasons, permit quotas, bag limits, and several other restrictions. Therefore, because society limits what farmers can do to protect their livelihood, it stands to reason that society and its empowered state wildlife agencies have the responsibility to assist farmers in reducing damage caused by wildlife.

State wildlife agencies must also protect the environment or endangered species from damage caused by overabundant wildlife. For example, deer overbrowsing in woodlands may eliminate sensitive or endangered plants. Excessive predation may threaten endangered colonial waterbird nesting sites. Wildlife agencies have a responsibility to maintain the long-term viability of rare or endangered species and enhance biodiversity.

History

Once called animal damage control and vertebrate pest control, experts in the field explored new terminology that would be more accurate, descriptive, and publically acceptable, leading to the contemporary terms of WDM (Cook 1991), human–wildlife conflicts, and ultimately human–wildlife coexistence. The first documented governmental act of WDM in the United States was conducted in 1683, when William Penn established a bounty on wolves to protect livestock owned by colonists (Lovich 1987). In 1886, the US Department of Agriculture (USDA) created the Division of Economic Ornithology and Mammalogy (renamed Division of Biological Survey in 1905) to address agricultural damage caused by mammals and birds. Their mission was codified in 1931 with the passage of the Animal Damage Control Act, which empowered the USDA to investigate, demonstrate, and control mammalian predators and rodent and bird pests (USDA 2015). In 1939, responsibilities were transferred to the Department of the Interior, US Fish and Wildlife Service's (USFWS) new Branch of Predator and Rodent Control (renamed Division of Wildlife Services in 1965). Throughout this period, most WDM was conducted by the lethal means of trapping, shooting, and poisoning. In the federal government, these activities were counter to the changing mission of the USFWS, and the Division of Wildlife Services soon fell out of favor with the agency. In 1985, responsibilities for WDM were shifted back to the USDA in the division that is today known as Wildlife Services (WS). The mission of WS is to provide federal leadership and expertise to resolve wildlife conflicts to allow people and wildlife to coexist.

Over time, some state agencies created programs that addressed wildlife damage, wildlife diseases, and nuisance wildlife. For example, from 1931 to 1980, the Wisconsin Department of Natural Resources (WDNR) administered a compensation program to pay landowners for damage to commercial crops and trees caused by deer and bear (Hygnstrom and Craven 1985). Sandhill cranes and waterfowl were later included in the program. In 1983 the program was tweaked and legislation created Wisconsin's Wildlife Damage Abatement and Claims Program (WDACP). The focus of this program was on damage compensation, with damage claims paid on a prorated basis. In 2013, 1,124 landowners voluntarily enrolled in the WDACP and requested damage abatement assistance on 255,702 acres of land, and the WDNR paid $1,394,577 on 325 wildlife damage claims (Koele et al. 2013). Eighty-three percent of the assessed losses were attributed to white-tailed deer. Such programs are not common, as currently only 13 state wildlife agencies pay compensation for wildlife damage.

Today, the field of WDM is recognized as an integral part of contemporary wildlife management. Evidence of this is seen in The Wildlife Society, an organization of wildlife professionals, which charters a Wildlife Damage Management Working Group with over 200 members. The Working Group supports a biennial Wildlife Damage Management Conference and associated proceedings that are national in scope. In addition, the Vertebrate Pest Council in California has hosted a biennial Vertebrate Pest Conference and associated proceedings since 1963 that are national/international in scope. During the past decade, at least 18 states across the nation have utilized University Extension Specialists with a focus in WDM. Products of their efforts and state and federal wildlife agency personnel include the Internet Center for Wildlife Damage Management, which is a clearinghouse of online information that at last count entertains 1.5 million visitors from all 50 states and 245 countries (Hygnstrom et al. 2015). A two-volume, 863-page book entitled *Prevention and Control of Wildlife Damage* (Hygnstrom et al. 1994) includes information on problem species, from alligators to polar bears, in North America. Two textbooks on the topic of WDM have been produced, *Resolving Human–Wildlife Conflicts: The Science of Wildlife Damage Management* (Conover 2002) and *Wildlife Damage Management: Prevention, Problem Solving, and Conflict Resolution* (Reidinger and Miller 2013), and are used in university courses across the nation.

The Role of States in Wildlife Damage Management

All states within the United States have developed laws and regulations to address various aspects of WDM and conflict management. Programs designed to assist citizens and communities often include technical assistance, investigation, compensation, land-use planning, and implementation of direct WDM practices. As populations of some species increase, or human land uses change, demand for services increases. Increasing WDM functions can burden state wildlife agencies and prevent fulfillment of other mission-related duties.

We conducted a survey of all 50 states and seven US territories in 2015 to determine their levels of involvement in WDM over the past five years. Initially, we examined all state wildlife agency websites and searched for information on wildlife damage, nuisance wildlife, and compensation. Then we contacted coordinators of WDM programs, or personnel most actively involved in WDM assistance. We sought responses for a series of questions, and the number of states answering each question is shown in parentheses.

We found considerable variability in the level of engagement by state wildlife agencies in assisting the public with human–wildlife conflicts. The number of species for which people are eligible for assistance by state ranged from "none" (12) to "all" (2). Most states focused efforts on game species (36), predators (21), furbearers (17), Canada geese (17), mesopredators (12), rodents (12), feral swine (8), bats (8), wild turkeys (6), and invasive species (6). In addition, problem types for which assistance could be obtained from state wildlife agencies varied from "none" (34) to "all" (10), and more specifically, growing crops (12), livestock (11), apiaries (6), stored crops (5), orchards (5), nurseries (4), garden crops (3), personal property (2), fences (2), forage crops (2), and irrigation equipment (1). Technical assistance and information were provided on 44 websites. Several had web pages that focused on "Wildlife Damage," "Living with Wildlife," and "Nuisance Wildlife." Technical assistance included species summary information, fact sheets on a wide range of species (up to 25), links to online resources, annual program reports, depredation regulations, lists of licensed wildlife control operators and trappers, and toll-free help lines.

Hands-on assistance was provided by 12 agencies, and 17 states provided cost-share programs for abatement materials, mostly fences for deer and bear. Thirteen states reported providing compensation for damage caused primarily by deer, elk, bear, wolf, and Canada geese (from $9,000 to $1.9 million per year). Depredation, shooting, or kill permits that allow property owners to react to damage were provided by 35 states, mostly for big game species such as deer (11), elk (3), bear (3), moose (2), pronghorn (2), and feral swine (2), but also including coyotes, mesopredators, squirrels, Canada geese, wild turkey, and wolves. Several states indicated that WDM is part of the responsibilities of all their wildlife field staff. Eight states reported employing 1–27 full-time staff dedicated to resolving human–wildlife conflicts. Twenty states reported managing annual budgets to support WDM programs, with funding levels ranging from $40,000 to $2.9 million per year. Annual budgets were supported by a variety of sources, including state game cash funds, hunter license fees, surcharges on deer and elk hunting licenses, sale of antlerless deer tags, state general revenue funds, Pittman–Robertson funds, grants, contracts, organizational funds, and interest in endowment funds.

Thirty-two state wildlife agencies provided oversight of the private wildlife control industry, in which 22 states required training. Nearly all state wildlife agencies restricted which species could be handled and managed by private wildlife control operators. State wildlife agencies collaborated with a wide range of agencies, organizations, industries, and individuals in implementing their WDM programs. Most notable were WS (21), the USFWS (10), and University Extension (7). Effectiveness of these collaborations was rated as high (9), medium (2), and low (0).

Federal Role in Assisting States

Wildlife do not abide by political boundaries, and thus cooperation is required for successful prevention and resolution of human–wildlife conflicts. As a result of shared authority, complexity, high costs, availability of expertise, or shared vision, states routinely work with federal, county, and nongovernmental land management agencies and organizations, as well as interest groups and individuals, to achieve goals.

Most states share WDM responsibility with federal WS for some species. Most WS programs are based on a state's need for assistance and expertise and work directly with state agriculture or natural resources agencies. In some cases, federal resources can augment state cost-share programs, allowing for greater service to those seeking relief. The division of duties typically is detailed in cooperative service agreements between agencies. Additional agreements exist among WS and county, township, and municipal governments; industry; and individuals for the provision of WDM services.

Legal Issues

Local, state, and federal laws and regulations are designed to manage wildlife, reduce human–wildlife conflict, and protect the public. We address several federal laws below that are applied across all states and territories. State and local regulations frequently are more restrictive than federal regulations and are too numerous to be addressed here. Wildlife control operators, pesticide applicators, hunters, trappers, wildlife rehabilitators, and those who manage wildlife populations must be aware that federal, state, and local laws and regulations all apply.

The Endangered Species Act (ESA) was passed in 1973 to protect imperiled plant and animal species. The ESA requires that an endangered or threatened species not be injured or harassed by wildlife control activities. Endangered and threatened species cannot be killed, harmed, or collected except under carefully described circumstances and only with appropriate federal and state permits. The presence of endangered or threatened species can affect how WDM activities occur by restricting use of traps, toxicants, and other control methods.

The Migratory Bird Treaty Act (MBTA) of 1918 protects all migratory birds in North America. Migratory birds and their nests and eggs cannot be taken, possessed, or transported without a federal permit. This does not include pigeons, house sparrows, or European starlings, which are non-native invasive species. Before attempting to control a migratory bird (e.g., woodpeckers, raptors, and waterfowl), landowners must obtain a 50 CFR Bird Depredation Permit. The permit allows the taking of migratory birds that destroy public or private property, threaten public health or welfare, and are a nuisance. The permit states the conditions under which the birds may be controlled and the methods that may be used. Permit holders may control migratory birds that are causing or are about to cause serious damage to crops, nursery stocks, or fish in hatcheries. An exception in the MBTA (50 CFR 21.43) is that "a federal permit shall not be required to control red-winged and Brewer's blackbirds; cowbirds; all grackles, crows, and magpies; when found committing or about to commit depredations upon ornamental or shade trees, agricultural crops, livestock, or wildlife, or when concentrated in such numbers and manner as to constitute a health hazard or other nuisance." Some states also have obtained a federal General Depredation Order for controlling Canada geese, gulls, and cormorants that are causing conflicts, inflicting property damage, or threatening endangered wildlife. A recent exception to the MBTA allows wildlife control operators to rescue migratory birds trapped inside buildings, provided that the birds are released unharmed and on-site.

In addition to the MBTA, the Bald and Golden Eagle Protection Act (BGEPA) of 1940 (amended in 1962 to include golden eagles) provides further protection for these two species, regardless of status under the ESA. The BGEPA prohibits "the take, possession, sale, purchase, barter, offer to sell, purchase or barter, transport, export or import, of any bald or golden eagle, alive or dead, including any part, nest or egg, unless allowed by permit" (16 U.S.C. 668 (a); 50CFR 22). Therefore, if WDM is needed for eagles that are jeopardizing human health and safety, or depredating livestock, additional permitting is required for the WDM entity.

The Federal Insecticide, Fungicide and Rodenticide Act (FIFRA), originally passed in 1947, established federal control of the distribution, sale, and use of pesticides. It has been amended several times and regulates the availability and use of all pesticides, including repellents and toxicants, including those used in WDM. It also mandates that the EPA provide oversight of research, registration, certification, sale, and use of pesticides to protect human health and the environment.

The National Environmental Policy Act (NEPA), enacted in 1970, promotes enhancement of the environment. The most significant outcome of the NEPA was the development of a process by which all executive federal agencies prepare environmental assessments (EAs) and environmental impact statements (EISs) that document the potential environmental effects of proposed projects in which a federal agency provides any portion of financing for the project, including WDM projects. The act does not apply to state actions where there is a complete absence of federal influence or funding.

Wildlife species that are not regulated by the federal government fall under state jurisdiction. In most cases where federal laws do apply, state laws and regulations add restrictions to those federal laws. They cannot be less restrictive. Under state law some problem species are unprotected and have no restrictions on their take. For example, many western states allow the unlimited take of coyotes and pocket gophers year-round. In eastern states, however, coyotes often are listed as a game animal with closed seasons and limited methods of harvest. States typically classify wildlife in the following ways: (1) "game species," which may be legally hunted; (2) "furbearer species," which are captured for fur, usually through trapping; (3) "nongame species," which are protected and for which no open seasons are available for their harvest; and (4) "unprotected species," which typically are non-native invasive species, or species that are very abundant. State and local ordinances may further restrict and define control activities. Local regulations may limit the techniques that can be used in controlling birds.

Linking Research, Practice, and Theory in Managing Human–Wildlife Conflicts

Conflict between humans and wildlife is increasing across American landscapes owing to urban and suburban expansion into new areas, changes in land-use practices, changes in resource extraction and production regimes, and shifts in wildlife management policy. To address the increasing conflicts, wildlife professionals build on basic ecological knowledge of population dynamics, animal behaviors, and landscape ecology to practice a form of applied ecology that exploits what we know about species to avert conflict in manners that align with long-held American beliefs, including the "greatest good for the greatest number in the long run" and Leopold's "land ethic." We draw on the overarching paradigm of the North American Model of Wildlife Conservation (NAMWC), which is covered elsewhere in this volume, when discussing WDM. All of the NAMWC components apply well to WDM, with the possible exception of "Non-Frivolous Use," which states that one can "legally kill certain wildlife for legitimate purposes" (Organ et al. 2012). This phrase is problematic because it assumes that the legitimacy of killing is not on a sliding scale based on individual circumstances, desires, and ethics. Regardless, the NAMWC is generally applicable to WDM and is being continually refined.

WDM programs can be thought of as having four parts: (1) problem definition, (2) ecology of the problem species, (3) management methods application, and (4) evaluation of management effort. Problem definition refers to determining the species and numbers of animals causing the problem, the amount of loss or nature of the conflict, the human role in the conflict, and other biological and social factors related to the problem. Ecology of the problem species refers to understanding the life history of the species, especially in relation to the conflict. Management methods application refers to taking the information gained from parts 1 and 2 to develop an appropriate management action to reduce or alleviate the conflict. Evaluation of management effort permits an assessment of the reduction in damage in relation to costs and impact of the management effort on target and nontarget populations. Emphasis often is placed on an integrated WDM approach (VerCauteren et al. 2012a).

Tools Used by State Wildlife Agencies to Address Human–Wildlife Conflicts

The specific methods used in integrated WDM often are categorized as lethal and nonlethal.

Lethal Strategies

Offending individual animals, such as gulls at airports, must sometimes be removed to protect human health, safety, and economic resources. Also, when populations of some species, such as white-tailed deer, become overabundant, damage becomes density dependent, and landowners turn to state agencies for relief. In general, three forms of lethal control are used to manage wildlife: shooting, trapping, and toxicants. As free-ranging populations are dynamic, lethal control often must be repeated to be effective. Relative to game species, this cycle of continually growing and harvesting animals at levels acceptable to various publics is a primary goal of state wildlife management agencies.

SHOOTING

State agencies have used regulated recreational hunting as their primary tool for managing game species for decades. In certain situations hunting is used to keep populations of some species from becoming too abundant and causing too much damage to agricultural and other resources. With most game species, agencies study and evaluate populations each year and determine what, if any, changes should be made to harvest levels in subsequent years. Besides being the most practical management tool for many species, hunting also can have social, economic, and ecological benefits. Especially in rural areas across the country, hunting provides many benefits to landowners, hunters, communities, and local economies.

Although regulated hunting often is the most practical and effective tool for managing populations, many situations occur in which it cannot be implemented or would not be effective in curtailing human–wildlife conflicts. An example is when a disease that is transmissible among deer and livestock is detected in a local population of deer. Immediate response is needed, and agency staff or other professionals may be called upon to perform culling to quickly and selectively decrease the population, with the intent of quickly reducing the potential for disease transmission. Similarly, disease or damage depredation permits may be issued to landowners to address local problems quickly. Strategies like this can be very effective because they target the specific population of individuals that are causing damage. Depredation permits differ from recreational hunting permits in that they allow landowners to cull animals that are damaging resources outside of hunting seasons and to use additional tools (lights at night, bait, etc.). In other instances professional shooters can be more effective, such as in culling feral swine or coyotes from aircraft and employing professional sharpshooters to cull deer in urban and suburban settings.

TRAPPING

Recreational trapping and cable restraints are tools used by agencies to manage furbearers in most states. Recreational trappers are an inefficient management tool for reducing damage when fur prices are low, however, because interest in recreational trapping often is driven by market value of furs. In these cases, agencies such as WS may trap nuisance furbearers. Examples of species and types of damage relative to furbearers include beavers building dams and flooding agricultural fields and roads, raccoons damaging sweet corn and killing poultry, and coyotes killing lambs. Lessons learned by generations of trappers and contemporary trap designs have led to significant advances relative to humaneness and species specificity of trapping and cable restraints. Commonly used tools for furbearers and nuisance species include cage traps, foothold traps, body-gripping traps, and foot-encapsulating traps. All but body-gripping traps can be used for nonlethal purposes as well, but if the intent is to euthanize the captured animal, humane methods such as a properly placed gunshot or asphyxiation by carbon dioxide must be employed when using nonlethal traps.

State agencies sometimes use trapping to manage common or invasive species of birds that are impacting other species of conservation concern. For example, populations of endangered Kirtland's warbler are being depressed in large part because common brown-headed cowbirds lay their eggs in warbler nests and warblers then raise the cowbird chicks instead of their own. Large cage traps are used to capture cowbirds in warbler habitat. The cowbirds are then euthanized humanely by cervical dislocation or asphyxiation.

State wildlife agencies also manage and regulate trapping done by private wildlife control operators or agents. As noted earlier, 22 states now require training for industry professionals who handle problem wildlife. A standardized, online curriculum has been developed, which can be modified for use in any state or province as a basic training program (Curtis et al. 2015). There is a trend toward increasing oversight and regulation of this industry by state wildlife agencies.

TOXICANTS

Toxicants are chemical compounds registered by the EPA that kill target animals through various physiological modes of action, such as coagulation response inhibition, disruption of metabolic processes, and inhibition of nerve impulses. Great care must be exercised in their use to minimize potential risks to humans, pets, livestock, and nontarget wildlife. Restricted use pesticides can only be applied by individuals certified

by the EPA, which in WDM typically includes WS personnel and certified pesticide applicators. Many firms that handle residential or commercial wildlife control hire pesticide applicators. While an important tool for rodent and bird control, toxicants seldom are used by state wildlife agencies. However, state agencies, often departments of either agriculture or wildlife, regulate use of toxicants for wildlife control in collaboration with the federal EPA.

In all cases, be it recreational hunting to manage populations on a large scale, application of a toxicant to control an invasive species, or selective trapping to remove a problem individual, lethal methods must be implemented responsibly and as part of a science-based strategy to achieve management and conservation goals. Wildlife researchers continually endeavor to develop strategies that are tailored to the target species, context of the conflict, and economics of the situation, while minimizing any negative impacts.

Nonlethal Strategies

Whenever possible and especially in small-scale local instances, nonlethal strategies are implemented by WDM personnel. They often suggest strategies and even provide labor and materials to help landowners address conflicts. Conover (2002) pointed out that nonlethal methods may result in the deaths of displaced, relocated, and excluded animals. For example, overpopulated animals that are fenced from a resource may suffer if alternative food sources are not available. Similarly, animals that are translocated from areas where they are causing damage to another area may not survive for a variety of reasons, including an inability to assimilate with individuals of the same species already present in the area and having no knowledge of their new landscape, its resources, and its dangers.

In general, nonlethal management strategies can be categorized as either physical or psychological strategies, or a combination of the two. These strategies usually include various forms of exclusion, habitat modification, frightening devices, repellents, reproductive control, and translocation. It must be realized that seldom are these options perfect fixes. Some have better utility than others, and the best tools for one situation may not perform as well in another.

When large animals become accustomed to human foods, dumpsters and other sources must often be fenced in the urban–wildland interface. *Photo courtesy of the USDA APHIS Wildlife Services.*

EXCLUSION

Exclusionary fencing is the most common method for physically separating wildlife from a resource they could damage. A wide variety of fence options are available, and the type used will depend on the level of protection desired, seasonality of the resource being protected, physical ability of the target species, motivation to breach, behavioral characteristics, costs associated with construction and maintenance, longevity of the building materials, and possible negative effects (VerCauteren et al. 2006). For example, a woven-wire fence is expensive upfront but could last 30 years and be virtually 100 percent effective, and thus it may be an excellent investment to keep deer from damaging high-dollar fruit trees. For an annual crop of lesser value, such as corn, a relatively inexpensive and easy-to-erect electric polytape fence may reduce damage even though it is not 100 percent effective in keeping deer out. Thus, while fences can prevent or eliminate agricultural damage, the costs associated with installation, materials, and maintenance can outweigh the economic benefits based on crop values. Fences often are used to exclude ungulates such as deer, elk, and feral swine from high-value crops and predators such as coyotes from small livestock pastures. In addition to fences, other exclusionary protection techniques include bird netting and lines, rodent-proof construction, wire mesh, cylinders, wraps, and bud caps.

HABITAT MODIFICATION

Habitat sometimes can be altered when exclusion is not an option owing to the nature of the resource, prohibitive costs, or environmental concerns. Habitat modification includes altering the biotic and abiotic components of the habitat or changing the management and maintenance of the resource in a manner that alters the carrying capacity for the target species or lessens the desirability of location (Reidinger and Miller 2013). Alteration of habitat can be done from landscape scale down to simply altering a stand of trees in a backyard to prevent birds from roosting. The landscape surrounding an airport can be manipulated to make it less attractive to birds and prey species to minimize bird strikes. For rodents, such as mice and voles, habitat modification can consist simply of mowing or removal of woodpiles, brush, and other habitat. Water levels can be raised and lowered to make habitat conditions inhospitable to beavers and muskrats. In cases where the resource being preyed upon is an agricultural crop, a switch to an unpalatable variety is effective, or altering the planting and harvesting timing can avoid the coinciding of bird migrations with crop vulnerabilities. Habitat modification can be cost prohibitive, owing to the cost of permitting processes, time, and labor; however, strategic modification can alter carrying capacity or desirability of the location for longer periods than other methods (Conover 2002; Reidinger and Miller 2013). Unfortunately, habitat modification often lacks target specificity, and several other species can be impacted. For example, to resolve a problem of deer causing damage to flower beds in a park, an adjacent woodlot where the deer live could be cut down, burned, and bulldozed, but the procedure would be expensive and impacts to coinhabiting squirrels, raccoons, songbirds, and associated park goers could be undesirable.

FRIGHTENING DEVICES

The goal of frightening devices is to influence the behavior of problem animals and move them from areas where they can cause damage. They fall into four categories: visual, auditory, audiovisual, and biological. The effectiveness of most frightening devices diminishes after initial success, as the animals become habituated to the frightening devices with repeated exposure. With the diminishing returns of frightening devices, it is important to consider strategic timing of use, visual or auditory range, and integration of multiple sensory stimuli. Visual frightening devices work by mimicking a predator's shape, sound, or movement to scare the target species or by exposing them to novel visual stimuli. Common visual frightening devices include plastic owls and coyotes, inflatable moving scarecrows, fence ribbons, flags, lasers for dispersing birds, strobe lights, and balloons. Auditory devices emit sounds within the audible range of the target species delivered through systems that are either recordings, such as alarm and distress calls, or controlled explosions, such as propane cannons. Audiovisual devices incorporate both stimuli, such as pyrotechnics, including bird bangers, bird screamers, and cracker shells. Biological frightening devices emulate natural factors that influence the behavior of problem animals. Livestock protection animals such as dogs, llamas, and donkeys are one of the oldest forms of WDM and can be used to reduce predation on livestock caused by canids, felids, and bears. Dogs also have been used to protect livestock from disease by providing a buffer from wildlife species (VerCauteren et al. 2012b) and to protect agricultural crops from damage by deer and other species (VerCauteren et al. 2005). Falcons and falconers have been used at airports to deter birds from using the area.

REPELLENTS

Repellents are chemical compounds registered by the EPA that disperse animals from an area or resource through various olfactory or taste senses associated with pain, displeasure, fear, conditioned aversion, or tactile response. Capsaicin, the chemical compound in chili peppers, induces pain and thus avoidance of treated plants by deer and other herbivores. Predator odors, such as coyote urine, induce fear in many herbivores and can be used as an area repellent. Compounds such as anthraquinone, when applied to turf grass and ingested by geese, induce illness and subsequent avoidance. In addition, polybutenes are sticky, tacky gel-like compounds that when applied to ledges will repel pigeons and squirrels from the area of application. Effectiveness of repellents is highly dependent on the motivation of the animals. If animals are food stressed, repellents likely will not be effective. Effectiveness also is affected by weathering, alternative food sources, and acclimation.

REPRODUCTIVE CONTROL

In some situations, altering the dynamics of the population is the best approach to mitigating the conflict or resource damage. The natality of a population can be changed through reproductive or fertility control, which reduces human–wildlife conflict if damage is density dependent. Reproduction in birds, such as

Overabundant species, like these blackbirds, can cause major conflict with humans and livestock. They are often controlled using repellents. *Photo courtesy of the USDA APHIS Wildlife Services.*

Canada geese, can be reduced during the nesting season by egg removal, but geese often will mitigate this technique by renesting. In addition, eggs can be oiled, addled, or punctured to inhibit hatching, but in most cases over 75 percent of the nests must be treated for several years to have a measurable effect on the population. For long-lived species such as geese and deer, lethal control often has two to three times the effect on population growth as inhibiting reproduction. Generally, fertility control is a long-term and expensive management approach rarely conducted by state and wildlife agency staff.

Strategies to control wildlife fertility include endocrine disruption, immunocontraception, intrauterine devices, surgical procedures, and chemosterilization. Since the late 1950s, research has been conducted on several species, including wild horses, white-tailed deer, prairie dogs, Canada geese, elephants, and bison. Contraceptive methods can only be implemented in specific situations as a result of a combination of needs that may include (1) reversibility, (2) suitable for field delivery, (3) effective with a single dose, (4) no hazard to nontarget species, (5) no harmful side effects, and (6) no effect on the social behavior of the animals (Conover 2002). Currently, few field-deployable single-dose contraceptive methods are available. Gonadotropin releasing hormone and porcine zona pellucida have shown some promise, but as with all chemical contraceptive methods, they have their limitations. While fertility control methods are gaining popularity with the public, they are not yet stand-alone methods for most situations, owing to a lack of applicability with large populations, effectiveness, and field readiness, as well as prohibitive costs. Most fertility control applications are considered experimental by state agency staff, and a research permit may be needed to use such methods. Very few products (e.g., GonaCon and OvoControl) have a current EPA registration for use on wildlife.

RELOCATION AND TRANSLOCATION

Animals that are captured can be relocated, translocated, placed in captivity, or dispatched by humane methods. Relocation is the release of a captured animal within its original home range, typically not far from the capture site. For example, raccoons that

Large carnivores, including black and grizzly bears and mountain lions, are frequently trapped and relocated when they intrude on the urban setting. *Photo courtesy of the USDA APHIS Wildlife Services.*

are denning within the chimney of a house can be captured with a cage trap, the chimney can be capped to prevent reentry, and the animals can be released at the base of the chimney. With any luck, raccoons will move away and use a natural cavity for denning, but often they go off in search of another chimney. Translocation is the release of a captured animal outside of its original home range, typically far from the capture site. For example, the same raccoons could be taken 25 miles away to a state wildlife area and released. Homeowners and the public often appreciate the perceived humane treatment of these problem animals. Unfortunately, raccoons and many other species of wildlife have strong homing abilities, and the animals may simply return after a short time and continue to cause problems. Raccoons also are highly territorial, so translocation may cause intraspecific strife with resident raccoons at the release site. Strife may result in injury, death, or disease transmission to the translocated and resident raccoons. In addition, if the translocated raccoons do survive, there are no guarantees that they will not cause problems in the new area. For these reasons, state wildlife agencies often prohibit the translocation of wildlife without specific permits. Three situations may warrant translocation: (1) when the animal is so valuable that euthanasia is not an option, (2) when the population is below carrying capacity at the release site, and (3) when public relations takes precedence over the other two (Conover 2002). Any animal that is a potential threat to human safety should not be translocated and should be removed from the wild.

Human Dimensions of Wildlife Conflict

Human dimensions of wildlife include methods and theory from a variety of disciplines, such as anthropology, sociology, economics, geography, and political science, among others. Human dimensions have become increasingly important, as citizen stakeholders are interested in and exert influence on wildlife policy.

There are three ways in which the American public perceives wildlife: positively affected by wildlife, negatively affected by wildlife, or not affected. The first two choices often sponsor opposing views regarding the need for action, and it may be difficult to reach an agreeable solution that pleases either viewpoint. Those not affected can help develop a course of action involving compromise, as they can look at the situation

Because overabundant wildlife is often very popular with the public, human dimensions research is often necessary to help address the inevitable conflicts that arise in urban settings. *Photo courtesy of the USDA APHIS Wildlife Services.*

objectively; however, it may be difficult to attract and hold their interest.

Decision-making for WDM policy is a public process that involves input by managers, stakeholders, and the general public. Within the bounds of local, state, and federal regulations, the decision-maker on privately owned land is often the landowner. For publicly owned lands, a public official typically is authorized to manage the property, but management must meet the legal mandates, and consider the interests and goals, of the public. Stakeholders often are categorized by their position as agricultural producers, rural landowners, urban dwellers, activists, advocates, consumptive resource users (e.g., hunters), and nonconsumptive users (e.g., bird-watchers). Regardless of their categorization, divisions typically align with positively affected, negatively affected, or nonaffected positions on the issue. Information provided by wildlife professionals and community input are crucial for local decision-makers to make informed choices. Several publications are available to assist community leaders and wildlife agency staff when dealing with the human dimensions of overabundant wildlife (Decker et al. 2002, 2004).

Changes in the Future

The field of wildlife management has changed greatly over the past century. Unchanging, however, are the facts that valued resources are damaged by wildlife, conflict between humans and wildlife exists, and wildlife itself is a valued resource. Our landscape is changing, and human-altered environments in some cases may lead to the decimation of habitat and decline of some wildlife populations, while some cases lead to increased diversity of habitats and overabundance of certain species that do well in developed landscapes. Public attitudes toward wildlife are changing, and the changes in public opinion and wildlife management policy never seem to slow or reduce in amplitude. Overabundant native species, invasive alien species, and infectious zoonotic diseases are just a few of the emerging issues that must be faced by wildlife professionals. Resolution of human–wildlife conflict continues to evolve and is a growing area of specialization for future generations of wildlife professionals to make a meaningful impact on wildlife conservation, wildlife management, and society at large. Highly trained wild-life professionals are needed at the local, state, and federal levels to apply scientific research and practice to the dynamic field of human–wildlife conflict management.

Invasive Species

Human–wildlife conflicts occur with both native and alien invasive species. In an ever more globalized world, there is an increase in exotic/alien species exploiting new ecological niches in the United States (Conover 2002). Federal and state agencies are required to act in the detection and control of invasive species. These species have been introduced either purposefully for hunting and aesthetics or accidentally by escaped animals from agricultural applications, the pet trade, freight, or ballast. Invasive species can displace or eradicate endemic species, damage crops, and cause economic hardship. Among the many established vertebrate invasive species in the United States and its territories are Burmese pythons, brown tree snakes, European starlings, rock pigeons, house sparrows, feral swine, Norway rats, house mice, and nutria. Each of these invasive species provides different challenges in management and control. For example, in the past decades feral swine have been expanding in abundance and distribution, causing an estimated $1.5 billion in annual damages. Feral swine exemplify all major facets of invasive species management and human–wildlife conflict in that they are niche generalists, carry zoonotic diseases, are involved in vehicle–wildlife collisions, damage agricultural crops, kill livestock, damage personal property, alter plant communities, contaminate waterways, and prey on endemic and endangered species.

Overabundant and Urban Deer

Deer populations have responded favorably to management and have adapted well to urban sprawl, and for the past few decades they have been the primary species responsible for several types of damage, including consumption of crops, vehicle accidents, and transmission of diseases to livestock and humans. Over 30 million deer currently occupy the United States (Ver-

Cauteren and Hygnstrom 2011). Although they cause millions of dollars of damage each year, as the most popular game species in the country, they are also a huge positive economic resource. The deer-hunting "industry" impacts the country's economy on multiple scales, from the sale of hunting gear and licenses to supporting local businesses and landowners.

State wildlife management agencies work diligently to use regulated, recreational hunting to manage deer populations at levels that provide a balance between positive ecological attributes (hunting and viewing opportunities, intrinsic values) and negative impacts (deer–vehicle collisions, impacts on plant communities, crop damage; VerCauteren et al. 2011). Unfortunately, in many locales it is difficult to reduce deer numbers to goal densities, even with extremely liberal hunting regulations and bag limits. Two of the largest impediments to increasing hunter harvest are that individual hunters only have the willingness and need to harvest so many deer a year and that hunters have difficulty obtaining access to private land that acts as refugia for deer. Across much of white-tailed deer range, the impact of hunting on deer populations is not great enough to reduce deer numbers to meet population goals.

In response to increased deer numbers in urban areas, much has been done in the past 25 years to assist communities with damage caused by deer, and many jurisdictions have implemented plans that include feeding restrictions, making properties less attractive (e.g., reducing cover), barriers, harassment, translocation, lethal removal, and fertility control. The technical guide *Managing White-Tailed Deer in Suburban Environments* (DeNicola et al. 2000) was developed to provide options for persons or communities to consider when experiencing conflicts. Most importantly, state agencies recently have worked cooperatively with municipalities to implement hunting in settings where it has not traditionally been allowed, owing to perceived safety concerns. As a result, urban hunts have become an important tool for managing deer in an increasing number of areas. In both rural and urban landscapes, wildlife professionals are tasked with continuing to develop creative strategies to use hunters and other tools in their efforts to keep deer populations at levels that do not unduly impact their habitats and human neighbors. Thus, by necessity deer management needs to be dynamic, and managers must work diligently to maintain deer numbers and distribution using a variety of management options.

Nonmigratory (Resident) Canada Geese

Populations of Canada geese declined significantly during the late 1800s and early 1900s primarily as a result of unregulated hunting and egg collecting. It was believed that the giant subspecies of Canada goose (*Branta canadensis maxima*) had actually gone extinct owing to overexploitation. However, a remnant population of giant Canada geese was discovered wintering in Rochester, Minnesota, in 1962. Canada geese typically nest in Canada and migrate significant distances in the fall to winter in moderate climates, but these giant Canada geese avoided migration by staying in an area that provided open water to roost and crops to feed on during the winter. These birds were used to restock areas throughout their former range. This highly successful wildlife restoration project was supported by wildlife agencies, hunters, and bird enthusiasts alike, but they did not anticipate the conflicts that loomed on the horizon. Geese started overwintering in many nontraditional areas of the eastern United States, including parks, golf courses, sewage treatment ponds, and other urban areas that provide open water and food during the winter. These nonmigratory or "resident" Canada geese have benefited from the way humans have altered landscapes, in the form of readily available agricultural fields, turfgrass, and other anthropogenic food sources. Nonmigratory goose populations often thrive because of protections provided by municipal ordinances, lack of predators, or expanses of mowed grass where predators can be observed easily. As geese congregate, they can make green space less attractive for recreational use by the accumulation of fecal deposits. Flocks of resident Canada geese have resulted in beach closings, reduced water quality, erosion, safety concerns at airports, and unsanitary conditions in parks, cemeteries, and yards and on sidewalks near businesses, hospitals, and schools. Adult nesting geese can be protective of their nests and young and become aggressive when an unsuspecting person gets too close to nests. Occasionally, people have been injured during these interactions. Communities that face these concerns often institute a public education effort to persuade people to stop feeding geese. Additional strategies may include the use of herding dogs, repellents, harassment and barrier devices, egg and nest destruction, juvenile translocation, and lethal removal. The technical guide *Managing Canada Geese in Urban Environments* (Smith et al. 1999) was developed to provide options for managers and the public to consider when experiencing conflicts.

LITERATURE CITED

Batcheller, G. R., M. C. Bambery, L. Bies, T. Decker, S. Dyke, D. Guynn, M. McEnroe, M. O'Brien, J. F. Organ, S. J. Riley, and G. Roehm. 2010. The public trust doctrine: Implications for wildlife management and conservation in the United States and Canada. Technical Review 10-01. The Wildlife Society, Bethesda, Maryland, USA.

Conover, M. R. 2002. Resolving human–wildlife conflicts: The science of wildlife damage management. CRC Press, Boca Raton, Florida, USA.

Cook, R. S. 1991. What's in a name? Proceedings of the Great Plains Wildlife Damage Control Workshop 10:165–168.

Curtis, P. D., R. Smith, and S. Hygnstrom. 2015. The National Wildlife Control Training Program: An evolution in wildlife damage management education for industry professionals. Human–Wildlife Interactions 9(2):166–170.

Decker, D. J., T. B. Lauber, and W. F. Siemer. 2002. Human–wildlife conflict management: A practitioner's guide. Northeast Wildlife Damage Management Research and Outreach Cooperative and Human Dimensions Research Unit, Cornell University, Ithaca, New York, USA.

Decker, D. J., D. B. Raik, and W. F. Siemer. 2004. Community-based deer management: A practitioner's guide. Northeast Wildlife Damage Management Research and Outreach Cooperative and Human Dimensions Research Unit, Cornell University, Ithaca, New York, USA.

DeNicola, A. J., K. C. VerCauteren, P. D. Curtis, and S. E. Hygnstrom. 2000. Managing white-tailed deer in suburban environments. Cornell Cooperative Extension, Ithaca, New York, USA. https://ecommons.cornell.edu/handle/1813/65.

Ehrlich, P. R., and J. P. Holdren. 1971. Impact of population growth. Science 171:1212–1217.

Hygnstrom, S. E., and S. R. Craven. 1985. State-funded wildlife damage programs: The Wisconsin experience. Proceedings of the Eastern Wildlife Damage Control Conference 2:234–242.

Hygnstrom, S. E., R. M. Timm, and G. E. Larson. 1994. Prevention and control of wildlife damage. University of Nebraska–Lincoln Extension, Lincoln, Nebraska, USA.

Hygnstrom, S. E., S. M. Vantassel, P. D. Curtis, and R. Smith. 2015. Internet center for wildlife damage management.

Proceedings of the Vertebrate Pest Conference 26:440–442.

Koele, B., D. Hirchert, and N. Balgooyen. 2013. Wildlife damage abatement and claims program. Wisconsin Department of Natural Resources. http://dnr.wi.gov/topic/WildlifeHabitat/documents/reports/damabate.pdf.

Lovich, J. E. 1987. Mountain nightingales: The story of wolves in western Pennsylvania. Mountain Journal 15:3–7.

McShea, W. J., B. H. Underwood, and J. H. Rappole. 1997. The science of overabundance: Deer ecology and population management. Smithsonian Institution Press, Washington, DC, USA.

Organ, J. F., V. Geist, S. P. Mahoney, S. Williams, P. R. Krausman, G. R. Batcheller, T. A. Decker, R. Carmichael, P. Nanjappa, R. Regan, R. A. Medellin, R. Cantu, R. E. McCabe, S. Craven, G. M. Vecellio, and D. J. Decker. 2012. The North American Model of Wildlife Conservation. Technical Review 12-04. The Wildlife Society, Bethesda, Maryland, USA.

Reidinger, R. E., Jr., and J. E. Miller. 2013. Wildlife damage management: Prevention, problem solving, and conflict resolution. Johns Hopkins University Press, Baltimore, Maryland, USA.

Seidl, I., and C. A. Tisdell. 1999. Carrying capacity reconsidered: From Malthus' population theory to cultural carrying capacity. Ecological Economics 31:395–408.

Smith, A. E, S. R. Craven, and P. D. Curtis. 1999. Managing Canada geese in urban environments. Jack Berryman Institute Publication 16, and Cornell University Cooperative Extension, Ithaca, New York, USA. https://ecommons.cornell.edu/handle/1813/66.

USDA (US Department of Agriculture). 2015. Wildlife Services Enabling Legislation. www.aphis.usda.gov/aphis/ourfocus/wildlifedamage/sa_program_overview/ct_legislation.

VerCauteren, K., C. Anderson, T. VanDeelen, D. Drake, W. D. Walter, S. Vantassel, and S. Hygnstrom. 2011. Regulated commercial harvest to manage overabundant white-tailed deer: An idea to consider? Wildlife Society Bulletin 35:185–194.

VerCauteren, K., R. Dolbeer, and E. Gese. 2012a. Identification and management of wildlife damage. Pages 232–269 in N. J. Silvy, editor. The wildlife techniques manual, 7th edition. Vol. 1. Johns Hopkins University Press, Baltimore, Maryland, USA.

VerCauteren, K., and S. Hygnstrom. 2011. Managing white-tailed deer: Midwest North America. Pages 501–535 in D. G. Hewitt, editor. Biology and management of white-tailed deer. CRC Press, Boca Raton, Florida, USA.

VerCauteren, K., M. Lavelle, T. Gehring, and J. Landry. 2012b. Cow dogs: Use of livestock protection dogs for reducing predation and transmission of pathogens from wildlife to cattle. Applied Animal Behaviour Science 140:128–136.

VerCauteren, K., M. Lavelle, and S. Hygnstrom. 2006. Fences and deer damage management: A review of design and efficacy. Wildlife Society Bulletin 34:191–200.

VerCauteren, K., N. Seward, D. Hirchert, M. Jones, and S. Beckerman. 2005. Dogs for reducing wildlife damage to organic crops: A case study. Proceedings of the Wildlife Damage Management Conference 11:286–293.

12

Colin M. Gillin
and John R. Fischer

State Management of Wildlife Disease

As described in chapters 1 and 2, regulating harvest of state wildlife resources was the principle reason state wildlife agencies were created over the past 150 years. Today, state agency responsibilities have expanded, ranging from protection of nongame and sensitive species to conservation of habitats and multiple non-harvest-related recreational programs. Along with these other responsibilities, management of disease has increasingly taken a more significant role in state agency programs, expenditures, and personnel efforts. Impacts of disease on wildlife populations, agency and community economics, and state policy have resulted in a need for states to contain, understand, and manage the interconnected influences of disease on the environment, humans, and wildlife.

As states improve wildlife health and disease management capabilities by developing new programs and broadening and building existing programs, the wildlife health field continues to change as well. Traditionally, wildlife health specialists and veterinarians mimicked the cultural makeup of hunting constituents within states, namely, a profession dominated by Caucasian males with a primary interest in game species management and disease issues affecting recreational hunting. Today, however, many highly trained and educated professionals are very conservation oriented and well versed in the management of both game and nongame species and their diseases. Many current disease specialists possess graduate research degrees with

veterinary educations, and several have gained board certification by the American College of Zoological Medicine. Diversity of the wildlife health profession is also increasing. Currently, 34 states have full-time staff in wildlife health programs, employing a minimum of 45 wildlife health professionals, including 35 wildlife veterinarians and another 10 well-trained wildlife disease biologists and coordinators.

Improvements in agency staff and state resource capabilities have evolved out of necessity owing to epic disease outbreaks, including chronic wasting disease (CWD), hemorrhagic disease, brucellosis, bovine tuberculosis (BTB), and highly pathogenic avian influenza. Evolving agency programs increasingly integrate wildlife health expertise into more developed and comprehensive state programs required to deliver a variety of ever-changing services to an engaged public and diverse cadre of stakeholders of wildlife resources.

History of Wildlife Disease Management

The evolution of today's wildlife disease management programs can be linked back to the historical role diseases in wildlife exhibited in interactions between humans and wild animals. This relationship has occurred since the early human occupation of North America, but it has increased as our population's footprint has expanded. Historically, disease always has played a role in wildlife population fluctuations, affecting the weak and old or acting to reduce inflated wild animal densities resulting from favorable resources of food, shelter, and water (Charles 1931).

After European settlers arrived and began to colonize North America in the 1500s, millions of Native Americans perished from small pox and other diseases to which they had no immunity (Thornton 1987). By the 1700s, Spanish immigrants in the southwestern United States and Florida introduced livestock that competed with wildlife and carried diseases against which wild animals similarly had no immunity or resistance (Brown and Wurman 2009). As described in previous chapters of this book, bison (*Bison bison*) herds once numbering in the millions were reduced to less than 1,000 animals in the Yellowstone region of Montana as a result of several factors. By the early 1900s, through presumed contact with infected cattle, this remnant herd of bison became a persistent source of bovine brucellosis, one of the most controversial and emotionally polarized livestock–wildlife disease outbreaks that challenge wildlife managers and animal agriculture officials in multiple states and agencies to this day (Mohler 1917; Kilpatrick et al. 2009).

In the early twentieth century, disease in wildlife was not considered an important factor in the management of wildlife (Baughman and Fischer 2005). However, since the 1930s, diseases affecting wildlife, including many species important to hunters, have become much more pervasive. As a result, the Michigan Department of Conservation established the first wildlife disease laboratory in the United States in 1933. The lab's initial role focused on starvation, nutrition, and diseases of Michigan wildlife, and by the early 1950s, it turned its diagnostic, research, and disease management resources to outbreaks of epizootic hemorrhagic disease (EHD) in white-tailed deer (*Odocoileus virginianus*; Fay et al. 1956).

The California Department of Fish and Wildlife's Wildlife Investigations Lab was established in 1941. This lab initiated a program to conduct wildlife disease investigations and monitor/manage population health issues in California's wildlife. The Wyoming Game and Fish Commission next established a Game and Fish Research Laboratory in 1947, recognizing wildlife disease as an important factor in management of populations.

The commission added the Sybille Wildlife Research Unit in 1952 to conduct field studies of wildlife diseases (Thorne et al. 1982).

Recurring EHD outbreaks that threatened white-tailed deer restoration efforts in the Southeast led to the formation of the Southeastern Cooperative Deer Disease Study in 1957 by the Southeastern Association of Game and Fish Commissioners (Thorne et al. 2005). In partnership with the University of Georgia's College of Veterinary Medicine and 11 southeastern state fish and wildlife management agencies, the organization soon broadened its work, was renamed the Southeastern Cooperative Wildlife Disease Study (SCWDS), and became a leader in the diagnosis, research, and management of wildlife disease. Currently, the SCWDS serves 19 state natural resource agencies, and funds they provide are leveraged with funds from several federal agencies within the US Department of the Interior (DOI), US Department of Agriculture (USDA), and granting organizations such as the National Institutes of Health and National Science Foundation. The SCWDS conducts wildlife disease research, diagnostics, surveillance, training, and consultation on a regional, national, and international basis with the primary objectives of defining population impacts of wildlife diseases and parasites and identifying the role of wildlife in the epidemiology of diseases in humans and domestic animals (Nettles and Davidson 1996).

Today, 34 of 50 states have dedicated programs supporting full-time wildlife health staff of veterinarians or disease specialists and coordinators. Several states with dedicated staff and health programs also utilize the services of cooperative programs like the SCWDS; however, at least eight states rely on the use of regional cooperative programs for disease investigation services. Of the remaining eight states with no formal wildlife disease programs, one is currently developing a program in multiyear stages. Six of these states are located in the northeastern United States, where veterinary expertise is provided through Cornell University College of Veterinary Medicine and Tufts Cummings Schools of Veterinary Medicine. States with no defined wildlife health program have personnel assigned to the tasks of surveillance or coordination of disease outbreak response as part of their broader job responsibilities.

Other universities and veterinary schools also have the capacity to provide various levels of disease services within their states and regions, including the University of California at Davis Wildlife Health Center, North Carolina State College of Veterinary Medicine, Texas A&M College of Veterinary Medicine and Biomedical Sciences, and others. Depending on disease type, species affected, and response required, all states also work cooperatively with federal partners at the USDA's Animal and Plant Health Inspection Service (APHIS) and the DOI to conduct surveillance and respond to fish and wildlife disease emergencies.

Today's state natural resource and wildlife disease professionals provide state management agencies with a variety of expertise, service, training, and resources to assist management personnel in evaluations of individual animals and populations. Examples of some of the responsibilities of these state programs include diagnosis and response to disease events, biological sampling, wildlife capture and necropsy training, wildlife rehabilitation, surveillance of disease, determining pesticide and toxin impacts on wildlife, health and condition of wildlife populations, prevention of zoonotic diseases, and investigations involving public safety and wildlife. Many wildlife health staff also assist their agency administrators in policy development, input on administrative rules, development of biological sampling standards, service and consultation on animal care and use in research, and research involving wildlife health and disease. The specialized knowledge and expertise states gain from hiring dedicated, trained staff in wildlife disease programs allow managers and administrators to utilize these programs as important tools in the management of wildlife. As political and social challenges influence and demand more comprehensive and adaptive wildlife management strategies, the importance of integrating the health and welfare of free-ranging wildlife into those strategies becomes more apparent.

Management Authority of State Wildlife Agencies

In the United States, management authority for most endemic wildlife is the responsibility of state wildlife agencies (Thorne et al. 2005), with several exceptions.

As described in chapter 1, states hold broad statutory authority of public resources, including the conservation of fish and wildlife by state fish and wildlife management agencies. As part of this authority, they are also entrusted with responding to and managing diseases that affect wildlife and respond to public expectation to address wildlife disease issues. Many states conduct annual disease surveillance through hunter field checks, or as part of an interagency surveillance effort like the Highly Pathogenic Avian Influenza (HPAI) program, or in response to localized wildlife die-offs.

Congress has affirmed federal authority of the secretary of agriculture (Animal Health Protection Act of 2002) to dispose of animals, inclusive of wildlife. This act stipulates that the secretary will take action "after review and consultation with the Governor or an appropriate animal health official of the state and will consult with the officials of the state agency having authority for protection and management of such wildlife."

In many states, statutes and laws authorize the chief animal health authority, namely, the state department of agriculture state veterinarian, to protect the state's livestock and agricultural commodities from reportable foreign animal diseases. This authority often directs the department of agriculture to act and take all measures necessary to control or eradicate disease threats depending on the particular disease risk to livestock, contagiousness, commodity groups affected, and expectations from agency administration, elected officials, and the public. Because of the transmissibility of many diseases between livestock/poultry and wild species in similar taxa, it is important for state wildlife management personnel and state agriculture health officials to plan ahead to facilitate coordination and collaboration between agencies and provide a clear understanding of management authorities, knowledge, resources, and capabilities before a disease outbreak occurs. It is also very important that all involved agencies understand the limits and scope of authorities as provided through the state's statutes and rules specifically affecting each agency.

Diseases occurring in wildlife can also fall under the purview of state public health authorities if the disease poses a zoonotic threat to the public, wildlife rehabilitators, or hunters. Diseases such as highly pathogenic

H5N1 avian influenza, West Nile virus, tuberculosis (TB), rabies, and other zoonotic diseases occur in wildlife and can affect how wildlife is managed as a result of the risk to public health. State and federal public health officials will often be involved in disease issues in cooperation with state wildlife management agencies and agricultural animal health officials, depending on disease threat and type, risk of infection to humans, and ability of public health authorities to provide biosecurity measures and guidance. Public health authorities may have broad decision-making authority if the disease causes severe illness or death in humans. Similar to the case for diseases affecting wildlife and livestock, effective planning and communication are essential to avoid conflicts and misunderstandings and to determine the appropriate management action with potential contradictory statutes, regulations, and policies between agencies.

Thorne et al. (2005) describe management conflicts of authority between agencies and the need for multiagency cooperation when addressing and managing persistent, large-scale, established disease outbreaks such as brucellosis, TB, and CWD. These diseases require extensive resources to manage, often more than any single agency can provide in funding and agency resource and personnel capacity. Many states fund major portions of their wildlife health programs with support from dollars obtained through the Pittman–Robertson Act of 1937. More recently, other state/federal cooperative programs have resulted in more effective and efficient surveillance and management of wildlife disease of national importance. States received federal funds through cooperative agreements with APHIS for CWD surveillance and management in wild and captive ungulates from 2002 to 2011 and for HPAI surveillance in wild birds from APHIS and the US Fish and Wildlife Service (USFWS). Other areas of cooperation between states and federal partners resulted in research and increasing state and federal understanding of TB in wildlife. Funding from the USFWS through state capacity grants has provided the foundation for interagency research geared to understand, manage, and respond to the devastating population effects of white-nose syndrome (WNS) as this disease moves westward, impacting federally threatened/endangered and state-managed bat species.

Wildlife Disease Management

Wildlife disease can be categorized several ways based on transmission pathways and species' impacts. Of greatest concern to wildlife agencies are diseases that specifically affect and diminish wildlife populations regardless of whether the disease is passed between domestic animals and humans. These types of diseases may only affect wildlife, lacking the ability to cross species barriers between other wildlife species, livestock, pets, or humans. When spillover occurs to domestic animals and humans, animal agriculture and public health agencies become involved in decision-making and response, thus adding to the complexity of disease management.

Wildlife managers consider several important criteria prior to enacting management actions to control wildlife diseases. These include the ability of the disease to affect the health of humans or their domestic animals, or its ability to reduce or limit wildlife populations. Wildlife diseases that reduce vast numbers of ecologically important wildlife species have the potential to affect the function of entire ecosystems. Wildlife diseases are important to wildlife managers because they can affect populations and abundance by decreasing reproduction and animal survival. Thus, the presence of disease in wildlife populations can lead to not only significant economic impacts through monitoring and control efforts but also loss of hunting opportunity and public trust.

From a wildlife management standpoint, factors most influencing disease in free-ranging wildlife populations include population density, environmental changes, movement of pathogens, land-use changes, interactions among humans and their domestic animals with wildlife, social pressures affecting disease management, wildlife privatization, feeding and baiting, and other highly artificial management activities that greatly enhance risks for disease introduction and establishment. Fischer and Davidson (2005) describe many of the disease issues affecting wildlife as being associated with highly artificial activities. Of the factors listed, the movement of pathogens and their animal hosts may be the activity that spreads disease the quickest and disseminates it the farthest in distance and distribution. There is always risk of moving pathogens anytime animals are moved because some pathogens may not be recognized (Wobeser 2002), or they may be missed during testing depending on the stage of infection and specificity or sensitivity of the test. Movement of disease through animal translocation has been described by many authors (Nielsen and Brown 1988; Davidson and Nettles 1992; Ballou 1993; Griffith et al. 1993; Wobeser 1994; Cunningham 1996; Hess 1996; Corn and Nettles 2001). Corn and Nettles (2001) pointed out that not only can disease affect the receiving population when diseased animals are introduced, but also the reverse can occur where translocations fail as a result of moving naive animals into a diseased population. It is important for managers to understand the disease status of the source and receiving site populations. Davidson and Nettles (1992) used the term "biological package" to describe the animal and its compliment of internal and external parasites, bacteria, viruses, prions, and other potential pathogens.

When moving wildlife from wild or captive environments, there are methodologies to reduce the risk of disease introduction (Corn and Nettles 2001). These include

1. evaluation of health status of the source population;
2. quarantine of translocated animals;
3. physical examination;
4. restriction on translocation from known disease areas; and
5. prophylactic treatment of translocated animals.

Surveillance of important emerging, reemerging, or novel diseases in wildlife provides managers with reasonable, but not unequivocal, assurance that wildlife populations are healthy, and it can act as an early warning system if disease is present. Managers use many methods of surveillance, including serology of antibodies and identification of antigens in blood products, the culture of pathogens, and microscopic survey of histopathological preparations, blood smears, and fecal and urine samples.

However, detection of disease in a wildlife population can be difficult (Thorne et al. 2000). Sample sizes of tested populations provide a level of confidence for detecting a disease but also indicate the amount of error, such that the disease could be missed if it occurs at a low prevalence. Sampling large numbers of animals can be expensive, labor and staff intensive, and time-consuming. Usually animals that have a borderline or weak positive test are not available for retesting unless they are radio-collared or marked for retrieval. Further, many serological tests used in wildlife were developed in domestic counterparts of similar families (Bovidae, Ovidae, etc.) and have not been validated for sensitivity or specificity in wild species. So the test itself may be a poor measure of the presence of disease in a particular species. As a result of these difficulties and the vast expanses wildlife occupy, disease may spread undetected to near-epidemic proportions before managers are able to identify it and develop an appropriate management strategy.

The goal of wildlife disease management is driven by knowledge of the disease pathophysiology, epidemiology, and transmission, which is critical for developing effective strategies. Strategies routinely used by wildlife managers to address wildlife disease issues have been categorized by Wobeser (1994).

Prevention

Prevention should be the first and most critical strategy. It is a strategy designed to ensure that disease is not transported or introduced into wildlife populations. Prevention measures might include banning the import or movement of animals or animal parts or conducting disease surveillance sampling of animals before transport. This can be one of the most difficult strategies to plan for and implement. In the case of CWD, managers may test dead animals to determine prevalence of the disease in a population. However, without an effective and sensitive test on live animals, it would be unknown whether animals being transported have the disease. This reduces the effectiveness of the prevention strategy and is a key reason wildlife managers fear that CWD may be transported between states in the farmed or captive cervid industry. When wild animals are to be relocated, they may be treated prophylactically for pathogens and parasites during captivity, including application of anthelmintics for parasites, long-acting antibiotics for capture injuries, injectable minerals and vitamins for deficiencies and capture myopathy, and vaccinations for clostridial and other potential infections.

Among all strategies, proactive prevention measures are the most effective, efficient, and economical. Managers that focus on prevention to avoid introduction of disease into susceptible populations must aggressively limit the chance that disease can enter a population by using animal movement restrictions, decontamination, and biosecurity measures and may employ physical space, structure, or immunological barriers ranging from fencing and species separation to vaccines. Prevention strategies require an effective communication plan and effort to educate and inform the public and stakeholders about the risks and costs of disease introduction and the measures undertaken to prevent infection of naive wildlife populations. Effective communication requires stakeholder buy-in and public participation with acknowledgement of public wildlife ownership and active disease prevention activities to gain support for regulations and policies to protect the health of wildlife. It can be especially challenging to maintain support and vigilance by stakeholders over a long period when the success of prevention has been rewarded by the lack of disease introduction. Often, without the apparent threat of disease over time, stakeholders become fatigued and less engaged. This produces ongoing communication challenges for agencies that need to keep their constituencies engaged and supportive of the actions required for prevention in the face of limited funding and resources.

Control

Controlling a disease once it is established is the most common strategy practiced by agencies because once a disease is present and being transmitted throughout a free-ranging population, it is very difficult to completely eliminate. Methods for control strategies include population density reduction, banning of feeding and baiting to keep wildlife from congregating unnaturally and spreading disease more rapidly, and habitat improvements and manipulations to maintain animal distribution, health, and vigor. Once a disease is introduced, state wildlife managers must assess risk to the state's wildlife and whether management efforts can be effectively conducted with available funding.

Management efforts in controlling or eradicating an introduced disease require increased staff time, possible staff reassignments, new permanent/temporary staff, and increased expenditure of funding. These are the principle reasons prevention through planned risk reduction is far more prudent and less expensive than reacting to disease introductions.

Management of disease in a wildlife population is complex, with multiple options employed to gain an effective response in either controlling the transmission or spread of the disease or eradication of the disease from the population. Controlling the disease agent or a vector carrying the disease is one type of management that was effective in eradicating screwworm from deer and livestock in the United States and other countries. Population management can also be effective by removing infected or exposed animals. Removal of an entire wildlife population is an extreme measure that can be considered when faced with widespread and rapid expansion of a zoonotic disease. It is very challenging to conduct successfully, however, owing to the difficulty in removing all individuals, the expense of locating and removing animals, and public opposition to depopulation.

Control of disease in a population is more often used, in conjunction with harvest management, to reduce animal density and distribution, thereby decreasing the probability that uninfected animals will come in contact with infected ones. Population control through hunting by increasing opportunity and hunter success with extended seasons and bag limits is the most cost-effective method to reduce population density and is generally considered acceptable by the public. The main limitation with this control method is "hunter harvest fatigue," where hunters, offered unlimited opportunity to remove deer from a defined area, lose interest in high and prolonged harvest. States often follow hunter harvest with the use of agency sharpshooters to further reduce densities. Another selective method of disease control is to capture, test, and slaughter animals testing positive for pathogen antibodies or antigens. This method has been used in bison and elk (*Cervus elaphus*) to test for exposure to the bacterium that causes bovine brucellosis. Test and slaughter can be effective, although it is limited by high costs. Additionally, many serological tests have been validated in selected livestock species, and they may have lower sensitivity and specificity in wildlife species.

Rehabilitative, Vaccination, and Environmental Treatments

More invasive and involved disease control methods have been used to directly treat individual animals in order to save populations. An example is the treatment of California condors (*Gymnogyps californianus*) for lead poisoning. In this instance, condors are captured and brought into a captive facility for extensive rehabilitative treatment. A similar scenario of wildlife rehabilitation used to offset losses from disease may occur during an outbreak of avian botulism in waterfowl.

When wildlife occurs in discreet populations with little emigration, vaccination for pathogens can be an effective treatment for disease. Vaccination is considered an effective tool under these limited and prescriptive conditions, as it can require multiple applications to individual animals, a stable compound that can withstand ambient environmental temperatures, an effective delivery system, and the ability to be administered to a significant proportion of the population, to ensure effectiveness. Vaccination has been successfully applied to wildlife in outbreaks of rabies by limiting spread of the disease to uninfected portions of a population. Prairie dogs (*Cynomys* spp.), a critical food source for black-footed ferrets (*Mustela nigripes*), have been vaccinated for plague (*Yersinia pestis*) with an oral vaccine to protect them from outbreaks. Similarly, elk have been injected remotely with a "bio-bullet" on winter feedgrounds to reduce prevalence of brucellosis in western Wyoming.

Environmental treatments have also been used by managers to decrease incidence of disease caused by contaminants and toxic effects of heavy metals from consumed lead. Manipulation of the environment may include removal of the contaminant, removal of substrate containing the contaminant, or burying the site with additional substrate. Environmental treatments to circumvent disease conditions in wildlife are quite expensive and cannot be applied over large areas of habitat.

Eradication of an established disease is the desired strategy for wildlife managers; however, this outcome is the most difficult to attain. Eradication of disease in free-ranging wildlife has been attempted on numerous occasions with the use of aggressive population reduction, vaccination and drug application, and test and cull management actions. Disease eradication as a strategy has limited success owing to the difficulty in locating and then treating or culling all infected animals. This applies to nearly all wildlife populations that are not closed, and immigration/emigration, along with animal movements within and between diverse habitats, makes complete animal treatment or removal of disease carriers nearly impractical.

No Management

Wobeser (2002) also identifies "no-action" or laissez-faire as a commonly used management action in many situations. A no-action strategy might be considered when feasibility and probability of successfully controlling or eradicating a disease are considered to be low, available funding and staffing are limited, or other priorities and problems are elevated by the management agency.

Disease management at nearly any level is difficult, time-consuming, and expensive and often results in public opposition from some stakeholders in reaction to depopulation or aggressive reduction in wildlife numbers (Fischer and Davidson 2005).

Diseases of Management Significance

There are many known diseases affecting the more than 400 mammalian, 900 avian, 300 amphibian, and 200 reptilian terrestrial wildlife species endemic to North America. However, most are of little significance to wildlife populations. The importance of a disease to a specific population is relative, depending on whether it causes population effects, reduces the health and value of domestic species, or causes economic loss. A wildlife management agency may determine the importance of a disease based on many factors, including the ability of a disease to extirpate populations; its effect on game species or highly visible species; or its effect on species that may interact with, share habitats with, and transmit diseases to livestock or people.

Several notable diseases currently are considered high priority and require aggressive management consideration, staff/department resources, and extensive funding by state wildlife management agencies. They are described below. Each state agency has different priorities, and these diseases may not be considered a high priority everywhere, particularly in states where the disease does not currently occur. Of all the diseases extant in wildlife, management agencies have spent considerable money and staff effort on those affecting free-ranging ungulates. Diseases of particular importance to the hunting and general public include brucellosis, the hemorrhagic diseases, BTB, and CWD in North American deer (*Odocoileus* spp.), elk, and moose (*Alces alces*). Disease prevention and response efforts have drawn significantly on limited state funding resources, in spite of nationally coordinated programs through APHIS, to prevent the spread of, detect, and control or eradicate these diseases. However, CWD in wild cervids may be considered the most expensive disease to manage in US and Canadian wildlife.

Chronic Wasting Disease

CWD became an important disease to wildlife managers as it emerged in free-ranging deer and elk populations on the North American landscape beginning in the early 1980s (Spraker et al. 1997; Miller and Kahn 1999). This transmissible spongiform encephalopathy (TSE) or "prion" disease of North American deer, elk, and moose was first diagnosed in captive research deer at a Colorado wildlife research facility as early as the late 1960s (Williams and Young 1980). It was diagnosed as a TSE through histopathological evaluation of brains from mule deer (*Odocoileus hemionus*) showing clinical signs of neurological disease and a physiological wasting (Miller et al. 2000). Affected elk also were diagnosed at the Colorado research site and at a Wyoming Game and Fish Department research facility (Williams and Young 1982). Although it has not been possible to retrospectively determine whether CWD first occurred in captive or free-ranging animals (Williams et al. 2002), the subsequent movement of captive deer and elk to private cervid farms and zoological collections facilitated the geographic expansion of the disease (Williams and Young 1992) and distrib-

uted infected animals to additional western states and locations in Canada. Exportation of captive elk from Saskatchewan to South Korea marked the first relocation of the disease outside of North America (Williams et al. 2002).

CWD continues to spread to additional North American states, primarily through live animal movement naturally in wild populations and via human-facilitated movement of infected captive cervids. This has led to the current distribution of CWD, spanning from Virginia and West Virginia in the east to Utah in the west, and from Texas in the south to Saskatchewan and Alberta in the north. Currently 24 states and two provinces have some history of CWD infection within their boundaries (CWD Alliance; www.cwd-info.org/).

The disease is important to wildlife managers because it is 100 percent fatal in infected animals. The CWD pathogen causes normal body prions occurring in neurological tissue of the brain to convert to the abnormal form. These prions accumulate in the brain and other tissues and eventually cause neurological disease, emaciation, and death. One complicating factor is a long incubation period (from 16–18 months to five years or longer for some genotypes of deer and elk) between acquiring the infection and showing clinical signs of CWD. The maximal incubation period is unknown. When CWD prions are present in a cervid, they are shed into the environment long before the animal appears sick (Tamgüney et al. 2009). For animals with genotypes that appear to prolong the period before disease is detected, the shedding period of the prions may be prolonged; however, the disease progresses and invariably causes death. With but one experimental exception (Williams et al. 2014), infection by CWD has been fatal to all North American deer, elk, and moose that contracted the disease experimentally, in captive settings or free-ranging environments.

CWD prions are extremely resistant to disinfectants, sunlight, heat, and other materials or processes that typically degrade proteins (Travis and Miller 2003). They can persist in the environment for years and remain infectious to susceptible animals introduced into that environment. Although there are gaps in current knowledge of its epidemiology, CWD is believed to be transmitted by direct contact between infected and susceptible animals and indirectly via consumption of or contact with materials contaminated with prions shed in the urine, saliva, feces, or decomposed carcasses of infected animals (Mathiason et al. 2009).

Research also has shown infectious prions binding to montmorillonite, a type of clay found in soil, and suggests that soil and soil minerals may facilitate CWD infectivity (Williams et al. 2002; Johnson et al. 2006). The length of time that infective prions remain viable in the soil is unknown; however, viability may be years to decades. Environmental contamination by CWD is currently a major field of management-based research affecting the captive cervid industry and free-ranging herds (Miller et al. 2004). Research has shown that certain plants are capable of uptake of the prion from contaminated substrate, indicating the potential for animals to ingest the pathogen through natural food uptake from the environment (Rasmussen et al. 2014). The prolonged incubation period, environmental persistence of CWD prions, and lack of a live animal test make CWD difficult for managers to detect and manage. For example, depopulation of an entire captive or wild herd may not eradicate CWD because of the persistence of prions in the environment. Subsequent reintroduction of susceptible animals can result in new infections.

There is no vaccine, treatment, or medical cure for CWD, and currently there is no practical free-ranging live animal test for the disease. Also, the infectious dose of CWD is not known, so determination of the level or degree of infectiousness cannot be determined. A disease agent that is 100 percent fatal, is difficult to detect in a live animal, is shed over time before clinical signs appear, and is widespread on a continental level is nothing less than a management nightmare.

CWD IN CAPTIVE CERVIDS

Live animal movement is the most frequently documented method for spread of CWD. Examples of this in wild animal movements include the spread of CWD from New Mexico to Texas, from West Virginia to Virginia, from Wisconsin to Iowa, and others. There also are many documented and circumstantial examples of CWD spread via live captive animal movement, including (1) the spread of CWD to 38 captive elk herds in Saskatchewan that received elk directly or indirectly

from a single infected herd, and (2) the spread of CWD to two captive elk herds in Colorado and one in Kansas when elk from a single infected facility in Colorado were shipped to 19 states and more than 40 other captive facilities within Colorado. Elk with CWD were shipped from Canada to South Korea in 2001, causing major international animal import trade concerns from the resulting epidemiological investigation. The disease reoccurred in a captive animal in the affected Korean area in 2004. As of November 2016, CWD has been found in 77 captive herds in 16 states.

To combat and control the disease in the captive cervid industry, CWD herd certification is required in order to ship animals across state lines, according to APHIS's National Herd Certification Program (HCP), which was implemented in 2012 (Code of Federal Register: 9 CFR 55 and 812014). Prior to implementation, states regulated the movement of captive cervids. The national HCP certifies herds in acceptable state CWD programs as being at low risk for having CWD. However, there is no "CWD-free" certification of captive cervid herds. The national HCP states that individual states may implement regulations more stringent than the national HCP and that their regulations will preempt the federal requirements, with one exception: states must allow transit of captive cervids through the state, even if they do not allow captive cervid farming in the state. From 2002 to 2012, federal funding was available to states for surveillance, monitoring, and management of CWD in wild and captive cervids and to the captive cervid industry for indemnity payments to owners if their herds became infected and required depopulation. This funding is no longer available, and the economic burden now falls solely on states.

Since the implementation of the HCP in 2012, CWD has been detected in additional captive cervid herds in the United States. Some of the herds had been certified, with animals being subsequently shipped across state borders. According to information provided by officials in affected states, all certified herds had been monitored for more than the five years required by the HCP before CWD was detected. Extensive animal movement in the captive cervid industry can make the epidemiological investigations challenging. Trace-back and trace-forward animal movements from infected or exposed herds often leave managers with major gaps in their epidemiological investigation. The buying and selling of animals and their movement through a facility can be frequent and result in high throughput.

To compound the threat of CWD entering wild cervid populations, escapes of animals from captive cervid facilities are not uncommon. An audit in Wisconsin in 2003 found that 432 deer that escaped between 2000 and 2002 remained unrecovered. Many escapes occurred because a gate was left open. In 2002, a previously captive Wisconsin deer was killed and tested positive for CWD six months after it had escaped from a captive facility that was known to be affected by CWD. This occurred again in 2015 when two animals from an affected Wisconsin facility tested positive for CWD months after their escape.

To date, CWD has not been detected in animals shipped across state lines from certified herds, although animals were shipped across state lines from certified herds that later were found to have CWD. And historically, CWD has been shipped unknowingly in infected, nonclinical animals from noncertified herds within a state, across state borders, and internationally. However, the point is that herd certification, whether it is regulated by the states or the USDA, does not mitigate all risks. There is concern that the HCP could create a false sense of security that CWD cannot be spread through movement of live animals from certified herds. Until science, veterinary medicine, and wildlife management progress beyond current levels of understanding and technology, the most prudent management action is to prohibit movement of live susceptible animals, a basic preventative disease principle. Without an effective and accurate live-animal test with high sensitivity and specificity across all age classes and conducted as a pre-shipment requirement, interstate movement of CWD in captive and wild cervids could continue to occur.

This issue is not without other problems and challenges. One particular challenge for state wildlife agencies is the movement of regulatory authority over captive cervids from wildlife management to animal agriculture agencies (see chap. 2 for a case study concerning the state of Vermont). Unlike state wildlife management agencies, the animal agriculture agencies typically do not have responsibility for the conservation of publicly owned, free-ranging wildlife.

CWD IN WILD CERVIDS

Aggressive management, including sustained population reduction, of wild cervids by hunters and sharpshooters in disease hot spots has maintained low prevalence and slowed the spread of CWD in some situations. This strategy appears to have eliminated CWD in Minnesota and New York when the disease occurred in a focal area and it was detected before large numbers of animals became affected. However, in regions or states where the disease is established, extensive funding and resources can be expended with varying success in prevalence reduction. This was attempted temporarily in Wisconsin at the cost of tens of millions of dollars, with the disease continuing to expand in distribution and prevalence.

Cost to states involving management and monitoring can be extreme. Wisconsin's Chronic Wasting Disease Response Plan: 2010–2025 (Wisconsin Department of Natural Resources) indicates that hunter surveys suggest that if prevalence of CWD is increased to 50 percent, nearly half would stop hunting. This revenue loss would be additive to expenditures used to manage/eradicate the disease. Costs often include indemnity payments to owners of affected captive herds, clean-up funds, surveillance and monitoring, hired sharpshooters, testing laboratories, field samplers, and loss of revenue from hunters and wildlife viewers related to direct (fees, licenses, etc.) and indirect (hotels, gas, restaurant) expenditures by consumptive and nonconsumptive users of the wildlife resource.

Actions designed to prevent disease in wildlife have been shown to be fiscally prudent and forward thinking as an investment by state agencies. We have described above that avoiding active movement of CWD over state borders begins with restrictive movement of potentially infectious tissues or animals. As with every important disease in wildlife, prevention is the only effective method to manage CWD in free-ranging wildlife, and, to date, no demonstrated agency action has been shown to effectively manage or remove CWD after it has become established in the wild. The increasing spread and infection rate of this disease have raised genuine concerns about long-term viability of affected wild cervid populations among wildlife managers and the citizens who hunt or otherwise appreciate wild deer, elk, and moose.

Hemorrhagic Disease

Many of the diseases of greatest concern to wildlife management agencies are those that affect deer. North American deer die-offs have been documented since the late 1800s (Nettles and Stallknecht 1992). Blue-tongue virus (BTV) and epizootic hemorrhagic disease virus (EHDV), closely related viruses in the Orbivirus family, are diagnosed in annual and seasonal North American deer die-offs (Nettles et al. 1992). Serotypes of these viruses occur worldwide, with 15 of 24 BTV serotypes and 3 of 10 EHDV serotypes currently occurring in the United States. Worldwide, hemorrhagic disease in free-ranging wildlife has been documented only in the United States and Canada, extending from the southeastern to the northwestern United States, and found sporadically in Alberta, British Columbia, and Saskatchewan. Bluetongue principally is a disease of domestic livestock, whereas EHD is considered a disease of wild ungulates, particularly white-tailed deer.

EHDV and BTV outbreaks generally occur in white-tailed deer, although mule deer and pronghorn (*Antilocapra americana*) have shown clinical signs and mortalities. In spite of the established occurrence of EHDV in the United States, long-term population declines of white-tailed deer have not occurred. Clinically, EHDV and BTV present identically in affected white-tailed deer, with the noticeable development of illness at about seven days postexposure. The disease generally has a rapid sudden onset, and progression is due to the virus's ability to damage the cells lining blood vessels, rendering them leaky or prone to clotting.

EHDV and BTV are transmitted by *Culicoides* spp. biting midges, not directly from one deer to another. Die-offs may involve large numbers of deer over a broad area, as infrequent, sporadic events, particularly in the more northern areas where hemorrhagic disease occurs. In contrast, deer may be minimally affected, if at all, in the more southern range of the disease where animals may be exposed to EHDV and/or BTV on an annual basis. Outbreaks of EHDV and BTV occur during late summer and early fall (August–October) and dissipate when cold weather reduces vector activity. Other wild ungulates and domestic livestock can be exposed and become infected, although they rarely are affected clinically. However, sheep may develop severe disease when infected with BTV. Humans, other species of wildlife, and non-ungulate pets are not susceptible to orbiviral hemorrhagic disease.

Effective treatment or control of the insect vector transmitting hemorrhagic disease has not been developed to date. There are vaccines for use in captive cervids that have shown highly variable effectiveness; however, population-level vaccination of free-ranging deer is not considered practical by wildlife managers.

Hemorrhagic disease currently is considered one of the most important viral diseases of white-tailed deer in the United States and can significantly reduce local deer numbers in a single season in populations with high mortality. Since many of the larger die-offs do not recur annually, deer numbers often rebound. Periods of drought can exasperate an outbreak by congregating deer near water sources, such as ponds and creek beds, which also serve as breeding habitat for the insect vector. There are no proven management actions that wildlife agencies can currently incorporate to alleviate the localized effects of this disease on deer populations.

Bovine Brucellosis

Bovine brucellosis is an important zoonotic disease occurring in wildlife and currently is a management issue in free-ranging elk and bison in the Greater Yellowstone Ecosystem (GYE). There are several species of *Brucella* bacteria; however, *Brucella abortus*, the species affecting cattle, bison, and elk, likely was introduced into wildlife in the GYE as early as 1917, when infected bison were brought to Yellowstone National Park from a brucellosis-infected cattle ranch (Mohler 1917; Meagher and Meyer 1994; Cheville et al. 1998). Subsequently, elk became infected and carriers of the disease around 1930 (Murie 1951), with the disease currently well established in bison and elk in GYE portions of Idaho, Montana, and Wyoming (Tunnicliff and Marsh 1935; Thorne et al. 1997; Aune et al. 2002; Etter and Drew 2006).

Brucellosis is an important disease because it is a reportable livestock disease and infection of cattle herds can cause the loss of a state's Brucellosis Class Free cattle status, which restricts animal movements, commerce, and exports of cattle and cattle products and increases herd owner expenses through additional testing and other management actions. Historically, affected cattle herds were depopulated, although other management actions, such as test and slaughter, currently may be considered.

When elk or bison are infected with brucellosis, many clinical manifestations can occur, including abortions, retained placentas, male reproductive tract lesions, arthritis, and bursitis (Rhyan 2013). Infection in female elk and bison generally causes abortion of the first calf, although some will also abort subsequent calves. Calves surviving to term may also be delivered stillborn or may fail to thrive if born alive (Moore 1947; Thorne et al. 1978a, 1978b; Rhyan et al. 1994, 2001; Meyer and Meagher 1995; Tessaro and Forbes 2004).

Transmission of the bacterium usually occurs orally when animals lick newborn calves or aborted fetuses or consume infected placentas or other fetal tissues (Thorne 2001). Persistent environmental contamination of the landscape through infected tissues can occur under moist and cool conditions for up to 81 days as determined by factors such as month, temperature, and exposure to sunlight (Aune et al. 2012). In highly contaminated areas on elk winter feeding grounds, animals are congregated during the peak of *Brucella* transmission from February through June, when abortions of infected fetuses are likely to occur (Roffe et al. 2004; Cross et al. 2007). Wyoming's 23 elk feedgrounds serve to separate elk from cattle and reduce the risk of interspecies disease transmission (Smith 2001), but this separation also facilitates maintenance of *Brucella* in elk populations (Thorne 2001). Bison typically herd together as part of their natural behavior, which also increases the likelihood of intraspecies disease transmission (Cheville et al. 1998).

Antibiotic treatments are not conducted in animals, and particularly not in free-ranging wildlife, owing to the need for repeated applications over time. In livestock, brucellosis is prevented by vaccination, and animals that react positively on testing are removed from the herd. Vaccination of elk has occurred on 22 of 23 of Wyoming's feedgrounds since 1985, with over 80,000 elk vaccinated (Scurlock and Edwards 2010). However, vaccination has not been shown to cause consistent declines in disease prevalence.

Brucellosis surveillance and monitoring are conducted by states surrounding the GYE and include testing of blood from elk on feedgrounds, from hunter-harvested animals, and from animals that have been captured for other projects. Because of the economic impact on the cattle industry, state and federal partners have worked cooperatively over the past 70 years to enact an eradication program. The brucellosis eradication program has greatly reduced disease incidence among domestic cattle; however, eliminating the disease in free-ranging bison and elk has been challenging and is not currently attainable with the diagnostic and treatment tools available or within the sociopolitical climate of diverse stakeholder interests in the GYE. Unfortunately, the persistence of brucellosis in wildlife in the GYE continues to result in occasional spillover to cattle herds in the area (Gillin et al. 2002; Kilpatrick et al. 2009).

An aggressive feedground vaccination program, coupled with extensive research and habitat management, has been practiced by multiple state wildlife and land management agencies, universities, and private nongovernment organizations and research institutions. In Wyoming, a Governor's Brucellosis Coordination Team has directed development of brucellosis management action plans for select feedground elk herd units and several bison herds. Plans are designed to reduce disease prevalence in wildlife and risk of transmission to livestock by relocating or closing feedgrounds when appropriate, reducing elk numbers, adjusting cattle operations, installing fences, testing and removing elk that react positively for brucellosis, vaccinating livestock and wildlife, enhancing elk habitat, and purchasing conservation easements and winter habitat.

Brucellosis has been on the North American landscape for more than a century and will require a cooperative multiagency approach to address complex issues and reach long-term management solutions. Managers of livestock health and state wildlife populations will need to continue to resolve issues surrounding elk feedgrounds by reducing the number of elk fed and strive to maintain separation of livestock and wildlife during the period of brucellosis transmission (Scurlock and Edwards 2010).

Bovine Tuberculosis

BTB is another zoonotic, contagious, and reportable livestock disease occurring in North American wildlife. The disease occurs in many parts of the world and historically was common in US cattle but was rarely observed in wildlife (fewer than 10 wild white-tailed or mule deer and several coyotes had been diagnosed with BTB; Rhyan et al. 1995; Schmitt et al. 1997). However, in 1994, a hunter-killed white-tailed deer was diagnosed with BTB in the northeastern portion of Michigan's Lower Peninsula, and this disease is now recognized as endemic in wild deer in this area. Since then, BTB has also been diagnosed in 66 cattle herds, four feedlots, four captive cervid herds, and in Michigan's wild elk and furbearers, including black bear, bobcat, coyote, opossum, raccoon, and red fox.

TB is a bacterial disease of the respiratory system. The tuberculosis complex includes human TB caused by *Mycobacterium tuberculosis*, BTB caused by *M. bovis*, and other forms of the disease. BTB is a contagious cattle disease capable of infecting many species of livestock and wildlife, as well as humans. Similar to brucellosis, BTB in cattle is important because infection of cattle herds can cause the loss of a state's Tuberculosis Accredited Free cattle status and restrict animal movements, commerce, and exports of cattle and cattle products. As with brucellosis, depopulation of affected herds was practiced historically, but other control methods currently may be employed.

BTB is transmitted primarily by respiratory secretions from infected to uninfected animals by coughing and sneezing or by ingestion of contaminated feed or other materials. Transmission usually occurs when animals congregate in close proximity. Increasing the density of animals per unit of habitat also increases the risk of disease transmission between animals. Survival of the bacterium in the environment primarily is affected by exposure to sunlight and temperature.

Development of BTB in wildlife is a chronic process often requiring years from initial infection to the first clinical sign of the disease. Once infected, many animals have shown the potential to act as reservoirs of infection for other wildlife species (de Lisle et al. 2002). White-tailed deer serve as maintenance hosts for BTB, facilitating the persistence of infection in wildlife and enabling the horizontal transmission of the pathogen between wildlife species. In deer, lymph nodes of the head are often enlarged and infected initially, and then disease progresses to the respiratory system and other organs. Most infected white-tailed deer with detectable BTB infection appear healthy. The disease in wild and domestic ungulates is chronic and progressive, causing pathological changes leading to debilitation, emaciation, poor body condition, and severe respiratory disease.

The tools available to state wildlife managers to effectively manage this chronic and debilitating disease are limited. Vaccine development and treatment options have yet to provide a viable alternative for mitigating the effects of BTB in free-ranging deer. However, in Michigan, the only state where the disease is persistent in wild deer populations, BTB surveillance and management activities have reduced prevalence and limited spread of the disease to new areas and populations. Eliminating unnatural deer congregation practices, such as feeding and baiting deer, reduces focal sites of transmission by allowing deer to disburse and distribute throughout available habitat for natural forage.

For the past 20 years, the Michigan Department of Natural Resources has attempted to eradicate BTB with aggressive hunter harvest (40% annual harvest of antlered deer and 16% of antlerless) and an attempt to ban deer baiting and feeding in core areas of infection, with vaccination trials added in recent years (Ramsey et al. 2014). This has all come at great expense, with a state expenditure of more than $150 million to date (O'Brien et al. 2011). Current models indicate that state management strategies focused on eradication likely are not practical. Aside from it being very difficult to find and remove affected deer in free-ranging populations, public support for continued high hunter harvest is often limited to the initial response period and can be unpalatable to public opinion over the prolonged period that might be needed to control, manage, or eradicate disease.

Similar issues with long-term public support of banning wildlife baiting and feeding also have developed (O'Brien et al. 2006, 2011; Rudolph et al. 2006) because these practices remain popular in spite of their negative impacts on wildlife (Brown and Cooper 2006).

Many areas where baiting was banned in 2008 saw a removal of the restriction by 2011 (Michigan Natural Resources Commission), with BTB-positive deer continuing to be found in areas where baiting previously was banned. Results from modeling predictions (Ramsey et al. 2014) indicate that BTB could be eradicated from Michigan deer if substantially increased control strategies could be sustained over a decade. Owing to the difficulty of attaining long-term public support, the authors suggested that more realistic goals might include controlling the disease spread and transmission to cattle.

Wildlife managers have no shortage of other important disease issues with regard to free-ranging wild mammals. Several high-profile and established diseases, like rabies and distemper viruses in wild canid and mustelid species, will circulate in North American wildlife for the foreseeable future (Gillin and Hunter 2010). Parasitic infestations of winter tick and meningeal worm are diseases affecting regional populations of moose. In spite of continued, reduced reproduction and survival of moose populations, a workable management option has not been identified.

In these cases and many others, chronic and rapidly emerging diseases in wildlife cross state, federal, and tribal land boundaries, requiring cooperation between multiple agencies and stakeholders. Current examples of this type of cooperation include management of the 2014–2015 incursion of HPAI into North America. The disease is considered a reportable foreign animal disease affecting domestic poultry and falls under the management purview of APHIS. However, wild birds are well-known reservoirs of avian influenza viruses of low pathogenicity, so efforts were made by state and federal natural resource agencies to determine whether wild birds, particularly waterfowl, were carrying HPAI viruses. With backyard and commercial domestic poultry operations holding highest economic and health risks, state agriculture departments were also at the forefront of cooperative planning. And stakeholders, from poultry owners and producers to hunters, falconers, wildlife rehabilitators, and the general public, were impacted by the reduction of poultry products and associated price increases.

Additional cooperative endeavors and coordinated responses have been implemented with more obscure but very important diseases affecting nongame species. WNS, caused by the fungus *Pseudogymnoascus destructans*, is currently spreading west, devastating millions of bats in more than 25 states and five Canadian provinces. The disease response by the USFWS is a coordinated effort with states that manage nonthreatened or endangered bat species and USFWS-managed listed bat species. Impacts of WNS can occur in resources ranging from managed caves on federal and state lands and parks to privately owned caves.

The Future of Fish and Wildlife Health

Michigan Department of Conservation's foresight in implementing the first dedicated wildlife health program in 1933 set the groundwork for the future of wildlife health and disease investigations. Wildlife health laboratories—including state programs, cooperatives such as the SCWDS (established in 1957), and the US Geological Survey (USGS) National Wildlife Health Center (established in 1975)—have elevated management and research capabilities and understanding of wildlife disease.

In 2008, the Fish and Wildlife Health Committee of the Association of Fish and Wildlife Agencies developed a National Fish and Wildlife Health Initiative toolkit for state fish and wildlife directors to help them assess their agency's fish and wildlife health preparedness. Goals of the initiative were to (1) facilitate establishment and enhancement of state, federal, and territorial fish and wildlife agency capability to effectively address health issues involving free-ranging fish and wildlife and (2) minimize negative impacts of health issues affecting free-ranging fish and wildlife through surveillance, management, and research. The initiative was developed by a National Fish and Wildlife Health Steering Committee composed of state directors and representatives from academia, the USDA, the USGS, the USFWS, and Native American fish and wildlife organizations. The resulting toolkit provided information to assist state directors in managing diseases in fish and wildlife and federal health policy as it relates to wildlife.

This document provided examples of position descriptions for fish and wildlife professionals and state wildlife health programs to guide states desiring to develop or enhance health programs to better address fish and wildlife disease issues. Disease outbreaks are expensive and drain agency resources in efforts to find an effective response to the disease. Although most state directors understand the need to balance resources while planning ahead for potential outbreaks, well-trained health staff, as well as understanding and trust among management authorities, are essential to effectively address fish and wildlife health management issues. This critical balance highlights the challenge of having the most experienced and competent decision-makers leading wildlife management agencies.

Siemer et al. (2013) conducted a human dimensions study to identify areas of capacity and key issues needed for development and implementation of a National Fish and Wildlife Health Plan. The study included 164 professional respondents from state fish and wildlife agencies (67%), federal agencies (22%), universities (5%), and nongovernmental organizations (4%) from 49 states and the District of Columbia. Respondents listed several factors as important, including limited agency capacity, funding, agency legitimation, and staffing. They also felt that priorities for action were (1) finding sources of additional funding and staffing, (2) fostering leadership, and (3) promoting interagency coordination. Finally, respondents felt that influencing agency leaders' views on the importance of disease management was a core component of fish and wildlife management programs and an important action priority.

This study pointed out the importance of wildlife health in management programs. The authors' conclusions included a series of actions that state fish and wildlife agencies could incorporate to enhance capabilities for early detection and coordinating their response to disease threats and incursions. These recommendations included (1) strengthening interagency relationships to provide effective coordinated responses through communication networks and interagency agreements; (2) acquiring resources and diverse funding for administering fish and wildlife health programs; (3) developing key components for a wildlife health program, from staff training and networking to incorporating strategic plans and ensuring adequate funding of existing and new programs; and (4) cultivating and maintaining public trust in state fish

and wildlife health programs through open communication with stakeholder groups. State wildlife agencies continue to develop programs that meet challenges of melding a free-ranging wildlife resource with the dynamic uses of the resource by the public and impacts caused by changing landscapes and habitats. State agencies will increasingly be challenged to provide staff and resources within an established program to address wildlife health and disease investigations in nearly all facets of wildlife management and conservation.

State wildlife management agencies have slowly and steadily increased their capacity to address wildlife diseases, generally in reaction to the incursion or nearby presence of an important disease. It is conceivable that at some point in the future all 50 states will maintain a state-managed wildlife health program. However, regional diagnostic and pathology programs will likely continue to be employed among southeastern and midwestern state agencies and to provide a sharing of collective expertise when dealing with wildlife diseases. And there will continue to be wildlife disease and policy challenges presented to agency personnel as new diseases continue to emerge. As part of the mission of a wildlife health program, veterinarians and wildlife health specialists provide an effective mechanism for specific technical support of all the agency programs. Demonstrated in the many wildlife health programs currently incorporated by state agencies, this combination of support and service has proven critical for field biologists, policy makers, and administrative staff in providing a comprehensively managed wildlife resource as part of the state's duty in expanding the public trust doctrine within the North American Model of Wildlife Conservation.

LITERATURE CITED

Aune, K., K. Alt, and T. Lemke. 2002. Managing wildlife habitat to control brucellosis in the Montana portion of the Greater Yellowstone Area. Pages 109–118 *in* T. J. Kreeger, editor. Brucellosis in elk and bison in the Greater Yellowstone Area. Wyoming Game and Fish Department, Cheyenne, Wyoming, USA.

Aune, K., J. C. Rhyan, R. Russell, T. J. Roffe, and B. Corso. 2012. Environmental persistence of *Brucella abortus* in the Greater Yellowstone Area. Journal of Wildlife Management 76:253–261.

Ballou, J. D. 1993. Assessing the risks of infectious diseases in captive breeding and reintroduction programs. Journal of Zoo and Wildlife Medicine 327–335.

Baughman, J., and J. R. Fischer. 2005. Programs for monitoring and managing diseases in free-ranging wildlife in the 21st century. Transactions of the North American Wildlife and Natural Resources Conference 70:346.

Brown, R. D., and S. M. Cooper. 2006. The nutritional, ecological, and ethical arguments against baiting and feeding white-tailed deer. Wildlife Society Bulletin 34(2):519–524.

Brown, R. D., and L. H. Wurman. 2009. A brief history of wildlife conservation and research in North America. Fair Chase Magazine. Boone and Crockett Club, Missoula, Montana, USA.

Charles, E. 1931. The study of epidemic diseases among wild animals. Journal of Hygiene 31:435–456. doi:10.1017/S0022172400017642.

Cheville, N. F., D. R. McCullough, and L. R. Paulson. 1998. Brucellosis in the Greater Yellowstone Area. National Academies Press, Washington, DC, USA.

Corn, J. L., and V. F. Nettles. 2001. Health protocol for translocation of free-ranging elk. Journal of Wildlife Diseases 37:413–426.

Cross, P. C., W. H. Edwards, B. M. Scurlock, E. J. Maichak, and J. D. Rogerson. 2007. Effects of management and climate on elk brucellosis in the Greater Yellowstone Ecosystem. Ecological Applications 17:957–964.

Cunningham, A. A. 1996. Disease risks of wildlife translocations. Conservation Biology 349–353.

Davidson, W. R., and V. F. Nettles. 1992. Relocation of wildlife: Identifying and evaluating disease risks. Transactions of the North American Wildlife and Natural Resources Conference 57:466–473.

de Lisle, G. W., R. G. Bengis, S. M. Schmitt, and D. J. O'Brien. 2002. Tuberculosis in free-ranging wildlife: Detection, diagnosis and management. Revue Scientifique et Technique de l'Office International des Epizooties 21:317–334.

Etter, R. P., and M. L. Drew. 2006. Brucellosis in elk of eastern Idaho. Journal of Wildlife Diseases 42:271–278.

Fay, L. D., A. P. Boyce, and W. G. Youatt. 1956. An epizootic in deer in Michigan. Transactions of the North American Wildlife Conference 21:173–184.

Fischer, J. R., and W. R. Davidson. 2005. Reducing risk factors for disease problems involving wildlife. Transactions of the North American Wildlife and Natural Resources Conference 70:289–309.

Gillin, C. M., and D. Hunter. 2010. Disease and translocation issues of gray wolves. Chapter 10 in R. P. Reading, B. Miller, A. L. Masching, R. Edward, and M. K. Phillips, editors. Awakening spirits: Wolves in the southern Rockies. Fulcrum, Golden, Colorado, USA.

Gillin, C. M., G. M. Tabor, and A. A. Aguirre. 2002. Ecological health and wildlife disease in National Parks. Chapter 19 in A. Aguirre, R. Ostfeld, G. Tabor, M. Pearl, and C. House, editors. Conservation medicine: Ecological health in practice. Oxford University Press, Oxford, UK.

Griffith, B. J., M. Scott, J. W. Carpenter, and C. Reed. 1993. Animal translocations and potential disease transmission. Journal of Zoo and Wildlife Medicine 24:231–236.

Hess, G. 1996. Disease in metapopulation models: Implications for conservation. Ecology 1617–1632.

Johnson, C. J., K. E. Phillips, P. T. Schramm, D. McKenzie, J. M. Aiken, and J. A. Pedersen. 2006. Prions adhere to soil minerals and remain infectious. PLoS Pathog 2(4):e32. doi:10.1371/journal.ppat.0020032.

Kilpatrick, A. M., C. M. Gillin, and P. Daszak. 2009. Wildlife–livestock conflict: The risk of pathogen transmission from bison to cattle outside Yellowstone National Park. Journal of Applied Ecology 46:476–485.

Mathiason, C. K., S. A. Hays, J. Powers, J. Hayes-Klug, J. Langenberg, and S. J. Dahmes. 2009. Infectious prions in preclinical deer and transmission of chronic wasting disease solely by environmental exposure. PLoS One 4(6):e5916. doi:10.1371/journal.pone.0005916.

Meagher, M., and M. E. Meyer. 1994. On the origin of brucellosis in bison of Yellowstone National Park: A review. Conservation Biology 8:645–653.

Meyer, M. E., and M. Meagher. 1995. Brucellosis in free-ranging bison (Bison bison) in Yellowstone, Grand Teton, and Wood Buffalo National Parks: A review. Journal of Wildlife Diseases 31:579–598.

Miller, M. W., and R. Kahn. 1999. Chronic wasting disease in Colorado deer and elk: Recommendations for statewide monitoring and experimental management planning. Colorado Division of Wildlife, Denver, Colorado, USA.

Miller, M. W., E. S. Williams, N. T. Hobbs, and L. L. Wolfe. 2004. Environmental sources of prion transmission in mule deer. Emerging Infectious Diseases 10:1003–1006.

Miller, M. W., E. S. Williams, C. W. McCarty, T. R. Spraker, T. J. Kreeger, C. T. Larsen, and E. T. Thorne. 2000. Epizootiology of chronic wasting disease in free-ranging cervids in Colorado and Wyoming. Journal of Wildlife Diseases 36:676–690.

Mohler, J. R. 1917. Abortion disease. Annual report of the Bureau of Animal Industry. Bureau of Animal Industry, Washington, DC, USA.

Moore, T. 1947. Brucella infection of buffalo and elk. Canadian Journal of Comparative Medicine and Veterinary Science 11:131.

Murie, O. J. 1951. The elk of North America. Teton Bookshop, Jackson, Wyoming, USA.

Nettles, V. F., and W. R. Davidson. 1996. Cooperative state action to address research needs—the experience of the Southeastern Cooperative Wildlife Disease Study. Transactions of the North American Wildlife and Natural Resources Conference 61:545–552.

Nettles, V. F., S. A. Hylton, D. E. Stallknecht, and W. R. Davidson. 1992. Epidemiology of epizootic hemorrhagic disease viruses in wildlife in the USA. Pages 238–248 *in* T. E. Walton and B. I. Osburn, editors. Bluetongue, African horse sickness, and related orbiviruses. CRC Press, Boca Raton, Florida, USA.

Nettles, V. F., and D. E. Stallknecht. 1992. History and progress in the study of hemorrhagic disease of deer. Transactions of the North American Wildlife and Natural Resources Conference 57:499–516.

Nielsen, L., and R. D. Brown, editors. 1988. Translocation of wild animals. Wisconsin Humane Society, Milwaukee, Wisconsin, USA.

O'Brien, D. J., S. M. Schmitt, S. D. Fitzgerald, and D. E. Berry. 2011. Management of bovine tuberculosis in Michigan wildlife: Current status and near term prospects. Veterinary Microbiology 151:179–187.

O'Brien, D. J., S. M. Schmitt, S. D. Fitzgerald, D. E. Berry, and G. J. Hickling. 2006. Managing the wildlife reservoir of *Mycobacterium bovis*. Michigan 112(2–4):313–323.

Ramsey, D. S., D. J. O'Brien, M. K. Cosgrove, B. A. Rudolph, A. B. Locher, and S. M. Schmitt. 2014. Forecasting eradication of bovine tuberculosis in Michigan white-tailed deer. Journal of Wildlife Management 78:240–254.

Rasmussen, J., B. H. Gilroyed, T. Reuter, S. Dudas, N. F. Neumann, A. Balachandran, N. N. V. Kav, C. Graham, S. Czub, and T. A. McAllister. 2014. Can plants serve as a vector for prions causing chronic wasting disease? Prion 8(1):136–142.

Rhyan, J. C. 2013. Pathogenesis and pathobiology of brucellosis in wildlife. Revue Scientifique et Technique (International Office of Epizootics) 32(1):127–136.

Rhyan, J. C., K. Aune, B. Hood, R. Clarke, J. Payeur, J. Jarnagin, and L. Stackhouse. 1995. Bovine tuberculosis in a free-ranging mule deer (*Odocoileus hemionus*) from Montana. Journal of Wildlife Diseases 31(3):432–435.

Rhyan, J. C., T. Gidlewski, T. J. Roffe, K. Aune, L. M. Philo, and D. R. Ewalt. 2001. Pathology of brucellosis in bison from Yellowstone National Park. Journal of Wildlife Diseases 37:101–109.

Rhyan, J. C., W. J. Quinn, L. S. Stackhouse, J. J. Henderson, D. R. Ewalt, J. B. Payeur, M. Johnson, and M. Meagher. 1994. Abortion caused by *Brucella abortus* Biovar 1 in a free-ranging bison (*Bison bison*) from Yellowstone National Park. Journal of Wildlife Diseases 30:445–446.

Roffe, T. J., L. C. Jones, K. Coffin, M. L. Drew, S. J. Sweeney, S. D. Hagius, and D. Davis. 2004. Efficacy of single calfhood vaccination of elk with *Brucella abortus* Strain 19. Journal of Wildlife Management 68:830–836.

Rudolph, B. A., S. J. Riley, G. J. Hickling, B. J. Frawley, M. S. Garner, and S. R. Winterstein. 2006. Regulating hunter baiting for white-tailed deer in Michigan: Biological and social considerations. Wildlife Society Bulletin 34:314–321.

Schmitt, S. M., S. D. Fitzgerald, T. M. Cooley, C. S. Bruning-Fann, L. Sullivan, D. Berry, and J. Sikarskie. 1997. Bo-

vine tuberculosis in free-ranging white-tailed deer from Michigan. Journal of Wildlife Diseases 33:749–758.

Scurlock, B. M., and W. H. Edwards. 2010. Status of brucellosis in free-ranging elk and bison in Wyoming. Journal of Wildlife Diseases 46:442–449.

Siemer, W. F., T. B. Lauber, D. J. Decker, and S. J. Riley. 2013. Agency capacities to detect and respond to disease threats: Professionals' views on limiting factors and action priorities. Human Dimensions Research Unit Series Publication 13-5. Department of Natural Resources, Cornell University, Ithaca, New York, USA.

Smith, B. L. 2001. Winter feeding of elk in western North America. Journal of Wildlife Management 65(2):173–190.

Spraker, T. R., M. W. Miller, E. S. Williams, D. M. Getzy, W. J. Adrian, G. G. Schoonveld, and P. A. Merz. 1997. Spongiform encephalopathy in free-ranging mule deer (*Odocoileus hemionus*), white-tailed deer (*Odocoileus virginianus*) and Rocky Mountain elk (*Cervus elaphus nelsoni*) in northcentral Colorado. Journal of Wildlife Diseases 33:1–6.

Tamgüney, G., M. W. Miller, L. L. Wolfe, T. M. Sirochman, D. V. Glidden, C. Palmer, A. Lemus, S. J. DeArmond, and S. B. Prusiner. 2009. Asymptomatic deer excrete infectious prions in faeces. Nature 461:529–532.

Tessaro, S. V., and L. B. Forbes. 2004. Experimental *Brucella abortus* infection in wolves. Journal of Wildlife Diseases 40:60–65.

Thorne, E. T. 2001. Brucellosis. Pages 372–395 *in* E. S. Williams and I. K. Barker, editors. Infectious diseases of wild mammals. Iowa State University Press, Ames, Iowa, USA.

Thorne, E. T., R. A. Humphries, D. J. O'Brien, and S. M. Schmitt. 2005. State wildlife management agency responsibility for managing diseases in free-ranging wildlife. Transactions of the North American Wildlife and Natural Resources Conference 70:3.

Thorne, E. T., N. Kingston, W. R. Jolley, and R. C. Bergstrom. 1982. Diseases of wildlife in Wyoming, 2nd edition. Wyoming Game and Fish Department, Cheyenne, Wyoming, USA.

Thorne, E. T., M. W. Miller, S. M. Schmitt, T. J. Kreeger, and E. S. Williams. 2000. Conflicts of authority and strategies to address wildlife diseases. Pages 123–137 *in* Proceedings of the 104th Annual Meeting of the United States Animal Health Association. Birmingham, Alabama, USA.

Thorne, E. T., J. K. Morton, F. M. Blunt, and H. A. Dawson. 1978a. Brucellosis in elk. II. Clinical effects and means of transmission as determined through artificial infections. Journal of Wildlife Diseases 14:280–291.

Thorne, E. T., J. K. Morton, and G. M. Thomas. 1978b. Brucellosis in elk. I. Serologic and bacteriologic survey in Wyoming. Journal of Wildlife Diseases 14:74–81.

Thorne, E. T., S. G. Smith, K. Aune, D. Hunter, and T. J. Roffe. 1997. Brucellosis: The disease in elk. Pages 33–44 in E. T. Thorne, M. S. Boyce, P. Nicolleti, and T. J. Kreeger, editors.

Brucellosis, bison, elk, and cattle in the greater Yellowstone area: Defining the problem, exploring solutions. Wyoming Game and Fish Department, Cheyenne, Wyoming, USA.

Thornton, R. 1987. American Indian holocaust and survival: A population history since 1492. University of Oklahoma Press, Norman, Oklahoma, USA.

Travis, D., and M. Miller. 2003. A short review of transmissible spongiform encephalopathies, and guidelines for managing risks associated with chronic wasting disease in captive cervids in zoos. Journal of Zoo and Wildlife Medicine 34(2):125–133.

Tunnicliff, E. A., and H. Marsh. 1935. Bang's disease in bison and elk in the Yellowstone National Park and on the National Bison Range. Journal of the American Veterinary Medical Association 86:745–752.

Williams, A. L., T. J. Kreeger, and B. A. Schumaker. 2014. Chronic wasting disease model of genetic selection favoring prolonged survival in Rocky Mountain elk (*Cervus elaphus*). Ecosphere 5(5):60. http://dx.doi.org/10.1890/ES14-00013.1.

Williams, E. S., M. W. Miller, T. J. Kreeger, R. H. Kahn, and E. T. Thorne. 2002. Chronic wasting disease of deer and elk: A review with recommendations for management. Journal of Wildlife Management 66(3):551–563.

Williams, E. S., and S. Young. 1980. Chronic wasting disease of captive mule deer: A spongiform encephalopathy. Journal of Wildlife Diseases 16:89–98.

———. 1982. Spongiform encephalopathy of Rocky Mountain elk. Journal of Wildlife Diseases 18:465–471.

———. 1992. Spongiform encephalopathies in Cervidae. Revue Scientifique et Technique (International Office of Epizootics) 11:551–567.

Wobeser, G. 1994. Investigation and management of disease in wild animals. Plenum Press, New York, New York, USA.

———. 2002. Disease management strategies for wildlife. Revue Scientifique et Technique (International Office of Epizootics) 21(1):159–178.

13

DANIEL J. DECKER,
WILLIAM F. SIEMER,
ANN B. FORSTCHEN,
AND CHRISTIAN SMITH

The Role of Human Dimensions in State Wildlife Management

Historically, wildlife management was narrowly described as protecting or regulating use of populations of certain wild animals. Species facing possible extirpation were protected, except predators and "vermin." Active wildlife management was focused on restoring or regulating use of "game" and "furbearer" species pursued by hunters or trappers and controlling numbers of predators for those valued species and livestock. Thus, from the inception of modern wildlife management the management foci and the actions they instigated were deeply values based. Indeed, the human dimensions of wildlife management are reflected in definitions of the field. For example, wildlife management has long been depicted as focusing on three interacting components: wildlife populations, wildlife habitats, and humans (Giles 1978; Decker et al. 2012a, 2013).

Contemporary wildlife management carried out by state wildlife agencies is an even more complex job than many previous depictions indicate. It requires understanding ever-changing, coupled human and natural systems and influencing such systems to achieve conservation of diverse wildlife resources. In current parlance, wildlife management systems (Decker et al. 2012c) are composed of ecological components, social components, and their interrelationships (i.e., they are social-ecological systems; Berkes et al. 2000). To be effective, state wildlife agencies need a comprehensive understanding of the wildlife management system they are working in. They also should know which

elements of it they can influence, how to exert that influence, and how much effect can be expected. But what outcomes should management strive to achieve? This is a societal values question, and general guidance toward an answer lies in the legal and philosophical underpinnings of public wildlife management in the United States. Determining objectives and selecting methods of wildlife management typically require specific human dimensions inquiry and stakeholder input, subjects of the remainder of this chapter.

Foundations of Wildlife Management and Role of Human Dimensions

Society's expression of the fundamental value of wildlife resources is reflected in the public trust doctrine. According to interpretations of this doctrine (e.g., Batcheller et al. 2010), wildlife in the United States belong to all citizens and are managed by state and federal governments as public trust resources for the benefit of both current and future generations (Smith 2011). The public trust doctrine establishes expectations for public wildlife managers to be knowledgeable about stakeholders (Decker et al. 2013; Organ et al. 2014); human dimensions research and stakeholder engagement support state agencies in fulfilling their public trust responsibilities (Decker et al. 2014a, 2014b; Forstchen and Smith 2014; Organ et al. 2014). Organ et al. (2014) articulate three key functions that public wildlife managers must perform under the public trust doctrine: sustaining trust assets (i.e., wildlife resources), developing trust assets, and distributing trust benefits. Human dimensions research and stakeholder engagement can play different roles as agencies perform these functions, including contributing to understanding of evolving societal values, norms, and expectations with respect to wildlife and its management.

In addition to the public trust doctrine, norms of good governance (e.g., participatory, transparent, accountable) also compel savvy public wildlife resource administrators to consider human dimensions. The ideas of public trust thinking and good governance are melded into a set of wildlife governance principles (Decker et al. 2016) that provide guidance for state agency wildlife professionals to improve conservation outcomes valued by stakeholders. Human dimensions research and stakeholder engagement ideas discussed in this chapter are necessary for application of wildlife governance principles. Wildlife management that applies the governance principles is expected to yield benefits for current and future generations.

Thus, state wildlife management is motivated by society's desire for valued outcomes or benefits that can be produced from managing public wildlife resources (Decker et al. 2001a). Expectations for such outcomes give rise to management aimed at (1) sustaining biodiversity and related ecological services, (2) providing renewable use and enjoyment of wildlife, and (3) minimizing negative consequences of human–wildlife interactions. In pursuit of these purposes, much of what is "managed" are human behaviors, typically humans' interactions with wildlife and habitats, or among each other regarding wildlife. Wildlife managers deal with individuals, informal and formal groups, and communities, all operating within social structures, cultural systems, and institutions. These components of wildlife management systems command a great deal of state agencies' time and attention. Consequently, the importance of human dimensions permeates all aspects of wildlife management.

Wildlife Management from a Human Dimensions Perspective

Human experiences with wildlife arise from many kinds of interactions between people and wildlife, as well as among people because of wildlife. Experiences occurring within wildlife management systems can be direct or indirect, vary in intensity and duration, occur at multiple scales (temporal, geospatial, jurisdictional) and levels (local, regional, statewide, multistate, continental, etc.), and be of many kinds. Effects of these experiences can be positive or negative and may take many forms (e.g., economic benefits or costs; threats to or enhancement of human health and safety; ecological services provided by wildlife and their habitats; physical, mental, and social benefits produced by recreational enjoyment of wildlife). The most important effects typically generate strong stakeholder reactions and prompt management attention; these effects are referred to as impacts (Riley et al. 2002). From this perspective, wildlife management can be defined as *the guidance of decision-making processes and implementation of practices that influence interactions between people, wildlife, and wildlife habitats, as well as among people about wildlife, to produce impacts valued by stakeholders* (Decker et al. 2012a). Understanding the reasons for and predicting outcomes of such interactions to inform management decisions are the primary purpose of human dimensions research.

Ultimately, under the public trust doctrine state wildlife management agencies are expected to produce benefits for society ("*impacts valued by stakeholders*"), where benefits are the outcomes (i.e., positive impacts created or negative impacts reduced) experienced directly or indirectly by citizens as a result of management actions. Wildlife managers attempt to enhance, regulate, or prohibit various experiences that people might have with wildlife for the purpose of influencing the nature and magnitude of impacts. Impacts of importance can arise from a wide range of management actions aimed at individual wild animals or populations of them, habitats, and people. Accordingly, state wildlife management programs often focus on protection or manipulation of wildlife populations and habitats, plus regulation of wildlife use (e.g., for hunting, trapping, and wildlife viewing). Wildlife management efforts tend to emphasize specific, on-the-ground activities such as manipulation of habitat through prescribed burns, influencing a wildlife population through hunting quotas, or relocating animals for purposes of establishing new populations. Activities of these kinds often are necessary to achieve many of the outcomes desired by society, but wildlife management as a whole enterprise includes a broader array of necessary activities (e.g., informative communication, negotiation, enforcement, development of strategic partnerships, decision-making). These have substantial human dimensions considerations, requiring wildlife managers to understand diverse stakeholders for wildlife management (Decker et al. 2013).

Wildlife Management Is a Process

Wildlife management is a process, but it seldom unfolds in a tidy linear or cyclic fashion (Decker et al. 2001b). Conserving a species or addressing a human–wildlife conflict is a multifaceted undertaking, with several process components: collecting and analyzing information (social and ecological research and monitoring); planning and decision-making (including working with stakeholders to set broad goals and specific objectives and then selecting actions to accomplish objectives); implementing various kinds of (preferably socially acceptable) actions directed at wildlife, habitat, and people; evaluating progress; and adjusting program components as needed (Decker et al. 2012c). Involving partners (e.g., other agencies, units of government, nongovernmental organizations [NGOs]) and other stakeholders in various aspects of management adds to the complexity of the process. Understanding the human dimensions of wildlife management systems is vital to the development of an informed wildlife management program.

Stakeholder Orientation

Stakeholders are central to why (toward what ends) wildlife management occurs and how it is conducted. A stakeholder is any person significantly affected by or significantly affecting wildlife or wildlife management decisions and actions (Decker et al. 1996). Stakeholders can have a variety of interests (i.e., stakes) in wildlife, human–wildlife interactions, and management interventions. Stakeholders may be well organized and represented by established NGOs; individuals joined together in ad hoc, situation-specific groups (grassroots); or simply individuals with an interest in

a management issue. Because wildlife are public trust resources, people do not need to be organized or even aware that they have a stake to be stakeholders in wildlife management.

Stakes in wildlife management reflect impacts of interest to people and typically take the form of recreational, cultural, psychological, social, economic, ecological, or health and safety impacts (Siemer and Decker 2006). A particular wildlife issue may involve a range of stakeholder-identified impacts, and a variety of these and other factors may influence stakeholder expectations of management (Carpenter et al. 2000). In white-tailed deer (*Odocoileus virginianus*) management, for example, the variety of stakeholders includes hunters, forest owners who experience suppressed regeneration of valuable tree species if deer browse too heavily on them, farmers who have crops consumed by deer, homeowners who have garden and landscape plants damaged by deer, motorists who face hazards of deer collisions, and public health officials concerned with tick-borne diseases. Sometimes people may not recognize their stakes in wildlife management decisions because they are unaware of benefits they receive from management (e.g., lower incidence of highway flooding because of beaver trapping) or do not anticipate impacts they will experience as a consequence of management (e.g., enjoyable sightings of rare raptors). The potential for citizens to become unknowing stakeholders is especially likely if management actions give rise to impacts they never previously experienced (e.g., total protection of black bears from hunting when density of bears is low and residents are not experiencing human–bear conflicts, followed by bear population expansion resulting in more urban nuisance situations, including attacks on pets and people) or they never have to experience negative impacts that would be likely in the absence of management (e.g., increase harvest of alligators to reduce predation on pets in certain areas, thus sparing numerous people from such tragedies).

Management of wildlife often occurs in response to stakeholders' expressed need for a state wildlife agency to influence impacts. Frequently, though, controversy about management of wildlife emerges and escalates when proposed management objectives and actions desired by some stakeholders are not acceptable to others

(Decker et al. 2004b; Siemer et al. 2007). For example, calls for management of suburban wildlife that cause negative impacts (e.g., deer, Canada geese, coyotes) often are initiated by those experiencing such impacts directly. These primary stakeholders seek relief from the problems, but others may enjoy their interactions with the wildlife in question or deplore any actions that would harm the animals. Managers' knowledge of variation in acceptance capacity among stakeholders for different impacts can improve management responses (Decker and Purdy 1988; Carpenter et al. 2000; Organ and Ellingwood 2000; Riley and Decker 2000a, 2000b; Lischka et al. 2008). Note that people opposed to specific management actions are a type of stakeholder *created by* the management effort itself.

Thus, two broad categories of stakeholders exist in many wildlife management situations: those who are most concerned about the impacts of human–wildlife interaction, and those most concerned about management methods. This means that even in situations where stakeholders can agree about desired outcomes of management, they may differ strenuously about which management actions are appropriate. These conflicts almost always reflect differences in values among stakeholders and are the primary reason why wildlife controversies tend to be what are referred to as "wicked problems" (Rittel and Weber 1973). Human dimensions research and public engagement activity help identify a priori the acceptability of various management methods and characteristics of human population segments likely to support or oppose particular actions (Lauber and Knuth 1997, 2000; Wittmann et al. 1998; Loker et al. 1999; Teel et al. 2002; Dougherty et al. 2003).

Diversity of stakeholder perspectives in wildlife management highlights the need for cooperation, collaboration, and coalition building among stakeholders, as well as between them and wildlife managers. Collaborative effort is often necessary to access needed funding and expertise, to overcome jurisdictional impediments, and to bridge chasms between values of various stakeholder groups (Yankelovich 1991a, 1991b; Wondolleck and Yaffee 2000; Beierle and Cayford 2002). Collaborative ventures can be challenging to create or coordinate, but knowledge of the values and motivations of collaborators, cooperators, and partners

helps build working relationships (Schusler et al. 2003; Decker et al. 2005). Human dimensions inquiry and public engagement activity can provide such knowledge (Lauber and Decker 2011) but do not guarantee avoidance of conflict.

Human Dimensions Research and Public Engagement

Put simply, most human dimensions research and public engagement activity focus on explaining what people think and do with respect to wildlife and incorporating that insight into wildlife management. Decker et al. (2004a:187) summarize agency uses of human dimensions inquiry in terms of improving understanding in four areas: (1) how people value wildlife, (2) what benefits people desire from wildlife management, (3) acceptability of management practices, and (4) how various stakeholders affect or are affected by wildlife and wildlife management decisions. They describe three interrelated domains of human dimensions inquiry: wildlife-related activity description (what people do), social psychological understanding (what people think and why), and management application (incorporating human dimensions insight into management). Modified descriptions of these domains are summarized as follows:

- *Wildlife-related activity description.*—Studies about participation in wildlife-related activities have two primary purposes: (1) count (or estimate), classify, and characterize users/viewers; and (2) monitor, describe, and predict trends in participation, including assessment of users' expenditures. Estimates of trends in various kinds of uses are valuable for planning where anticipation of demand for and supply of wildlife-use opportunities can be critical to species conservation and provision of benefits for people.
- *Social psychological aspects of human–wildlife coexistence.*—Inquiry in this domain addresses three questions of interest to wildlife agencies: (1) how different segments of society value and assess the presence of wildlife and associated interactions and consequences (impacts experienced); (2) which management actions are considered

acceptable for different situations; and (3) why people participate in various wildlife-dependent activities, reasons for degree of avidity, substitutability, and so on. These lines of inquiry may take various theoretical directions (e.g., focus on values, attitudes, norms, or motivations), depending on the information needed by managers. Decker et al. (2004a) provide more detail and examples of these theoretical realms.

- *Application of human dimensions knowledge in management.*—Human dimensions knowledge has aided state wildlife management by informing policies, practices, and education and communication strategies. This includes assessing the economic trade-offs associated with various management strategies.

In summary, wildlife managers who work on the front lines of conservation navigate an increasingly complex set of human dimensions considerations. To do their jobs well, they need the best research-based human dimensions insight they can get. In response, human dimensions research and application, including stakeholder engagement, have become more diverse and sophisticated to meet managers' needs for effective wildlife governance and management. Countless human dimensions insights regularly used today by state wildlife agencies as conventional wisdom at one time were areas of uncertainty about the "people aspects" of wildlife management and topics of active inquiry (Decker et al. 2004a).

Approaches to Stakeholder Input and Involvement

Wildlife managers are continuously looking for better ways to facilitate stakeholder input and involvement for decision-making (Decker and Chase 1997; Chase et al. 2000). Stakeholder input can be sought through research and by facilitated engagement processes. Rather than reviewing the many specific forms of stakeholder engagement used by state wildlife agencies, we briefly describe six general postures taken by state wildlife agencies with respect to stakeholder involvement: expert authority, passive-receptive, inquisitive, inter-

mediary, transactional, and co-managerial. The role of human dimensions research is noted in each.

Expert Authority Approach

This approach is the least attentive to stakeholder engagement. It reflects a belief that wildlife managers are experts who should make decisions and take actions without methodical stakeholder input. This approach was the norm when managers focused on a small set of stakeholders (e.g., hunters, trappers, farmers, ranchers, and forest owners) and were concerned about fewer management outcomes. Differences in values among these stakeholder groups were assumed to be well known by managers, but early human dimensions studies cast doubt on this assumption (Purdy 1987; Enck and Decker 1997). The expert authority approach is still used in some situations, perhaps most notably under emergency circumstances, for example, when a wildlife disease outbreak is discovered and an incident command system is triggered to deal with disease containment. Even then, state agencies may use human dimensions research or stakeholder engagement activity to gauge public reaction to their approach, especially if stakeholder cooperation is required for management to succeed (Brown et al. 2006).

Passive-Receptive Approach

The passive-receptive approach, where wildlife managers are receptive to input offered by stakeholders but do not seek their input systematically, traditionally has been the most common. When making decisions in this approach, wildlife managers rather than stakeholders themselves determine the relative weight to give concerns voiced by various stakeholders. Greater weight is placed on concerns of those stakeholders who make their views known. The functioning assumption is that if managers do not hear from stakeholders about their concerns, they must be disinterested in the outcome of management and therefore have little standing in decision-making. Active, organized stakeholders have the advantage in this approach because they both are prepared to give voice to their interests and concerns and have established access to managers and decision

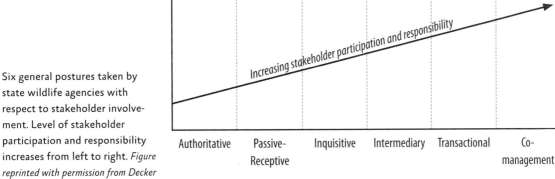

Six general postures taken by state wildlife agencies with respect to stakeholder involvement. Level of stakeholder participation and responsibility increases from left to right. *Figure reprinted with permission from Decker et al. (2012b).*

makers, sometimes to the exclusion of other legitimate stakeholders with less capacity to effectively represent their interests. The passive-receptive approach is common, often reflected in statements at the end of agency press releases and websites inviting input, such as, "If you have any comments or questions, please call, write, or visit your local Department of Natural Resources office." Prevalent as this approach may be, human dimensions research has revealed that many legitimate stakeholders simply do not prefer such means for making their attitudes and preferences known (Chase et al. 1999). Moreover, many citizens are "latent" stakeholders; they do not suspect that they are stakeholders in current decisions, realizing this only after they experience impacts arising from the decisions. The passive-receptive approach misses many of these concerns initially during decision-making, but it ultimately deals with them as collateral impacts afterward.

Inquisitive Approach

An inquisitive approach recognizes that reliance on unsolicited stakeholder input alone can lead to wildlife managers developing a biased perspective where importance of a subset of stakes can be exaggerated and some important stakes can be missed altogether. "Inquisitive" wildlife agencies actively seek information about stakeholders to inform management planning and implementation. They also seek stakeholder input to evaluate programs that are in place (i.e., to refine management policies, regulations, or activities). For

example, in southwest Florida, the population of the Florida panther (*Puma concolor coryi*), a federally listed endangered species, has been increasing—now estimated at 100–180 adults. Land stewardship practices of ranchers in the area significantly contribute to and improve panther habitat. However, state and federal agencies are receiving increasing reports of suspected calf depredation by panthers. A study was funded by the US Fish and Wildlife Service to understand ranchers' opinions about panthers and learn what ranchers might regard as acceptable compensation for calf predation. Using focus groups to elicit concerns, researchers learned of ranchers' low trust of government agencies and frustration about how agencies had managed efforts to protect and recover the Florida panther. This inquiry and stakeholder engagement activities help federal and state agencies working with private landowners in south Florida ensure that ranching continues on the landscape, while supporting a sustainable panther population. Human dimensions inquiry and application are essential to coexistence of ranchers and panthers in this situation.

Intermediary Approach

The intermediary approach emerged as wildlife managers recognized the value of dialogue with stakeholders (i.e., not just collecting data about stakeholders remotely in surveys). This approach encourages two-way communication between individual stakeholder groups and the wildlife management agency, but it does not emphasize dialogue among stakeholder groups with different concerns. Instead, managers act as intermediaries who shoulder responsibility for uncovering similarities and differences in stakeholder interests and positions, and then attempt to weigh and balance their concerns in decision-making. When dealing with a controversial topic such as wolf (*Canis lupus*) conservation, wildlife managers operating as intermediaries may choose to interact one-on-one with multiple stakeholder groups (e.g., ranchers, hunters, environmentalists) in an attempt to reveal common ground prior to convening a meeting of representatives from the various interests—essentially doing preparatory work. A difficulty for managers to avoid with this approach is assuming primary responsibility for creating compromise, when arguably that burden should be shared among the various stakeholder groups involved. Human dimensions inquiry can aid managers acting as intermediaries by revealing stakeholder beliefs, attitudes, and preferences about the issue and perceptions of other stakeholders.

Transactional Approach

A transactional approach aids wildlife managers when stakeholders need to be involved in prioritizing seemingly incompatible interests. When managers need to find objectives and choose actions that are acceptable to diverse stakeholders, they commonly use professionally facilitated processes where stakeholders interact with one another directly to articulate their interests (i.e., desired outcomes of management), as well as their preferences for management actions. Note that in this approach stakeholders describe their stakes and management preferences to each other, rather than through the manager-intermediary. They work together to rank desired outcomes and thereby produce objectives and means they can all live with. *These processes are not the equivalent of systematic inquiry*; thus, state wildlife agencies may also sponsor survey research to ensure that the incidence and salience of various citizen interests are documented in a generalizable way, rather than relying on a stakeholder committee to estimate prevalence of various interests in a wildlife resource. By learning about diverse perspectives on (or stakes

in) the issue directly from one another, conducting a survey, discussing viewpoints, debating the trade-offs, and compromising, stakeholder participants are more likely to reach consensus about appropriate objectives and courses of action (Nelson 1992).

A transactional approach was recently piloted for decisions about deer population management in local management zones in New York. An ad hoc stakeholder committee was established in a deer management zone for the purpose of identifying a mutually acceptable population management objective for white-tailed deer, couched in terms of impacts desired. An independent facilitator (not wildlife agency staff) assisted discussions among members of the input group. Stakeholder participants were chosen to represent a wide breadth of interests, but they were not officially representing organized interest groups, the intent being to avoid revisiting established arguments in deer management disputes between interest groups and the agency or among interest groups themselves. A systematic stakeholder survey was conducted in the management zone for the purpose of providing the input group and the state wildlife agency with information about the nature and extent of deer-related impacts being experienced by citizens in the management zone.

Co-managerial Approach

Co-management involves state wildlife agencies engaging other agencies, NGOs, and local communities in both making decisions and "sharing" responsibility for management. A fundamental distinction exists between the co-managerial approach and the other approaches discussed thus far. In co-management scenarios, the resources necessary for effective management derive from partnerships, i.e., they are not limited to the financial, land and water, and human resources of the state wildlife agency. Specifics of co-management partnerships are negotiated on a case-by-case basis. Because these partnerships are tailored to individual circumstances, they take many forms. For example, in community-based co-management, local communities (either governments or individuals and groups) have devised a variety of tactics to make their communities' desires for management of a wildlife resource possible (for a discussion of community-based co-management of white-tailed deer, see, e.g., Raik et al. 2005). The need still exists in co-management for accurate information about stakeholders: their experiences with wildlife of interest, their desired objectives for wildlife management in terms of impacts experienced, and relative social acceptability of various management actions.

In all of the approaches identified, insight about stakeholders' beliefs and attitudes, patterns of behavior, and expectations of wildlife management are needed for success. It is common for state wildlife agencies to combine elements of several of these approaches, and both social science inquiry and stakeholder engagement activities are relied on to support various kinds of management decisions. Below we present a case study that illustrates many of the human dimensions considerations, information-seeking activities, and application attributes outlined in this chapter.

- -

Case Study: Black Bear Management in New York

Many states have a long and at times contentious history of black bear (*Ursus americanus*) management. This case highlights how the New York State Department of Environmental Conservation (NYSDEC) has evolved in its approach to human dimensions research and stakeholder engagement for bear management, particularly in the Catskill Mountain region of southeastern New York (the Catskills; see table 13.1).

EARLY HUMAN DIMENSIONS APPLICATION IN BLACK BEAR MANAGEMENT

As mentioned earlier, wildlife management is an iterative process involving analysis of current and desired conditions, designing and taking actions to achieve desired conditions, making decisions to solve problems, and evaluating outcomes to learn how to improve future decisions and performance. Black bear management in New York has experienced several cycles.

Two management cycles occurred during the period from 1977 to 1988 that made efforts to incorporate human dimensions research findings

into management and policy decisions (Decker and O'Pezio 1989; Decker et al. 2001a). Earlier, managers' concern about a substantial decline in hunters' harvest of black bears in the Catskills between 1954 and 1969 led to an ecological study of the bear population during the 1970s. Findings from biological research and hunter harvest reports led state wildlife managers to conclude that hunting was the leading source of bear mortality in the Catskills, and by 1975 hunting pressure had stabilized the bear population well below biological carrying capacity (McCaffrey et al. 1976). Bear managers also determined that unsolicited citizen complaints about black bear nuisance or damage experiences were low, interpreted as indicating an acceptable level of human–bear conflict in the region. Consequently, in 1976 the state proposed a two-year hunting moratorium as a means of achieving a 60–80 percent increase in the northern Catskill bear population. During the moratorium, the state wildlife agency sponsored additional ecological and sociological research to assess response of the bear population and human tolerance of an expected increase in bear abundance. A pilot survey of Catskill landowners in 1976 (O'Pezio 1977) and a more comprehensive survey in 1978 (Brown et al. 1979; Decker et al. 1981) supported managers' assumption that the bear population at that time was well within the acceptance capacity of area residents, i.e., the level of problems residents experienced was socially tolerable (Decker and Purdy 1988).

Follow-up research indicated that the hunting moratorium achieved its bear population growth objective (managers documented an 80% increase in the bear population). Evaluation of the bears' physical condition suggested that the population increase had not exceeded biological carrying capacity. Additional survey research documented that little change in landowner attitudes occurred between 1978 and 1983, indicating that landowner acceptance capacity had not changed even though bear numbers and recorded human–bear encounters had increased (Decker et al. 1985). The combined ecological and sociological studies

Table 13.1 Broad categories, purposes, and applications of human dimensions research, with specific examples related to black bear management in New York State

Domains of HD research and application	Purposes	Specific examples from New York State
Wildlife-related activity description	Count, classify users; describe, predict use trends. Use harvest or sightings reports as indices of species abundance (surrogate biology)	Harvest survey data (combined with data on natural mortality) revealed hunting as a main factor limiting Catskill bear population (McCaffrey et al. 1976)
Social psychological understanding	Improve understanding of what people think, and why; how people value wildlife; what benefits people desire; acceptability of management practices; how stakeholders affect or are affected by wildlife or wildlife management decisions	Landowner surveys documented acceptance capacity was unchanged after a two-year hunting moratorium (Decker et al. 1981, 1985). Statewide survey documented high sensitivity to and expectations for agency response to interactions with bears in residential settings (Siemer and Decker 2003). Based on synthesis of information from HD research and engagement processes, managers defined 12 specific effects of human–bear interactions as management priorities (Siemer and Decker 2006)
Application of HD knowledge in management	Incorporating HD insights into management policies, management practices, education programs, communication strategies, stakeholder engagement	HD insights used to develop a planning framework (NYSDEC 2003), support proposals to expand bear-hunting zones and seasons (NYSDEC 2014, Appendix 1), develop and evaluate a problem-prevention education program (Gore et al. 2008), understand media coverage and risk perception after a bear-related human fatality (Gore et al. 2005), develop and issue education guide (Siemer et al. 2007)

gave managers more confidence in their understanding of primary and collateral effects of their decisions (i.e., primary effects on bear conservation and bear-hunting satisfactions; secondary effects on landowners who interact with black bears).

Stakeholders had limited input to setting bear management objectives or determining acceptability of bear management interventions in New York State prior to the 1970s. At that time, state agencies typically exercised expert authority and took a passive-receptive approach to stakeholder input on wildlife management issues. With the exception of human dimensions studies just described, the state wildlife agency employed a passive-receptive approach to stakeholder engagement in bear management into the late 1980s.

NEED TO UNDERSTAND AND ENGAGE STAKEHOLDERS INTENSIFIES AS MANAGEMENT ISSUES EMERGE

During the 1990s, the cultural, political, and ecological environment in which bear management occurred was changing across the country. In New York, stakeholder groups opposed to bear-hunting practices expressed their concerns directly to the governor, who created an executive order prohibiting bear hunting with dogs. For numerous reasons, the black bear population grew for approximately two decades, and the geographical distribution of bears expanded across the state, into areas historically closed entirely to bear hunting. Complaints about bears rose steadily statewide throughout the 1990s, with the Catskills being the epicenter of problems.

The state wildlife agency initiated a new management cycle (from situation analysis through monitoring action outcomes) as a series of linked activities between 2001 and 2008 (Decker et al. 2014c). The agency relied on human dimensions research during this management cycle. Five innovations distinguished this from previous bear management cycles: extensive situation analysis (including social science inquiry), an explicit focus on stakeholder-defined impacts, transactional stakeholder engagement, use of quantitative systems-thinking techniques, and a decision by the wildlife agency management team to approach the entire experience as a learning opportunity.

Charged to develop a statewide comprehensive bear management plan, the state's Bear Management Team collaborated with human dimensions specialists in 2002 to develop a new framework for black bear management planning. The agency's new bear management framework was prepared, approved, and released to the public in 2003 (NYSDEC 2003). Based on the precepts of impact management (Riley et al. 2002), with its strong emphasis on systematically documenting impacts of importance to stakeholders, the framework established a process for adapting New York's management program to changing social and environmental conditions. Comprehensiveness in stakeholder input and engagement, a focus on impacts, manager–stakeholder deliberation, and adaptive management were featured elements of this planning cycle.

Several activities conducted in this case (i.e., synthesis of public meeting data from 1992 to 1994, small group meetings held with stakeholders across the state, and a statewide survey of residents; Siemer and Decker 2003) represent consultative or inquisitive forms of public engagement (Rowe and Frewer 2005), where stakeholders convey information to policy makers through processes initiated by the policy-making body (the state wildlife agency in this case). The state wildlife agency also created regional committees of stakeholders, called stakeholder input groups, who met with wildlife managers two or more times to identify the impacts of greatest concern in their region

and to translate those interests and concerns into objectives for bear management in their region. Because this input process was designed to encourage multidirectional information exchange and deliberation focused on impacts, it provided opportunities to question both stakeholders' and managers' assumptions about the relationship between bear abundance and managing the impacts of greatest concern in each region of the state.

Three management priorities were generally identified by all input groups convened in the state: (1) minimally, a viable population of bears should be maintained in New York; (2) public education about bears should be expanded; and (3) negative impacts (e.g., agricultural and residential property damage) should be reduced (Schusler and Siemer 2004).

APPLICATION OF HUMAN DIMENSIONS INSIGHTS TO MANAGEMENT POLICIES AND PROGRAMS

Extensive public involvement created public support necessary to implement controversial management interventions (e.g., hunting regulation changes that expanded bear-hunting zones into areas previously closed to hunting; NYSDEC 2014, Appendix 1). Agency-sponsored research and outreach were implemented, including (1) studies to understand bear management stakeholders (Siemer and Decker 2003, 2006), (2) a study of media frames used to characterize black bear management in New York (Siemer et al. 2007), (3) pilot testing a prevention education program in one Catskill community (Gore et al. 2008), and (4) development of a guide to educate communities about bear management issues (Siemer et al. 2007). The state also improved the system it used to monitor citizen reports of interactions with bears. The management priorities, processes, and products developed during the 2001–2008 management cycle guided the bear program in the subsequent management cycle and provided a foundation for New York's 2014–2024 bear management plan (NYSDEC 2014).

Another, unusual use of human dimensions research occurred in the wake of a tragic event involving a black bear. During the summer of 2002, a black bear killed an infant in a small community in the Catskill region of New York. This was the first such case recorded in New York State. The details surrounding the infant's death need not be presented here, but the incident is notable for its testing of an assumption held by state bear managers about the effects of such an event on public acceptance of bear management policy in New York. Basically, in systems modeling that bear managers engaged in a few years earlier, where system effects were explored and scenarios considered, it had been assumed that if a human injury from a black bear was experienced, the public would pressure the state to reduce the bear population (reduce bear numbers markedly) and also invoke harsher treatment of any bear that appeared to be habituated or food conditioned. Immediately following the infant's death, which occurred only months after a human dimensions survey had been conducted in the Catskills, a new study was conducted to assess effects of the widely publicized death on policy-relevant public sentiment about black bears and bear management. The thought was that such information would be needed to guide black bear management policy and practice changes in the region. However, contrary to expectations, the study did not reveal public backlash against black bears or the need for a change in bear management in the Catskills (Gore et al. 2005). In this case, human dimensions research helped avoid unnecessary action that otherwise may have been undertaken based on a logical, yet incorrect, mental model of system reaction to the tragic incident.

The nearly four decades of Catskill black bear management described above cover three broad domains of human dimensions research and application: wildlife-related activity description, social psychological understanding, and application of human dimensions knowledge in management. Table 13.1 summarizes the purposes of the human dimensions work in each domain. Note that perhaps the most important contribution of human dimensions research was pointing out where assumptions were not valid and where

following common beliefs held by managers at the time might have sent management in the wrong direction.

Summary

"Wildlife resources are a public trust" (Organ et al. 2012:11). This simple, powerful tenet of wildlife management in the United States means that all citizens can legitimately seek benefits from public wildlife management. This idea can lead to controversy over allocation of benefits from wildlife management. Decision-making in wildlife management that follows a representative governance model but incorporates a more participatory approach might be the most effective means of attaining equity in the allocation of trust benefits when stakeholder interests differ or conflict. In practice, active participation by all beneficiaries is constrained by logistic and pragmatic factors. Systematic social science surveys of beneficiary populations present a surrogate for obtaining representative face-to-face input from a breadth of stakeholders. Coupling human dimensions inquiry with stakeholder engagement practices provides an inclusive strategy for capturing the breadth of stakeholder interests and building knowledge of relevant core values that relate to wildlife conservation.

Human dimensions research and stakeholder engagement are important in helping wildlife managers meet their responsibilities as public trust resource administrators in agreement with general wildlife governance principles (Organ et al. 2014; Decker et al. 2016). Identifying salient human values in wildlife management often can be critical to avoid or address the potential for conflict among stakeholders. The public's understanding of impacts of human–wildlife interactions experienced by all stakeholders, desired benefits from wildlife and its management, preferences for management action alternatives, and willingness to engage in governance/management are important information for wildlife managers. Moreover, the obligation of public trust managers to consider future generations calls for social valuation of wildlife to be forward looking. That is, while contemporary values are included in wildlife management decisions, they do not exclusively dictate decision outcomes, as they may not promote sustainability of wildlife. Consequently, comprehension of the ever-changing human dimensions of wildlife management is imperative for the administration of wildlife as trust resources shared commonly and in perpetuity.

LITERATURE CITED

Batcheller, G. R., M. C. Bambery, L. Bies, T. Decker, S. Dyke, D. Guynn, M. McEnroe, M. O'Brien, J. F. Organ, S. J. Riley, and G. Roehm. 2010. The public trust doctrine: Implications for wildlife management in the United States and Canada. Technical Review 10-01. The Wildlife Society, Bethesda, Maryland, USA.

Beierle, T. C., and J. Cayford. 2002. Democracy in practice: Public participation in environmental decisions. Resources for the Future, Washington, DC, USA.

Berkes, F., C. Folke, and J. Colding. 2000. Linking social and ecological systems: Management practices and social mechanisms for building resilience. Cambridge University Press, Cambridge, UK.

Brown, T. L., D. J. Decker, and D. L. Hustin. 1979. Public attitudes toward black bear in the Catskills. Cornell University, Department of Natural Resources, Outdoor Recreation Research Unit Publication 79-1. Ithaca, New York, USA.

Brown, T. L., D. J. Decker, J. T. Major, and W. H. Gordon. 2006. Assessment of the multi-agency approach to managing chronic wasting disease in Oneida County, New York: Perceptions of incident responders. Cornell University, Department of Natural Resources, Human Dimensions Research Unit Series Publication 06-3, Ithaca, New York, USA.

Carpenter, L. H., D. J. Decker, and J. F. Lipscomb. 2000. Stakeholder acceptance capacity in wildlife management. Human Dimensions of Wildlife 5:5–19.

Chase, L. C., T. M. Schusler, and D. J. Decker. 2000. Innovations in stakeholder involvement: What's the next step? Wildlife Society Bulletin 28:208–217.

Chase, L. C., W. F. Siemer, and D. J. Decker. 1999. Suburban deer management: A case study in the Village of Cayuga Heights, New York. Human Dimensions of Wildlife 4:59–60.

Decker, D. J., T. L. Brown, D. L. Hustin, S. H. Clarke, and J. O'Pezio. 1981. Public attitudes toward black bears in the Catskills. New York Fish and Game Journal 28:1–20.

Decker, D. J., T. L. Brown, and W. F. Siemer. 2001a. Human dimensions of wildlife management in North America. The Wildlife Society, Bethesda, Maryland, USA.

———. 2001b. Wildlife management as a process. Pages 77–90 in D. J. Decker, T. L. Brown, and W. F. Siemer, editors. Human dimensions of wildlife management in North America. The Wildlife Society, Bethesda, Maryland, USA.

Decker, D. J., T. L. Brown, J. J. Vaske, and M. J. Manfredo.

2004a. Human dimensions of wildlife management. Pages 187–198 in M. J. Manfredo, J. J. Vaske, B. L. Bruyere, D. R. Field, and P. Brown, editors. Society and natural resources: A summary of knowledge. Modern Litho, Jefferson, Missouri, USA.

Decker, D. J., and L. C. Chase. 1997. Human dimensions of living with wildlife—a management challenge for the 21st century. Wildlife Society Bulletin 25:788–795.

Decker, D. J., A. B. Forstchen, J. F. Organ, C. A. Smith, S. J. Riley, C. A. Jacobson, G. R. Batcheller, and W. F. Siemer. 2014a. Impacts management: An approach to fulfilling public trust responsibilities of wildlife agencies. Wildlife Society Bulletin 38:2–8.

Decker, D. J., A. B. Forstchen, E. F. Pomeranz, C. A. Smith, S. J. Riley, C. A. Jacobson, J. F. Organ, and G. R. Batcheller. 2014b. Stakeholder engagement in wildlife management: Does the public trust doctrine imply limits? Journal of Wildlife Management 79:174–179.

Decker, D. J., C. C. Krueger, R. A. Baer, Jr., B. A. Knuth, and M. E. Richmond. 1996. From clients to stakeholders: A philosophical shift for fish and wildlife management. Human Dimensions of Wildlife 1:70–82.

Decker, D. J., and J. O'Pezio. 1989. Consideration of bear–people conflicts in black bear management for the Catskill region of New York: Application of a comprehensive management model. Pages 181–187 in M. Bromley, editor. Bear–people conflicts: Proceedings of a symposium on management strategies. Northwest Territories Department of Renewable Resources, Yellowknife, Canada.

Decker, D. J., and K. G. Purdy. 1988. Toward a concept of wildlife acceptance capacity in wildlife management. Wildlife Society Bulletin 1:53–57.

Decker, D. J., D. B. Raik, L. H. Carpenter, J. F. Organ, and T. M. Schusler. 2005. Collaborations for community-based wildlife management. Urban Ecosystems 8:227–236.

Decker, D. J., D. B. Raik, and W. F. Siemer. 2004b. Community-based suburban deer management: A practitioner's guide. Northeast Wildlife Damage Management Research and Outreach Cooperative, Ithaca, New York, USA.

Decker, D. J., S. J. Riley, J. F. Organ, W. F. Siemer, and L. H. Carpenter. 2014c. Applying impact management: A practitioner's guide, 3rd edition. Cornell University, Department of Natural Resources, Human Dimensions Research Unit and Cornell Cooperative Extension, Ithaca, New York, USA.

Decker, D. J., S. J. Riley, and W. F. Siemer. 2012a. Human dimensions of wildlife management. Pages 43–57 in P. R. Krausman and J. W. Cain, editors. Wildlife management and conservation: Contemporary principles and practices. Johns Hopkins University Press, Baltimore, Maryland, USA.

———, editors. 2012b. Human dimensions of wildlife management, 2nd edition. Johns Hopkins University Press, Baltimore, Maryland, USA.

———. 2012c. Wildlife management as a process. Pages 87–100 in D. J. Decker, S. J. Riley, and W. F. Siemer, editors. Human dimensions of wildlife management, 2nd edition. Johns Hopkins University Press, Baltimore, Maryland, USA.

———. 2013. Human dimensions of wildlife management. Pages 34–50 in P. R. Krausman and J. W. Cain, editors. Wildlife management and conservation: Contemporary principles and practices. Johns Hopkins University Press, Baltimore, Maryland, USA.

Decker, D. J., C. Smith, A. Forstchen, D. Hare, E. Pomeranz, C. Doyle-Capitman, K. Schuler, and J. Organ. 2016. Governance principles for wildlife conservation in the 21st century. Conservation Letters 9:290–295.

Decker, D. J., R. A. Smolka, Jr., J. O'Pezio, and T. L. Brown. 1985. Social determinants of black bear management for the northern Catskill Mountains. Pages 239–247 in S. L. Beasom and S. F. Roberson, editors. Game harvest management. Caesar Kleberg Wildlife Institute, Kingsville, Texas, USA.

Dougherty, E. N., D. C. Fulton, and D. H. Anderson. 2003. The influence of gender on the relationship between wildlife value orientations, beliefs, and the acceptability of lethal deer control in Cuyahoga Valley National Park. Society and Natural Resources 16:603–623.

Enck, J. W., and D. J. Decker. 1997. Examining assumptions in wildlife management: A contribution of human dimensions inquiry. Human Dimensions of Wildlife 2:56–72.

Forstchen, A. B., and C. A. Smith. 2014. The essential role of human dimensions and stakeholder participation in states' fulfillment of public trust responsibilities. Human Dimensions of Wildlife 19:417–426.

Giles, R. H. 1978. Wildlife management. W. H. Freeman, San Francisco, California, USA.

Gore, M. L., B. A. Knuth, C. W. Scherer, and P. D. Curtis. 2008. Evaluating a conservation investment designed to reduce human–wildlife conflict. Conservation Letters 1(3):136–145.

Gore, M. L., W. F. Siemer, J. E. Shanahan, D. Schuefele, and D. J. Decker. 2005. Effects on risk perception of media coverage of a black bear-related human fatality. Wildlife Society Bulletin 33:507–516.

Lauber, T. B., and D. J. Decker. 2011. Developing adaptability: The promise and pitfalls of collaborative conservation. Human Dimensions of Wildlife 16:219–221.

Lauber, T. B., and B. A. Knuth. 1997. Fairness in moose management decision-making: The citizen's perspective. Wildlife Society Bulletin 25:776–787.

———. 2000. Suburban residents' criteria for evaluating contraception and other deer management techniques. Human Dimensions of Wildlife 5:1–17.

Lischka, S. A., S. J. Riley, and B. A. Rudolph. 2008. Effects of impact perception on acceptance capacity for white-tailed deer. Journal of Wildlife Management 72:502–509.

Loker, C. A., D. J. Decker, and S. J. Schwager. 1999. Social acceptability of wildlife management actions in suburban areas: 3 cases from New York. Wildlife Society Bulletin 27:152–159.

McCaffrey, E. R., G. B. Will, and A. S. Bergstrom. 1976. Preliminary management implications for black bears, Ursus americanus, in the Catskill region of New York State as a result of an ecological study. Pages 235–245 in H. R. Pelton, J. W. Lentfer, and G. E. Folk, Jr., editors. Bears: Their biology and management. IUCN Publication News Service 40, International Union for Conservation of Nature and Natural Resources, Morges, Switzerland.

Nelson, D. 1992. Citizen task forces on deer management: A case study. Northeast Wildlife 49:92–96.

NYSDEC (New York State Department of Environmental Conservation). 2003. A framework for black bear management in New York. Albany, New York, USA.

———. 2014. Black bear management plan for New York State 2014–2024. www.dec.ny.gov/docs/wildlife_pdf/bearplan2014.pdf.

O'Pezio, J. 1977. Public attitudes toward black bears in the Catskills. New York Federal Aid in Fish and Wildlife Restoration Project W-89-R-21. New York State Department of Environmental Conservation, Albany, New York, USA.

Organ, J. F., D. J. Decker, S. S. Stevens, T. M. Lama, and C. Doyle-Capitman. 2014. Public trust principles and trust administration functions in the North American model of wildlife conservation: Contributions of human dimensions research. Human Dimensions of Wildlife 19:407–416.

Organ, J. F., and M. R. Ellingwood. 2000. Wildlife stakeholder acceptance capacity for black bears, beavers, and other beasts in the East. Human Dimensions of Wildlife 5:63–75.

Organ, J. F., V. Geist, S. P. Mahoney, S. Williams, P. R. Krausman, G. R. Batcheller, T. A. Decker, R. Carmichael, P. Nanjappa, R. Regan, R. A. Medellin, R. Cantu, R. E. McCabe, S. Craven, G. M. Vecellio, and D. J. Decker. 2012. The North American Model of Wildlife Conservation. Technical Review 12-04. The Wildlife Society, Bethesda, Maryland, USA.

Purdy, K. G. 1987. Landowners' willingness to tolerate deer damage in New York: An overview of research and management response. Pages 371–375 in D. J. Decker and G. R. Goff, editors. Valuing wildlife: Economic and social perspectives. Westview Press, Boulder, Colorado, USA.

Raik, D. B., W. F. Siemer, and D. J. Decker. 2005. Intervention and capacity considerations in community-based deer management: The stakeholders' perspective. Human Dimensions of Wildlife 10:259–272.

Riley, S. J., and D. J. Decker. 2000a. Risk perception as a factor in wildlife acceptance capacity for cougars in Montana. Human Dimensions of Wildlife 5:50–62.

———. 2000b. Wildlife stakeholder acceptance capacity for cougars in Montana. Wildlife Society Bulletin 28:931–939.

Riley, S. J., D. J. Decker, L. H. Carpenter, J. F. Organ, W. F. Siemer, G. F. Mattfeld, and G. Parsons. 2002. The essence of wildlife management. Wildlife Society Bulletin 30:585–593.

Rittel, H. W. J., and M. M. Webber. 1973. Dilemmas in a general theory of planning. Policy Sciences 4:155–169.

Rowe, G., and L. J. Frewer. 2005. A typology of public engagement mechanisms. Science, Technology and Human Values 30:251–290.

Schusler, T. M., D. J. Decker, and M. J. Pfeffer. 2003. Social learning for collaborative natural resource management. Society and Natural Resources 15:309–326.

Schusler, T. M., and W. F. Siemer. 2004. Report on stakeholder input groups for black bear management in the Lower Catskills, Upper Catskills and Western New York, October 2003–January 2004. Cornell University, Department of Natural Resources, Cornell Cooperative Extension and Human Dimensions Research Unit, Ithaca, New York, USA.

Siemer, W. F., and D. J. Decker. 2003. 2002 New York State black bear management survey: Study overview and findings highlights. Cornell University, Department of Natural Resources, Human Dimensions Research Unit Series Publication 03-6, Ithaca, New York, USA.

———. 2006. An assessment of black bear impacts in New York. Cornell University, Department of Natural Resources, Human Dimensions Research Unit Series Publication 06-6, Ithaca, New York, USA.

Siemer, W. F., D. J. Decker, P. Otto, and M. L. Gore. 2007. Working through black bear management issues: A practitioners' guide. Northeast Wildlife Damage Management Research and Outreach Cooperative, Ithaca, New York, USA.

Smith, C. A. 2011. The role of state wildlife professionals under the public trust doctrine. Journal of Wildlife Management 75:1539–1543.

Teel, T. L., R. S. Krannich, and R. H. Schmidt. 2002. Utah stakeholders' attitudes toward selected cougar and black bear management practices. Wildlife Society Bulletin 30:2–15.

Wittmann, K., J. J. Vaske, M. J. Manfredo, and H. C. Zinn. 1998. Standards for lethal control of problem wildlife. Human Dimensions of Wildlife 3:29–48.

Wondolleck, J. M., and S. L. Yaffee. 2000. Making collaboration work: Lessons from innovation in natural resources management. Island Press, Washington, DC, USA.

Yankelovich, D. 1991a. Coming to public judgment. Syracuse University Press, Syracuse, New York, USA.

———. 1991b. The magic of dialogue: Transforming conflict into cooperation. Simon & Schuster, New York, New York, USA.

14

Chad J. Bishop
and Michael W. Hubbard

The Role of Field Research in State Wildlife Management

Research has been integral to the success of modern state fish and wildlife agencies in accomplishing their missions to conserve and manage fish and wildlife species within the public trust. Agencies use research, surveys, and monitoring to inform most aspects of their work. From human dimension surveys used to understand the needs and desires of constituents to demographic studies on both harvested and protected species, research plays a vital role.

Research not only informs biological decisions but also shapes the direction of an agency, both socially and politically. Agencies that use well-designed research and surveys as a basis for starting discussion on management-related questions always have a solid anchor point to refer back to as decisions are made and actions are implemented. Agency administrators are obligated to integrate political and social ramifications of their actions into the decision-making process, but research provides the foundation upon which justifiable, defendable decisions are made. Without research and survey efforts, political influences and outspoken constituencies can often guide decisions that are not in the best, long-term interests of the resource or society.

Structural Research Models within State Wildlife Agencies

The structure and size of fish and wildlife agencies vary widely from state to state, which influences the quantity of research and survey efforts undertaken, as well as the process used to initiate that work. State wildlife agency structures range from divisions within much bigger departments of natural resources to autonomous agencies directly responsible to citizens. Agency size ranges from a couple hundred permanent positions to a couple thousand employees. This variance in agency structure and size leads to several different structural research models and a variety of approaches for initiating and conducting research.

What may be considered the ideal structure is for an agency to have in-house research units with permanent scientists on staff. This system allows agencies to have direct access to scientists, and it sets up a situation where the scientists themselves are concerned about the mission of the agency and the natural resources they are charged to manage. This approach also creates a situation where scientists work directly with on-the-ground managers to understand the issues they face in their day-to-day activities, resulting in research and monitoring efforts that are designed to directly modify management activities. Most notably, in-house scientists develop an understanding of the issues that are or will be impacting the agency and will often initiate projects to provide insight into those issues before they become problem areas for the organization.

The specific roles and responsibilities of in-house research scientists vary among states. Some in-house research units contain scientists who focus almost exclusively on designing, conducting, and publishing original research. These units have greater capacity to address a number of different research needs simultaneously and to conduct longer-term, intensive research. Research scientists in these units typically do not have formal decision-making authority with respect to management issues, nor are they responsible for implementing management decisions. Rather, their role is to provide original biological information to managers, agency leadership, and wildlife commissions or boards that have formal decision-making authority and, in doing so, to advance wildlife science. This formal research role helps maintain agency credibility in these situations where science supporting regulatory decisions is produced from within. That is, since most wildlife management decisions are informed by political and socioeconomic considerations in addition to biology, this structure helps ensure that the underlying science is unbiased and can be relied on by all stakeholders, regardless of their perspectives on any given regulatory or management issue.

Other research units housed within state agencies contain staff members who are assigned a dual research and management role. This structure is more common in smaller agencies or where funding and personnel limitations necessitate such dual responsibilities. Research scientists in these units often conduct research in direct collaboration with Cooperative Fish and Wildlife Research Unit (CFWRU) scientists and faculty at universities. Most of the actual research is accomplished by graduate students, which has the additional benefit of supporting the training of future professional biologists. These direct associations with university scientists, who are independent of the state agency regulatory process, help maintain scientific credibility as the state agency research scientists simultaneously accomplish their management responsibilities. Under this model, the agency scientists typically function as wildlife species or system specialists, with broad responsibilities for the management of those species or systems. They are often charged with integrating biological science, human dimensions, and socioeconomic and political considerations to formulate regulatory and management recommendations for consideration by their agency's leadership.

The final state agency research model is one where the agency does not employ any research scientists, but instead relies on other entities to meet their research needs, most often the wildlife department at one of their state's universities. Nearly all land-grant universities in the United States have a wildlife biology or natural resources department that has historically worked collaboratively with the given state's wildlife agency. In many states, the wildlife agency collaborates with university faculty via a CFWRU housed within the university's wildlife department. The CFWRU allows the wildlife agency to fund research at the university while paying minimal or no overhead, which amounts to considerable cost savings for state agencies, considering that universities typically charge high overhead rates on research funds (e.g., ~45%). For states without a CFWRU, other agreements have been put in place to facilitate research collaboration between the state wildlife agency and university faculty.

For this state agency research model, universities and CFWRUs are fundamentally critical to meeting the biological science needs of state fish and wildlife agencies and therefore maintaining agency scientific credibility. Agency biologists identify research needs and coordinate with university faculty and/or CFWRU scientists to accomplish the work. Most often the agency provides most or all of the funding, and as with research units housed in state agencies, much of the research is accomplished by graduate students. The quantity of research may ebb and flow over time depending on an agency's competing needs at any given moment. The downside of this model is that the state wildlife agency is dependent on other institutions to meet its biological science needs and thus may be impacted when those institutions face budget or personnel reductions or evolve their missions over time. This model as currently structured will remain effective as long as universities maintain strong wildlife programs composed of faculty willing and interested to conduct research on issues of priority importance to state wildlife agencies. The model also depends on the continued support of agreements that restrict the amount of overhead charged by universities, such as those associated with the CFWRUs.

Types of Research

Research conducted by state wildlife agencies is almost always applied in nature, meaning that the research question has a direct tie to a management issue. This ensures that research results are of value to wildlife managers by informing species management strategies and decisions, wildlife monitoring protocols, and optimal resource allocation, among others. When research projects become less applied or fail to address the highest management priorities facing an agency, research funding may be redirected to other uses. Additionally, research staff and associated budgets within state wildlife agencies are rarely considered essential to routine agency operations, such that when revenues decline, research units may be especially vulnerable to personnel and budget reductions. For these reasons, it is imperative that research supported by state agencies has direct application for species management or conservation. Successful in-house research units have established respected track records within their agencies of doing long-term proactive research that directly improves management of hunted species, conservation of imperiled species, and human dimensions work and therefore are considered necessary for the efficient and effective operation of the agency as a whole.

Applied research is broad in scope, pertaining to anything that helps solve a management or conservation need. The *Journal of Wildlife Management*, *Wildlife Monographs*, and *Wildlife Society Bulletin* provide a historical accounting of applied research conducted by state agencies through time. State agency research results are also published in numerous other scientific journals and in-house agency publications. Generally speaking, state agency research programs have focused heavily on how best to monitor population size of species or species assemblages, understanding factors that limit species population growth, and evaluating effectiveness of various wildlife and habitat management actions. Agency researchers have placed a heavy focus on research questions pertaining to management of hunted species given the historic roles of state agencies in implementing and enforcing game regulations and setting license numbers and harvest quotas. There has also been an emphasis on research pertaining to human–wildlife conflicts, which includes minimizing ungulate impacts on farmers and ranchers and managing large carnivores in areas with expanding human populations. Understanding and minimizing effects of wildlife disease is another focal area of many programs.

More recently, state agency research programs have placed a heavy emphasis on conservation of declining species and human dimensions.

As part of conducting applied wildlife research, state agencies have a strong track record of developing and improving field techniques used within the wildlife profession. From the profession's earliest days, state agency researchers and biologists have been instrumental in advancing techniques used to sample, capture, mark, and monitor wildlife. They spend countless hours observing and interacting with wildlife in the field, leading to continually refined hypotheses about how to improve techniques. Agency researchers have worked in close collaboration with various industry partners, university faculty, and other scientists to make technique improvements. For example, radio collars and other types of tracking devices have evolved from rudimentary VHF instruments to sophisticated satellite transmitters capable of remotely tracking fine-scale movements of animals, including small mammals and birds. Population monitoring techniques that once required the repeated capture and handling of animals can now be accomplished noninvasively by obtaining DNA from animal hair, scat, and tissue samples. And as field techniques have improved, field researchers have worked in close collaboration with CFWRU and university faculty to advance statistical analysis techniques. Mathematical optimizations that could only be conceptualized two decades ago can now be performed in minutes on a standard laptop computer. These technique advancements over time have steadily improved the efficiency and effectiveness of wildlife management and research (Silvy 2012).

Research accomplished by state agencies has also led to theoretical advances in science. Occasionally, state agencies have research or management needs that cannot be adequately addressed given current tools and existing models, which leads to theoretical advances as part of achieving an applied need. A historic example is the advancement of statistical theory to support defensible population management of waterfowl in North America. During the mid-twentieth century, state and federal waterfowl research biologists began working collaboratively with Canadian biologists to implement a continental-scale waterfowl banding program to monitor waterfowl harvest. Harvest data were generated from hunters who harvested banded birds and reported the information. Defensible data were needed to inform harvest objectives and allocation across the provinces and states as ducks and geese migrated south during the fall. Harvest allocation can be highly controversial to say the least, particularly with numerous stakeholders and governmental entities involved, demanding credible information to support decision processes. Waterfowl researchers needed assistance analyzing the volumes of band recovery data. This need prompted a collaborative effort among federal scientists, university faculty, and waterfowl researchers to generate defensible estimates of harvest and survival rates, which ultimately advanced statistical theory (Brownie et al. 1978). Various other theoretical advances in statistics have occurred over time as state wildlife agency researchers have worked in collaboration with university and federal agency colleagues to obtain defensible survival and population estimates from field data for any number of species (White and Burnham 1999; Burnham and Anderson 2002). Similarly, state agency research and management needs have contributed to important theoretical advances in wildlife nutrition (Robbins 1993), ecology (Fryxell et al. 2014), conservation genetics (Allendorf et al. 2012), and disease etiology (Foreyt and Jessup 1982; Miller and Williams 2004), to name a few.

Unique Attributes of State Agency Research Programs

There are several unique aspects of state wildlife agency research units that distinguish them from most other research institutions. First, agency researchers have a history of conducting longer-term (e.g., 5–10 year) field studies addressing complex ecological questions. University faculty and federal research scientists typically have less flexibility to commit to single, long-term research projects given obligations to train graduate students, teach courses, secure research funding, and meet publication requirements. In most cases, state agency researchers design and implement these studies in collaboration with university faculty, CFWRUs, or other scientists. As a typical example, a state agency commits to funding a research project, with an agency researcher identified as the principal investigator (PI).

The agency PI commits a significant amount of his or her time to the study for the duration, focusing on one or several overarching research question(s). There are invariably additional important research questions or needs that could be addressed with minimal additional funding but which the PI lacks capacity to address, creating opportunities for collaboration with universities and other research institutions, most often in the form of graduate student projects. These state, federal, and university research partnerships have been instrumental in advancing wildlife science and management.

A second unique aspect of research within state wildlife agencies is that projects are mostly funded internally with state dollars and federal grants earmarked for state agencies (e.g., Pittman–Robertson Act), providing a reliable source of funding. This allows researchers more time to focus on designing and implementing research rather than writing grants, and it minimizes demands to meet requirements of diverse funding entities. With that said, state agency researchers have been increasingly successful in leveraging this base funding to secure millions of dollars in external grant funds. Such efforts to secure external funding have become increasingly important as the responsibilities of state wildlife agencies for species management and conservation have expanded at the same time that funding has stagnated or declined.

Finally, state agencies employ large numbers of field-based management biologists, which is a logistical asset for research. Agency biologists can provide significant field support at no additional cost during critical times of a research project, facilitate access to private lands, and provide field equipment. Biologists are spread thin with diverse responsibilities, but in most cases they will prioritize research assistance because the results will benefit them. In some cases, local biologists provide the impetus for a research study and therefore are personally vested in the outcome. Further, if the agency has made a commitment to a research study, most likely that commitment will involve at least some support from field staff.

Research Process

Budgets for research and monitoring efforts can vary between state wildlife agencies. However, it is not correct to assume that research conducted by those organizations with limited financial resources results in less rigorous studies. Agencies that can only conduct a few projects at a time must ensure that those projects are of the highest priority for their organization and have been well designed and reviewed by both internal and external experts.

State wildlife agencies that have divisions or units focused on research and monitoring often have highly rigorous project selection protocols. These processes may require months to take information needs and develop them into full-blown research or monitoring projects. These organizations often develop their research requests by having scientists work directly with managers, administrators, and other stakeholders. Research projects developed through these approaches often go through multiple steps to help determine project need, refine the research question, propose alternative models, determine achievable objectives, and develop an appropriate statistical design. These agencies also use outside reviewers and scientists to help develop research studies. At each step in the development of these projects, studies can be dropped if they cannot address the original need or fail to meet the statistical rigor necessary to modify management actions or inform policy decisions. In Missouri, for example, it is not uncommon for only 30–50 percent of the research ideas proposed to result in on-the-ground projects.

Agencies with limited research staff often have committees to work with managers and administrators on determining what studies need to be conducted in the upcoming years. These committees are often composed of individuals with different professional backgrounds that represent the various areas of the organization (aquatics, terrestrial, law enforcement, forestry, etc.), and they often use outside expertise from universities, CFWRUs, and consultants to develop research, monitoring, and human dimension survey efforts. While smaller wildlife organizations may not have long-term proposal development processes to determine their highest priority needs, this is often offset by their close interaction with agency leadership, both management and policy, which provides direct input on the issues that need additional information to help guide future decisions.

While all research conducted by state wildlife agencies should be reviewed and vetted through both internal and external experts related to design and analysis, it is also important to make sure that animal handling protocols are humane and in compliance with the Animal Welfare Act of 1966, as amended. At most universities and some state agencies, faculty and staff engaged in research projects that require the handling of live animals must have their methods and protocols approved by an Animal Care and Use Committee (ACUC). This step ensures that all capture and handling protocols are safe for both scientists and the animals they are studying. Even though approval from an ACUC is often required for research projects, wildlife management activities, including routine animal capture and handling, are not subject to the Animal Welfare Act and remain under authority of the state or federal wildlife agency with jurisdiction.

Regardless of the size or organizational structure of a wildlife agency, the need to engage statisticians and biometricians during the project design phase is paramount. Without the proper statistical design at the beginning of a project, data collected may not be of the proper quantity and quality to be useful in answering the question that the project is supposed to inform. In these situations, erroneous conclusions may be inferred, resulting in ill-advised management decisions. Consequently, regardless of the method that agencies undertake to develop research projects, statistical expertise should be obtained before the project is initiated. Many state wildlife agencies have their own biometricians for this purpose, while others rely on contracted services or collaborations with universities. Researchers will also often engage database managers and geographic information system experts prior to on-the-ground data collection activities, as these individuals can help with the organization and maintenance of huge data sets that often result from large-scale research projects. Given that numerous research projects last in excess of five years, database managers can also help establish the necessary metadata (information about the individual pieces of data) for large amounts of information that may need to be transferred from one scientist to another through time.

All scientists have professional obligations to present research results at conferences and publish those findings. The dissemination of information through these channels allows other wildlife agencies and staff to utilize and build on research findings in other areas. However, unlike in federal or academic research positions, where job advancement is partially tied to the publication of research results, state agency scientists have traditionally had less pressure to publish their findings. On the positive side of this approach, the lack of publication requirements provides agency researchers greater flexibility to focus on fewer, higher-quality long-term studies. On the negative side, agency research may not be published until long after the study is completed, and on occasion never published in an accessible journal or in-house publication series. As difficult as it can be to conduct quality research on wildlife resources, it is imperative that state agencies insist on the dissemination and publication of research findings supported by their organizations.

Advocacy

Earlier in the chapter we emphasized the importance of state agencies supporting defensible research and monitoring programs to inform management decisions. A state agency's scientific credibility is critical because it promulgates regulations for managing wildlife that directly impact individuals, businesses, and communities. If science underlying a particular decision is identified as flawed or insufficient, impacted constituencies will have a legitimate basis to challenge the agency's decision. While this is relatively straightforward, an agency's scientific credibility can also be called into question if it is perceived as biased. Most agency employees are passionate about their work and strongly motivated to protect and conserve wildlife and their habitat, which is generally in alignment with their agency's mission. However, this passion and value orientation toward wildlife sometimes allows stakeholders with differing values to question the agency's objectivity in policy deliberations. State agency researchers, in particular, can be questioned because they provide key data and professional judgment used by all stakeholders to work through a decision process. For this reason, state agency researchers must ensure that any advocacy is based on science alone. Advocacy is defined by the *Oxford Dictionary of English* as "public support for or recommendation of a particular cause or policy." Research scientists are expected to present data and advance management recommendations that are consistent with the results of their studies, irrespective of political pressures or their personal philosophies. Simply put, they must be trusted to provide objective data and information. Most governmental decision processes are inherently political and formally allow for socioeconomic impacts to be considered, leading to management or policy alternatives that are not always in the best interests of wildlife. When researchers go beyond their science to advocate for a particular policy position that aligns with their values, they can become marginalized by stakeholders, rendering them less effective. In sum, wildlife agencies are required to navigate complex political processes while simultaneously conducting and communicating objective science. This biopolitical framework demands professionalism, integrity, and close working relationships among researchers, managers, and administrators if the collective agency is to effectively conserve wildlife and their habitat (Thomas 1985).

Case Study 1: Monitoring Western Mule Deer Populations

Mule deer are an iconic western species valued by sportsmen and enjoyed by wildlife enthusiasts. Revenues generated from mule deer hunting have been critical for funding wildlife management for decades. Mule deer can also serve as an indicator of overall ecosystem health. Given the importance of mule deer in the West, generations of wildlife managers have placed a high priority on monitoring the size of mule deer populations to inform harvest and other management decisions. Dating back to the 1960s, state wildlife researchers have been instrumental in developing techniques for wildlife management biologists to effectively monitor deer populations.

Mule deer are distributed across large landscapes in the West, often spanning rugged terrain with relatively low road densities. Wildlife managers, therefore, have long depended on helicopters and to a lesser extent fixed-wing planes as a means to survey mule deer and other ungulate populations. However, the question of how to most effectively use aircraft to obtain suitably precise and unbiased population estimates is far from trivial. Managers have relied on state agency researchers and their university colleagues to develop and improve aerial monitoring techniques. The first challenge researchers faced was developing a suitable sampling approach. Research biologists divided deer winter ranges into rectangular or irregular-shaped sample units, often referred to as quadrats. A subset of quadrats were then selected, using simple or stratified random sampling schemes, for conducting aerial counts (Gill 1969; Kufeld et al. 1980). Within each sampled quadrat, aerial observers attempted to count all deer present. These counts could then be extrapolated to generate a population estimate.

The second challenge researchers faced was counting bias. Researchers recognized that it was inappropriate under most circumstances to assume that every animal in a quadrat was detected, given visibility barriers such as trees and tall shrubs (Caughley 1974). Researchers put forth considerable efforts to estimate visibility bias when counting deer. They captured deer and placed radio collars on them with visual markers attached, thereby establishing samples of marked deer that could be radio-tracked and visually identified by aerial observers. Aerial observers would count deer in sample units as part of aerial surveys, and at the same time researchers would keep track of the position of the radio-marked deer. Researchers could then quantify the number of times radio-marked deer were present

Western mule deer monitoring case study. *Photo courtesy of Colorado Parks and Wildlife.*

in a sample unit but not seen by the observers, allowing them to estimate the percentage of animals that were not detected (i.e., sightability correction factor; Bartmann et al. 1986). In some cases, they evaluated animal sightability as a function of habitat type, snow conditions, animal group size, and animal activity. For example, a lone animal in heavy cover is much more difficult to detect than a group of animals in open habitat. Researchers developed sightability models, which were used to correct raw animal counts during aerial surveys based on a series of sightability factors (Ackerman 1988; Steinhorst and Samuel 1989). Collectively, these research efforts provided wildlife managers with techniques for accurately estimating mule deer populations using aerial surveys.

Mule deer management improved once management biologists had the ability to estimate size of deer populations and monitor them over time. Significant management challenges remained, however. Namely, it is expensive to conduct sample-based aerial surveys, and it became cost prohibitive for states to rely exclusively on aerial surveys to monitor deer populations statewide (Gill et al. 1983). Additionally, there are a number of logistical challenges biologists must address to implement aerial population surveys effectively. Given these challenges, research biologists looked into the use of population models to monitor the size of deer populations, which became increasingly more practical as computers evolved. The challenge for researchers was to identify a set of model inputs that could be obtained annually, without being cost prohibitive, and could generate reasonable estimates of population size. They initially capitalized on three parameters that were routinely collected by management biologists: harvest, age ratio, and sex ratio. Harvest simply reflects the number of animals taken by hunters, which is typically estimated by sample-based post-hunt surveys of hunters. Early winter age ratios (i.e., number of fawns per 100 does) provide an estimate of annual reproduction and survival of fawns to winter. Post-hunt sex ratios (i.e., number of bucks per 100 does) indicate the proportion of the population composed of bucks after the hunting season. Harvest, in combination with observed post-hunt sex and age ratios, provides a basic framework for modeling annual changes in a deer population. Age and sex ratios can be adequately estimated with less flight time and cost than population size, although they are subject to the same potential sampling and visibility biases discussed above. Research scientists therefore developed protocols, sampling strategies, and analytical tools to help management biologists obtain unbiased, suitably precise ratio estimates (Bowden et al. 1984; Samuel et al. 1992). Population models with adequate model inputs provided biologists with an additional tool for monitoring deer population size to inform management decisions, which allowed more deer populations to be monitored per unit cost.

Unfortunately, models did not always perform well even with good harvest and ratio inputs, which created problems for management biologists when stakeholder groups were upset over decisions being made. A weak population model made it more difficult for biologists to develop and defend hunt recommendations. It became increasingly clear to research scientists that adult female and overwinter fawn survival estimates were needed for models to perform well (White and Bartmann 1998). However, it would not be feasible to obtain survival estimates for every deer population on an annual basis, given the costs necessary to capture, radio-mark, and monitor samples of deer. An understanding of how survival varied over time and among populations was needed to inform an optimal survival monitoring strategy. Fortunately, state agency researchers were conducting studies in a number of locations across the West to measure survival and better understand factors that limit populations. Thus, data were available to support analyses of survival variability among adjacent populations and regionally across the West. Researchers from Colorado, Idaho, and Montana came together and accomplished the first multistate mule deer survival analysis to better understand how survival varied temporally and spatially (Unsworth et al. 1999). They found that annual variation was significant, particularly in overwinter fawn survival, demonstrating the need to measure survival annually as part of an effective monitoring program. Spatial variation was less, demonstrating that populations tended to fluctuate similarly in spite of broad habitat differences among states, although notable survival differences were documented among states in one year of study. Bishop et al. (2005) further documented spatial variability in survival among adjacent populations, emphasizing the need to measure survival in multiple populations annually to account for spatial variation in survival. This interstate research effort led to the establishment of multiple, perpetual deer survival monitoring areas across multiple states, providing agency biologists with empirical survival rates for modeling populations. Additionally, researchers evaluated strategies to inform management biologists how best to allocate available funds toward survival and aerial monitoring efforts (Bowden et al. 2000). Finally, White and Lubow (2002) developed a spreadsheet-based modeling framework for agency biologists to incorporate survival, age and sex ratios, harvest, and occasional population size estimates into an integrated population model. These various research efforts dramatically improved the ability of state agencies to obtain accurate population estimates for informing deer harvest and other management decisions.

The survival monitoring areas mentioned above have been instrumental in tracking deer population changes and informing management decisions in multiple western states over the past 15 years. They have also allowed more comprehensive analyses of survival variability among populations, thereby expanding our understanding of deer population dynamics and how best to utilize survival rates from a set of core populations to model other populations (Lukacs et al. 2009).

In summary, research tied to deer population monitoring provides a good example of how state wildlife researchers have worked closely with peer scientists and management biologists to benefit wildlife management. Collection of defensible deer population data has been fundamentally important to state agencies over time for informing harvest and various other management decisions. These data are becoming increasingly important for helping inform present-day efforts to maintain deer numbers in light of considerable landscape change (e.g., development, increasing human densities, noxious weeds). These data may also have utility for broader wildlife management and conservation objectives, constituting one of the best data sets collected for any species. For example, modeling of deer population data has enhanced our general understanding of parameter sensitivity, assumptions about model inputs, environmental stochasticity, and monitoring design considerations (Bowden et al. 2000; White and Lubow 2002). Long-term mule deer population data sets may ultimately serve as an invaluable resource for better understanding environmental changes, to the extent that mule deer can be deemed an umbrella species for western shrubland and forested habitats. Such empirical information on wildlife response to landscape change could be beneficial for other species and help solve present-day and future challenges of monitoring and managing a diverse set of sensitive and declining species for which population data are limited.

Case Study 2: The Missouri Ozark Forest Ecosystem Project

All state wildlife agencies conduct habitat manipulations to enhance wildlife populations and the health of various ecosystems. Most habitat activities occur at fairly small scales and can range from wildlife openings that may be from two to five hectares up to 1,000-hectare controlled burns. The results from these management activities are often studied for specific wildlife populations or specific aspects of the ecosystem and often for only three to five years after the management action was taken. For specific information needs, these studies, when well designed, often provide solid information on which to base future management actions so long as the information is not extrapolated beyond the scope of the work. However, wildlife populations do not exist in a vacuum, and it is important to remember that any management activity will result in both positive and negative benefits for various aspects of the system. It is also important to remember that most populations are managed by state agencies at a landscape scale (county or unit within a state, watersheds, etc.). Consequently, to truly understand the impacts of various management actions, they must be studied at a scale that is appropriate, both temporally and geographically, for the systems and populations that are being impacted by that management.

The Missouri Ozark Forest Ecosystem Project (MOFEP) is a 100-plus-year experiment designed to look at the impacts of even-age, uneven-age, and no-harvest forest management techniques on a number of the biotic and abiotic attributes of this ecosystem

(Brookshire et al. 1997). From a state agency perspective, the Missouri Department of Conservation needs to understand the long-term impacts of habitat manipulations associated with forest management techniques.

Given the temporal (a minimum of 100 years) and the spatial (3,680 hectares) extent of the MOFEP, it took significant time to determine the appropriate design and analytical approach to this study. The study was first conceived in 1989, but pre-treatment data collection was not initiated until 1991, with the first set of treatments (even- and uneven-age harvests) not occurring until 1996 (Sheriff and He 1997). The design of this adaptive experiment was selected to allow for a high degree of flexibility for managers (Sheriff and He 1997) while also allowing the results obtained throughout the study to be used to modify the ongoing treatments. The overall emphasis of the MOFEP was to use an adaptive learning approach while conducting on-the-ground management activities (Olson 2015).

The temporal and spatial scale of the MOFEP has allowed scientists from state, federal, nonprofit, and academic organizations to study various aspects of the ecosystem since the study's inception. The project has five main areas of focus and includes the impacts of

Typical sampling site used to determine impacts of forest management practices on ecology in the Missouri Ozark Forest Ecosystem Project. *Photo courtesy of Missouri Department of Conservation.*

forest management on birds, small mammals, herpe-tofauna, forest composition, and ground flora (Olson 2015). There are also numerous other studies being conducted on a wide range of variables from forest entomology to mycology (Knapp et al. 2014). During the first 21 years of the study, over 65 peer-reviewed journal publications have been developed from data collected during the MOFEP, as well as hundreds of proceeding reports, presentations, and technical papers (Knapp et al. 2014).

The Missouri Department of Conservation has modified the way it conducts uneven-aged forest management throughout the state based on results from the MOFEP. Actual management treatments being conducted on the project related to uneven-aged harvest have also been modified based on results from the ongoing project (Knapp et al. 2014). These modifications are reflective of the adaptive nature of the project design and, given the temporal scale of the project, are absolutely necessary for a state agency to justify the long-term investment. As information is acquired, management approaches across the state are altered, resulting in the need to modify the actual treatments being evaluated. Without this type of an approach, it would be unrealistic for a management agency to invest in a long-term research program with a design that does not allow the treatments to be modified in an adaptive fashion.

Future Challenges

Research supported by state wildlife agencies has figured prominently in the history of wildlife management in North America. The future role of research will depend heavily on how state agency research units and CFWRUs respond to several key challenges on the horizon. Perhaps the most important challenge is securing sufficient funding to sustain strong research programs. State wildlife agencies routinely deal with fiscal constraints and have competing demands for their available funds. Although this funding challenge is familiar, state agencies are currently struggling to meet a broadening conservation mandate without sufficient additional funding. Also, research programs are now competing for a limited set of funds with a greater number of internal support units as agencies diversify to remain relevant and competitive. Examples of expanding work units that were less prominent historically include information technology, marketing, outreach, and education. To complicate the picture, agencies are being required to direct greater funding toward maintenance of aging infrastructure. These fiscal and operational realities make it increasingly difficult for agencies to justify engaging in large-scale research efforts. The budget process that most agencies must follow also makes it difficult to enter into multiyear projects where funding is legislatively allocated on annual or biennial cycles. When budget reductions must be made, as many agencies have experienced in recent years, research is often considered a programmatic area where cuts can be absorbed without compromising an agency's ability to meet basic operating requirements such as issuing hunting and fishing licenses, enforcing laws, and providing customer service.

Now more than ever, research units need to demonstrate their utility by conducting applied science to simultaneously address near-term and long-term problems facing agencies. Agency leadership has become increasingly political, where it is now common for a director to remain in place for only a few years, which is less than the duration of a typical research project. The near-term or acute problems confronting an agency director will necessarily take precedence over various other longer-term conservation challenges. However, to meaningfully address the prevailing conservation problems, research must have been initiated years prior, often under previous directors when the issues were less acute. The challenge for research units is to constantly provide meaningful research results on acute problems facing an agency at any given moment, while at the same time securing financial support to initiate research on issues that are anticipated to be major conservation challenges in the future. Ultimately, state-supported research programs will need to be increasingly strategic to sustain funding as competing demands within agencies intensify and leadership is increasingly dynamic.

A notable technical challenge facing state research programs, or perhaps an opportunity in light of the above discussion, is the need to develop field and analytical techniques for monitoring and managing a diverse array of sensitive or declining species. Historically, many state agencies placed a heavier emphasis on monitoring a relatively small set of principal game species. Researchers, in coordination with management biologists, were able to develop and evolve monitoring strategies over time for these species. In contrast, agencies are now being tasked with monitoring a comparatively large number of nonhunted species, with priorities changing from one year to the next. For example, a biologist may be required to conduct a population survey on a bat species one year and one on a toad species the next. Researchers are being asked to come up with solutions, in terms of both field and statistical techniques. Fortunately, many state agencies have supported nongame programs over time, resulting in a wealth of institutional knowledge on a diverse array of species. Also, considerable progress has been made in recent years to monitor species using animal DNA and remote cameras. However, significant and numerous challenges remain for agencies to implement defensible monitoring programs, particularly for widely distributed species in low abundance. Pressures are arguably greater than in the past because decisions regarding conservation of potentially threatened or endangered species are routinely litigated, and therefore the science is intensely scrutinized. These pressures demand that state agency science and monitoring strategies are sound, which is an opportunity for state research programs to receive additional funding and support.

Finally, one of the areas that is critical to the success of any wildlife agency is the development of an understanding of their constituents' expectations and desires (see chap. 12). Developing this information requires human dimensions research that, owing to changes in society, should be repeated at regular intervals to remain relevant. Finding ways to support additional human dimensions research without compromising other research priorities remains a significant challenge for the future.

LITERATURE CITED

Ackerman, B. B. 1988. Visibility bias of mule deer aerial census procedures in southeast Idaho. PhD diss., University of Idaho, Moscow, Idaho, USA.

Allendorf, F. W., G. H. Luikart, and S. N. Aitken. 2012. Conservation and the genetics of populations, 2nd edition. John Wiley and Sons, Chichester, West Sussex, UK.

Bartmann, R. M., L. H. Carpenter, R. A. Garrott, and D. C. Bowden. 1986. Accuracy of helicopter counts of mule deer in pinyon-juniper woodland. Wildlife Society Bulletin 14:356–363.

Bishop, C. J., J. W. Unsworth, and E. O. Garton. 2005. Mule deer survival among adjacent populations in southwest Idaho. Journal of Wildlife Management 69:311–321.

Bowden, D. C., A. E. Anderson, and D. E. Medin. 1984. Sampling plans for mule deer sex and age ratios. Journal of Wildlife Management 48:500–509.

Bowden, D. C., G. C. White, and R. M. Bartmann. 2000. Optimal allocation of sampling effort for monitoring a harvested mule deer population. Journal of Wildlife Management 64:1013–1024.

Brookshire, B., L. R. Jensen, and D. C. Dey. 1997. The Missouri Ozark Ecosystem Project: Past, present, and future. Pages 1–25 in B. L. Brookshire and S. R. Shifely, editors. Proceedings of the Missouri Ozark forest ecosystem project symposium: An experimental approach to landscape research. US Department of Agriculture, Forest Service, North Central Forest Experiment Station, General Technical Report NC-193, St. Paul, Minnesota, USA.

Brownie, C., D. R. Anderson, K. P. Burnham, and D. S. Robson. 1978. Statistical inference from band-recovery data—a handbook. US Fish and Wildlife Service Resource Publication 131, Washington, DC, USA.

Burnham, K. P., and D. R. Anderson. 2002. Model selection and multimodel inference: A practical information-theoretic approach, 2nd edition. Springer, New York, New York, USA.

Caughley, G. 1974. Bias in aerial survey. Journal of Wildlife Management 38:921–933.

Foreyt, W. J., and D. A. Jessup. 1982. Fatal pneumonia of bighorn sheep following association with domestic sheep. Journal of Wildlife Diseases 18:163–168.

Fryxell, J. M., A. R. E. Sinclair, and G. Caughley. 2014. Wildlife ecology, conservation, and management, 3rd edition. John Wiley and Sons, Chichester, West Sussex, UK.

Gill, R. B. 1969. A quadrat count system for estimating game population. Colorado Division of Game, Fish, and Parks, Game Information Leaflet 76, Fort Collins, Colorado, USA.

Gill, R. B., L. H. Carpenter, and D. C. Bowden. 1983. Monitoring large animal populations: The Colorado experience. Transactions of the North American Wildlife and Natural Resources Conference 48:330–341.

Knapp, B. O., M. G. Olson, D. R. Larsen, J. M. Kabrick, and R. G. Jensen. 2014. Missouri Ozark forest ecosystem project: A long-term, landscape-scale, collaborative forest management research project. Journal of Forestry 112:513–524.

Kufeld, R. C., J. H. Olterman, and D. C. Bowden. 1980. A helicopter quadrat census for mule deer on Uncompahgre Plateau, Colorado. Journal of Wildlife Management 44:632–639.

Lukacs, P. M., G. C. White, B. E. Watkins, R. H. Kahn, B. A. Banulis, D. J. Finley, A. A. Holland, J. A. Martens, and J. Vayhinger. 2009. Separating components of variation in survival of mule deer in Colorado. Journal of Wildlife Management 73:817–826.

Miller, M. W., and E. S. Williams. 2004. Chronic wasting disease of cervids. Current Topics in Microbiology and Immunology 284:193–214.

Olson, M. G. 2015. Strategic plan for the Missouri Ozark forest ecosystem project. Missouri Department of Conservation, Jefferson City, Missouri, USA.

Robbins, C. T. 1993. Wildlife feeding and nutrition, 2nd edition. Academic Press, San Diego, California, USA.

Samuel, M. D., R. K. Steinhorst, E. O. Garton, and J. W. Unsworth. 1992. Estimation of wildlife population ratios incorporating survey design and visibility bias. Journal of Wildlife Management 56:718–725.

Sheriff, S. L., and Z. He. 1997. The experimental design of the Missouri Ozark forest ecosystem project. Pages 26–40 in B. L. Brookshire and S. R. Shifely, editors. Proceedings of the Missouri Ozark forest ecosystem project symposium: An experimental approach to landscape research. US Department of Agriculture, Forest Service, North Central Forest Experiment Station, General Technical Report NC-193, St. Paul, Minnesota, USA.

Silvy, N. J., editor. 2012. The wildlife techniques manual, 7th edition. 2 vols. Johns Hopkins University Press, Baltimore, Maryland, USA.

Steinhorst, R. K., and M. D. Samuel. 1989. Sightability adjustment methods for aerial surveys of wildlife populations. Biometrics 45:415–425.

Thomas, J. W. 1985. Professionalism—commitment beyond employment. Transactions of the Western Section of the Wildlife Society 21:1–10.

Unsworth, J. W., D. F. Pac, G. C. White, and R. M. Bartmann. 1999. Mule deer survival in Colorado, Idaho, and Montana. Journal of Wildlife Management 63:315–326.

White, G. C., and R. M. Bartmann. 1998. Mule deer management—what should be monitored? Pages 104–118 in J. C. deVos, Jr., editor. Proceedings of the 1997 deer and elk workshop. Arizona Game and Fish Department, Rio Rico, Arizona, USA.

White, G. C., and K. P. Burnham. 1999. Program MARK: Survival estimation from populations of marked animals. Bird Study 46(Supplement):120–139.

White, G. C., and B. C. Lubow. 2002. Fitting population models to multiple sources of observed data. Journal of Wildlife Management 66:300–309.

15

Jonathan W. Gassett

Future Needs and Challenges for State Wildlife Agencies

In the mid- to late 1800s, many wildlife species on the North American continent had reached their penultimate lows. Unregulated killing of wildlife for meat, fur, feathers, and other products was the primary cause of these uncontrollable reductions in population numbers. The immigration of Europeans into the Eastern Seaboard and their subsequent westward expansion drove demand for these products (Riess 2013), and the associated slaughter for market caused the drastic reduction, extirpation, and even extinction of many species across the continent. The creation of state fish and wildlife agencies stemmed the tide of unregulated harvest and began the long, slow process of continental wildlife restoration. The development and implementation of the North American Model of Wildlife Conservation was instrumental in the recovery of many wildlife species across the continent (Geist 1995; Geist et al. 2001).

Early efforts by state fish and wildlife agencies to recover imperiled North American fauna largely focused on species desirable for hunting, food, clothing, fur, and ornamentation. These efforts were bolstered by the conservation agenda of President Theodore Roosevelt, which precipitated the formation of the Boone and Crockett Club, The Wildlife Society, the American Game Protective and Propagation Association (later called the Wildlife Management Institute), and the Cooperative Wildlife Research Units, as well as the beginning of conservation curricula at the University of Wisconsin and the University of Michigan (Leopold 1930). The corresponding federalization of certain wildlife protection efforts, including implementation of the Lacey Act and Migratory Bird Treaty Act and Roosevelt's push to protect lands critical to wildlife conservation through federal appropriation (Brinkley 2009), resulted in the birth of the North American Model of Wildlife Conservation, which uniquely fit the needs and demands necessary for the successful recovery of numerous important species.

As many state fish and wildlife agencies reach or surpass their centennial anniversaries, they face unprecedented needs and challenges to their continued success. Agencies have achieved many accomplishments since their establishment around the turn of the twentieth century. Wildlife populations that were scarce to almost nonexistent at the time have now recovered to healthy numbers or, in some cases, to the point of superabundance. Early wildlife managers faced the daunting challenges of recovering such iconic game species as white-tailed deer, mule deer, pronghorn antelope, black bear, elk, turkey, bison, and others. Many species that were decimated by the proliferation of unregulated market hunting have now been recovered. States now face the challenge of recovering and protecting the much larger group of more than 12,000 species classified as at-risk, as well as implementing significant landscape-level habitat changes that impact all wildlife species. Agencies face these challenges with numerous obstacles ahead, most notably the lack of adequate, stable funding for substantive continental conservation efforts.

Most state agencies initially implemented user fees in the form of hunting licenses and permits to pay for conservation, but substantially more funding was needed to underwrite the expensive projects necessary for restoring many species. In 1937, Nevada senator Key Pittman and Virginia congressman Absalom Willis Robertson authored legislation that redirected a general excise tax on firearms and ammunition to specifically fund wildlife restoration efforts of the state fish and wildlife agencies, while simultaneously protecting state license dollars from diversion into other state coffers. This much-needed infusion of funds, which to date has cumulatively provided more than $9 billion to wildlife restoration, built the financial foundation for conservation across the United States (USFWS 2015).

Later actions by Congress supported by state and federal fish and wildlife agencies greatly contributed to the success and sustainability of the conservation effort across our country. The passage of the Duck Stamp Act, the Dingell–Johnson Act and its Wallop–Breaux Amendment (companions to the Pittman–Robertson Act), the Clean Water and Clean Air Acts, the National Environmental Policy Act, and the Endangered Species Act (and its companion—the Canadian Species at Risk Act) and creation of the Wildlife Conservation and Reinvestment Program and the State Wildlife Grant (SWG) program were all efforts to help ensure that our native natural resources received adequate protection and funding.

With many iconic game species now thriving, today's state fish and wildlife agencies face a whole new slate of challenges and essential needs that may limit their success and effectiveness into the future. Human population growth and the resulting increased urbanization; the introduction of invasive species; habitat degradation and loss; communication, collaboration, and partnerships; climate change; long-term, sustainable funding for conservation; fair chase, ethics, and public perception; and increasing hunter and angler recruitment, retention, and reactivation represent some of the greatest needs facing today's state fish and wildlife agencies.

Human Population Growth

The human population in the United States has grown from 76 million in 1900, to 152 million in 1950, to 275 million in 2000 and is expected to reach 399 million by 2050 (Day 1996). Increased demand for space, clean water, food, and other resources will grow exponentially with the expanding human population, undoubtedly placing significant pressure on native fish and wildlife resources. Human–wildlife conflicts, the introduction of invasive exotic species, habitat loss, degradation, fragmentation, and increasing controversy regarding the appropriate use of science in wildlife management are by-products of an increasingly urban, high-density human population that is further disconnected from nature (Kellert and Wilson 1995; Louv 2008).

As society becomes increasingly urbanized, wildlife species able to successfully coexist with man often result in wildlife population levels that become a nuisance or are otherwise sociologically unacceptable. State fish and wildlife agencies are seeing an unprecedented rise in the number of complaints associated with human–wildlife conflicts. Faced with an increasingly disconnected and urban populace, state agencies now expend considerable resources dealing with problem animals by way of actions that rarely have any impact at the population level. Much of the funding diverted to nuisance wildlife issues was previously used for more ecologically significant wildlife management programs, but it now serves a primarily sociological function.

Introduction of Invasive Species

Invasive species, especially those introduced by a burgeoning human population, pose another significant challenge to the future of state fish and wildlife agencies. Exotic species introduced into favorable habitat often have fewer biological checks and balances and thus may thrive to the point of superabundance. These introductions often result in displacement of or negative impact on native flora and fauna. Species such as common reed (*Phragmites* spp.), reed canary grass (*Phalaris arundinacia*), zebra mussels (*Dreissena polymorpha*), quagga mussels (*D. bugensis*), kudzu (*Puer-*

Massive invasive Burmese python (*Python molarus*) caught in the Florida Everglades. *Photo courtesy of the Florida Fish and Wildlife Conservation Commission.*

aria lobata), Asian carp (Cyprinidae), Burmese python (Python molarus), spotted knapweed (*Centaurea maculosa*), and feral swine (*Sus scrofa*) are examples of exotics introduced by humans that have swamped or otherwise damaged native populations and habitats, in some cases to the point of localized extirpation.

Case Study: Burmese Python
(with permission of the Florida Fish and Wildlife Conservation Commission)

The Burmese python is one of the largest snakes in the world. Adult Burmese pythons caught in Florida average between 1.8 and 2.7 meters (6–9 feet); the largest Burmese captured in Florida measured over 5 meters (17 feet) in length. Burmese pythons are semiaquatic and often found near or in water. They are also excellent climbers and can be found in trees. Often cited as having a docile nature, Burmese pythons are popular in the pet trade. However, they are currently listed as a conditional species in Florida, which are species that may be dangerous to the ecology and/or health and welfare of the public, and therefore can no longer be acquired as pets in the state. They are also federally listed by the US Fish and Wildlife Service (USFWS) as an injurious species under the Lacey Act, which prevents the importation of pythons into

the United States and also prohibits snakes from being transported across state lines. Their native range includes India, lower China, the Malay Peninsula, and some islands of the East Indies.

FLORIDA DISTRIBUTION
A population of Burmese pythons is now established in South Florida, mainly within the Florida Everglades. Individual snakes have been found near Naples, suggesting that the population may be moving northwest. Python observations outside of South Florida are typically escaped or released pets.

INTRODUCTION HISTORY
Burmese pythons have been reported from the saline glades and mangroves at the south end of Everglades National Park since the 1980s. The actual mechanism of introduction is not known; however, it is likely that Burmese pythons escaped from a breeding facility that was destroyed during Hurricane Andrew in 1992. It is also likely that pet pythons have been released in and around the Everglades.

CONCERNS
In Florida, Burmese pythons have been found to prey on a variety of mammals, birds, and even alli-

also consume threatened or endangered native species. One python that was caught on Key Largo had eaten an endangered Key Largo wood rat. Burmese pythons also can pose a threat to human safety and may prey on pets such as cats and dogs. There is potential for the population to spread west toward Naples, and more research is required to determine how far pythons can survive outside of South Florida. These invasive snakes pose significant environmental, economic, and social concerns.

WHAT IS THE FLORIDA FISH AND WILDLIFE CONSERVATION COMMISSION DOING?

The Florida Fish and Wildlife Conservation Commission (FFWCC) works with several agencies and organizations to manage Burmese pythons that are established in and around the Everglades. They coordinate their management activities and objectives with other field offices and agencies, tribes, universities and researchers, and nongovernmental organizations (NGOs) so that their collective efforts and projects are complimentary. FFWCC staff continually survey and monitor pythons in South Florida. They track sightings and respond rapidly to potential new populations of all exotic constrictors.

As part of the FFWCC's program to reduce the population of Burmese pythons in South Florida, the FFWCC allows snake experts to remove these non-native constrictors from wildlife management areas and several properties managed by the South Florida Water Management District. These volunteers not only remove Burmese pythons but also provide valuable data to the FFWCC on the locations and sizes of the snakes. This will help to contain the population from spreading north from the Everglades. The FFWCC also allows licensed hunters to kill Burmese pythons they encounter in several South Florida Wildlife Management Areas.

To prevent future invasions, the FFWCC sponsors Exotic Pet Amnesty events where pet owners can surrender pythons and other non-native animals rather than release them in Florida's woods and waters. People can also report Burmese

Current range of Burmese python (*Python molarus*) from the introduction and subsequent dispersal throughout South Florida.

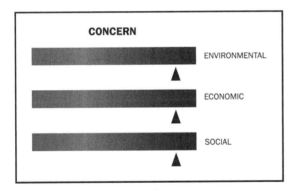

Degree of environmental, economic, and social concern over Burmese python (*Python molarus*) impacts to South Florida.

gators. Because of their large size, adult Burmese pythons have few predators, with alligators and humans being the exceptions. They prey on native species and may reduce their populations locally. Research is under way to ascertain the impacts pythons have on native mammal species. While pythons will eat common native species and exotic species such as Norway rats, they can python sightings to the FFWCC's Exotic Species Hotline. Callers reporting a live snake are routed to a hotline operator or to an FFWCC dispatcher. Although it may be impossible to eradicate Burmese pythons from South Florida, the FFWCC has learned a lot about their habits and is optimistic that it will be able to contain this population and reduce its impacts on native wildlife.

PYTHON REMOVAL PROGRAM

The Burmese Python Removal Program is a management tool that allows people to remove Burmese pythons and other non-native reptiles from state lands. It is not a python-hunting program or a recreational program. While the primary focus of the program has historically been Burmese pythons, the intent of the program is to allow properly trained and permitted people to remove all invasive exotic reptiles that are encountered during collection trips. Permit holders must turn in all captured pythons, but they can request to have the carcasses returned to them. They are allowed to sell skin and meat, thus providing a type of compensation (note: Burmese pythons from Everglades National Park have been found to have very high levels of mercury and may not be recommended for human consumption). All non-native reptiles, including conditional reptiles, may be taken under this permit. Snakes can be captured by hand or by using handheld equipment (snake tongs, snake hooks, etc.). Pythons may be euthanized on-site by legal and humane methods, or dropped off live at a site designated by the FF-WCC. Burmese python removal permits also allow for the live transport of other conditional reptiles. This option is for Python Patrol responders, Cooperative Invasive Species Management Area cooperators, nuisance trappers, and government employees who need to transport conditional reptiles as part of their job duties.

Habitat Degradation and Loss

With the unprecedented growth and expansion of the human population comes the associated loss and fragmentation of wildlife habitat. Development, improved farming techniques, energy development and transmission routes, highway construction, increased greenhouse gas emissions, and coastline development frequently result in the loss of secure, high-quality habitat necessary for many species of wildlife. Particularly in the eastern United States, along the West Coast, and in areas of intensive energy production, a loss of habitat resulting from urban sprawl, a dense network of roads, and a changing climate is compromising the connectivity of habitat and wildlife populations. The challenge for state fish and wildlife agencies, as well as other land management agencies, will be to delineate a landscape that, if protected and appropriately managed, will restore and maintain gene flow between viable breeding populations of native wildlife species, reduce direct mortality along transportation corridors, and ensure long-term existence of wildlife species throughout these areas of dense human habitation. State fish and wildlife agencies must begin to focus more at the landscape scale in order to provide the necessary habitat and connectivity that will be required to sustain our native resources. Agencies must critically evaluate their state and strategically identify specific priority locations and landscapes in need of protection.

Communication, Collaboration, and Partnerships

Communication with the public on issues of critical conservation challenges is also of paramount importance for state agencies. The failure of state fish and wildlife agencies to accurately and clearly communicate the priorities for sound natural resource management is becoming more pervasive as they transition to a more diverse constituency and increasing public interest as it relates to all fish and wildlife species. Agencies must strive to openly communicate their issues and priorities so that the public understands not only why these factors are critical but also that they are the highest priorities for ensuring abundant fish and wildlife into the future.

Given issues facing natural resource management in an increasingly urban setting and at larger geographic scales, new approaches will also be required that foster interdisciplinary collaboration and provision of training opportunities that meet the information needs of regional and continental natural resource issues. In general, state agencies are facing an ever-shrinking workforce, so establishing or building partnerships with land-grant universities, research institutions, US Geological Service Cooperative Research Units, and NGOs will be instrumental in addressing contemporary conservation issues and developing new and innovative strategies for effective wildlife conservation.

Agencies must continue to make collaboration and partnerships a priority, building a better cohesive system of conservation with public agencies at all levels and with conservation NGOs. State agencies will not be able to accomplish all necessary conservation objectives without first building conservation capacity and a conservation ethic in the broader community. In some early adopter states, this is already occurring with the advent of legacy constitutional dollars, primarily in the form of legislatively dedicated sales tax funds, distributed across government and NGOs that pursue a conservation mission. This will lead to increased public support beyond what any individual agency could hope to accomplish.

The proper role and use of science is another significant communication need important to the future of state agencies. This issue ranges from what disciplines are actually considered science (from a complete lack of scientific method to the peer review process) to the use and/or abuse of data and science in wildlife management. Within the life span of many state fish and wildlife agencies, we have witnessed the birth of, maturity of, and significant decline in the use of science in wildlife management. At the turn of the twentieth century, little was known about relationships between predators and prey, wildlife and habitat, or wildlife and exigent environmental factors. From the 1930s through the 1980s, sound science took on an increasingly necessary role for the restoration, recovery, and stabilization of many of our wildlife species.

Within the past 30 years, science has begun to take a back seat to social pressure, politics, anecdotal references, and other nonecological factors. Agencies are witnessing an exponential increase in opinion pieces that carry the same weight as peer-reviewed and published scientific papers, particularly in Endangered Species Act listings, Clean Water and Clean Air Act

rules, and other controversial decisions. Additionally, the potential and realized spread of disease from ill-advised private wildlife translocations, the widespread habitat loss resulting from urban and suburban sprawl, and the continuing battle for accessible lands and waters for hunters are all having significant negative economic impacts on local communities. These trends must be reversed to most efficiently and effectively manage our native resources.

Climate Change

Climate change will likely prove to be one of the most significant challenges to state fish and wildlife agencies in the future. As the political debate over mankind's contribution to climate change rages on, it is widely accepted by the scientific community that the earth is undergoing increasingly rapid periods of climatic shifts that are anthropogenic in nature. A failure to take climate change into account during the decision-making process will likely result in failure to reach wildlife management objectives in the future (Inkley et al. 2004).

As climatic conditions continue to shift, we will see landscape-level habitat alterations and associated changes to the species of fish and wildlife that occupy them. Vulnerable species will likely become threatened to the point of regional extirpation or extinction, while other opportunistic species will thrive and become dominant on landscapes, rivers, streams, and oceans where they have never been found before. Extreme floods and droughts will regulate plant and animal communities more so than ever within our lifetimes. Changes to aquatic systems, including base flow rates and maximum and minimum air and water temperatures, will necessitate shifting of both aquatic and terrestrial species guilds to compensate for these changes driven by climate change. State agencies must begin to balance management of rare species versus abundant and common fish and wildlife resources. Indeed, the conservation tent must be expanded and become more inclusive—beyond sport fish and hunted wildlife species—if agencies intend to survive, prosper, and succeed. They must also begin planning for the new normal as climate continues to shift and species guilds respond. Many species, by necessity, will adapt to these new conditions, and state fish and wildlife agencies and the public they serve must adapt as well.

Effective adaptive management approaches will be necessary to protect and enhance fish and wildlife populations and their habitats in the face of climate, land-use, invasive species, and human population trends. Climate change and development of land for food, housing, and other uses by a burgeoning human population will continue to stress many types of habitat and fish and wildlife populations. Climate change will also likely allow invasive species to increasingly stress and change aquatic and terrestrial communities.

One of the most significant recent tools created to address conservation of fish and wildlife populations are State Wildlife Action Plans (SWAPs). These plans were developed by each state fish and wildlife agency and designed to address the needs of more than 12,000 species of greatest conservation need. SWAPs are scientifically based, strategic, and cost-effective plans designed to preserve our wildlife resources for the future. Recovery of species that have reached threatened or endangered status is typically more costly than preventative actions that keep species populations from reaching such declines. Proactive management actions identified in the SWAPs, including those that address climate change impacts, are intended to prevent species from becoming threatened or endangered.

Case Study: Sea Level Rise in Coastal North Carolina

(with permission of the Wildlife Management Institute)

Sea level rise has been identified as a significant climate change issue along coastal North Carolina, which is the third-lowest-lying state in the United States. Much of the land is just above sea level (below 1 meter) and is currently experiencing significant levels of erosion (Poulter 2005; Feldman et al. 2009). Climate change contributes to sea level rise increases through thermal expansion of ocean waters and ice field melting. During the twentieth century, the average global (eustatic) sea level rose by about 0.17 meters (6.7 inches), at an average rate of 0.017 meters (0.07 inches) per year. This was 10 times faster than the average rate of sea level rise during the past 3,000 years (Parry et al.

2007). Sea level rise is projected to increase by between 0.19 and 0.59 meters (7–23 inches) by the end of the century (2090–2099; Parry et al. 2007; Titus et al. 2009). However, these projections do not account for recent changes in ice flows in Greenland and Antarctica, meaning that these values likely underestimate future global rates of sea level rise (Union of Concerned Scientists 2006; Titus et al. 2009).

North Carolina has experienced increased rates of relative sea level rise, including the global rate, as well as localized factors such as land subsidence, based on geological data and tide gauge data. Several studies indicate that, on average, 1 millimeter of relative sea level rise likely occurred per year for the past 2,000 years up until the twentieth century. More recently, rates have ranged between 3.0 and 3.3 millimeters (Zervas 2004; Kemp 2009; Kemp et al. 2009). The North Carolina Coastal Resources Commission Science Panel on Coastal Hazards used these estimates to determine that North Carolina will likely see an increase of 0.4–1.4 meters of sea level rise by the end of the century. The panel recommended adopting a 1-meter rise scenario by 2100 for policy and decision-making purposes in the state (NCCRC 2010). The panel also notes that 2 meters of rise is possible but unlikely, unless there are accelerated rates of ice sheet melting and warming (NCCRC 2010).

Various tools exist to visualize what sea level rise may look like for the Albemarle-Pamlico Sound. Screenshots from the National Oceanic and Atmospheric Administration's Digital Coast's Sea-Level Rise and Coastal Flooding Impacts Viewer (www.csc.noaa.gov/digitalcoast/) allow a user to see what a range of sea level increases might look like for the peninsula. The tool allows a user to look at 1, 2, 3, 4, or 5 feet of sea level rise. The tool uses a simple bathtub sea level rise model that shows areas vulnerable to inundation from sea level rise based solely on elevation projection. It does not take into account factors such as erosion, marsh migration potential, or hydrology that could influence how sea level rise may occur on the landscape.

IMPACT OF SEA LEVEL RISE ON ENDANGERED RED WOLVES

The red wolf (*Canis rufus*) is an iconic native predator of the southeastern United States that

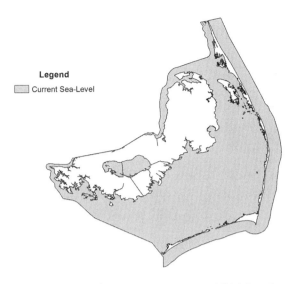

Albemarle Peninsula, current status. *Source: NOAA Digital Coast Sea-Level Rise and Coastal Flooding Impacts Viewer.*

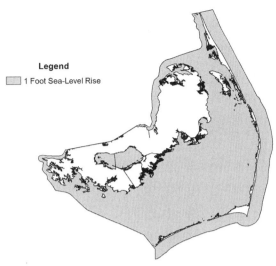

Albemarle Peninsula, with 1 foot of sea level rise. *Source: NOAA Digital Coast Sea-Level Rise and Coastal Flooding Impacts Viewer.*

Albemarle Peninsula, with 3 feet of sea level rise. *Source: NOAA Digital Coast Sea-Level Rise and Coastal Flooding Impacts Viewer.*

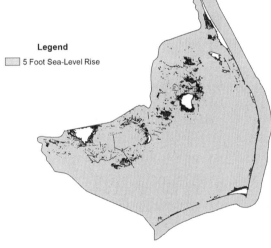

Albemarle Peninsula, with 5 feet of sea level rise. *Source: NOAA Digital Coast Sea-Level Rise and Coastal Flooding Impacts Viewer.*

was persecuted throughout the past 200 years until the last remnant population was removed from coastal Louisiana and Texas in the 1970s and placed into captivity by the USFWS. A recovery attempt was initiated in the 1980s with a target of establishing populations in the Great Smoky Mountains National Park in eastern Tennessee and the Alligator and Pocosin Lakes National Wildlife Refuges in coastal North Carolina. While the Tennessee population failed fairly rapidly as a result of an insufficient prey base, the North Carolina population persists at a level of around 100 individuals, making this one of the most endangered vertebrate species in the world.

Red wolf restoration efforts have already seen significant resistance from neighboring private landowners, who are concerned for their livestock, wildlife, and pets. Wolves currently move freely between federally owned national wildlife refuges and neighboring private lands. With a sea level rise of only 1 foot, much of the federal land would become inundated, thus pushing the wolves further onto private lands. With a rise of 3–5 feet, much of the available red wolf habitat on both public and private land would be lost, potentially jeopardizing restoration efforts and predisposing the red wolf to extinction in the wild once again.

Conservation Funding

Funding for state fish and wildlife agencies comes from a variety of sources. Licenses, permit fees, and registrations for hunting, fishing, and trapping constitute the primary state-based funding mechanism, although specialized items such as license plates, state tax check-offs, and voluntary donations are used by some states to provide funds for specific projects and activities. Some states also receive general fund appropriations from their legislatures, although the annual competition for limited dollars makes this a less-than-desirable option. States such as Missouri, Arkansas, Minnesota, Virginia, and Iowa possess unique funding mechanisms where a dedicated portion of state sales tax is earmarked for conservation efforts. These states utilize a public-pay/public-benefit system, where all citizens and taxpayers support the costs of fish and wildlife management.

Participant Funding through Licenses, Permits, and Federal Excise Taxes

The majority of state fish and wildlife agencies still receive most of their conservation funding through the sale of licenses, permits, and federal excise taxes on firearms and ammunition, archery equipment, and fishing tackle. License and permit revenue and the associated federal excise tax still remain the single largest source of revenue to state agencies, averaging over 70 percent of their annual budgets.

In 1937, Congress passed the Pittman–Robertson (PR) Act, which redirected an 11 percent excise tax on firearms and ammunition from the general treasury to the USFWS, with a further directive that these funds be apportioned to the states to provide funding for wildlife restoration. The Dingell–Johnson (DJ) Act, passed in 1950, provided similar support for game fish restoration and management by assessing a tax on angling equipment. In 1984, the Wallop–Breaux Amendment to DJ included a percentage of the national fuel tax that provides funding to develop and maintain motorboat access. Collectively, these funds, along with the SWG program, are managed by the USFWS through their Wildlife and Sport Fish Restoration (WSFR) program and are apportioned to the states through a formula based on the landmass of the state and the number of licenses sold annually. The WSFR program, through PR, DJ, and the SWG program, has cumulatively provided more than $18 billion for fish and wildlife conservation in this country since 1937 (USFWS 2015).

Future funding models for state fish and wildlife agencies must leverage fiscal, economic, and advocacy support of hunters and anglers and build on that support to leverage additional resources from all beneficiaries. Many state agencies are still highly reliant on user fees from hunters and anglers. However, current trends indicate a significant demographic change in the next 10–20 years.

Nontraditional Funding

Several avenues for external long-term funding for wildlife conservation have been approached in the recent past. In 2001, Congress established the Wildlife Conservation and Restoration Program (WCRP), nested within PR, to provide funding for wildlife diversity conservation, but funding was only appropriated for the first year. After that, Congress failed to appropriate funding under this program. In the past 10 years, there has been a significant push by the conservation community to provide stable funding for wildlife diversity using an excise tax on other outdoor products, such as binoculars, camping gear, and other outdoor equipment. This initiative, however, called the Conservation and Reinvestment Act (CARA), failed to pass Congress and become law. As a substitute, the SWG program was

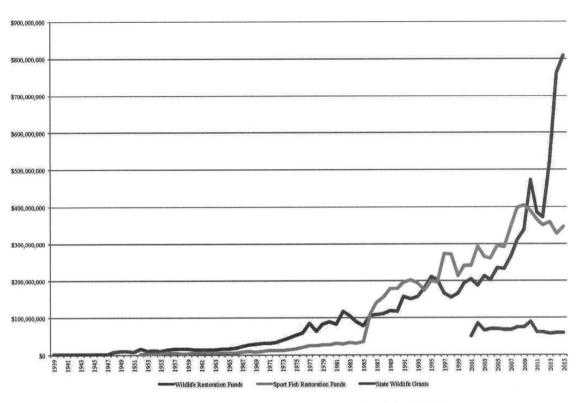

Funding levels of Wildlife and Sport Fish Restoration programs and State and Tribal Wildlife Grant programs.

created with funding from an offshore oil and gas tax that was directed to the states to meet conservation objectives for species of greatest conservation need. The SWG program has been in place since 2002, and funding has ranged from $58 million in 2013 to $90 million in 2010. Because the SWG program is appropriated annually during the budget-making process of Congress, states are not provided with the stability and consistency of funding needed to support long-term conservation initiatives. States are appropriated with SWG funds using the same formula as used for PR and DJ. The estimated funding requirement for comprehensive wildlife diversity conservation is $1.3 billion (to reach 75% implementation of SWAPs; Southwick Associates 2012a), leaving even the best funding year for the SWG program short by $1.2 billion.

Blue Ribbon Panel

In 2014, a Blue Ribbon Panel on Wildlife Diversity Funding was formed for the purpose of developing a twenty-first-century model for wildlife conservation funding. This panel consists of leaders across the spectrum of conservation, outdoor recreation, energy, retail sales, and government. The outcome will be to produce a recommendation and model legislation to resolve the long-term funding issues, including the annual deficit of funding for at-risk species of more than $1.3 billion. The panel will further evaluate the mechanism of using the WCRP account under PR as the vehicle to appropriate and transfer these funds to the agencies with authority to manage fish and wildlife resources throughout the United States, with the goal of adequate and stable funding for the more than 12,000 species and their habitats in need of protection and conservation.

Fair Chase, Ethics, and Public Perception

The ability for hunting to continue in North America largely depends on participants, lawmakers, and the public better understanding the North American

Model of Wildlife Conservation, which is a set of principles that is the foundation of conservation success stories throughout the United States and Canada (Organ et al. 2012). The model's seven principles are outlined in chapter 1.

Strict adherence to these principles has produced tremendous success stories in the recovery of multitudes of game species directly, through restoration and responsible management, and of many nongame species through the acquisition, protection, and improvement of habitat (chap. 2). The model has also set the stage for evaluation and introspection by state fish and wildlife agencies. Not only have they learned from both successes and failures, but they have also evaluated challenges to various fish and wildlife management approaches based on how those approaches coincide with or differ from the model.

While the North American model is still the most successful model for wildlife conservation in the world, it is currently facing unprecedented challenges. States will need to determine whether they can continue to work effectively with the primary funding source being generated from the user-pay/public-benefit model. If they cannot, they need to begin to think about a new model for conservation, such as a public-pay/public-benefit model that diversifies funding, since the work that they do provides goods and services for many more people than just anglers, hunters, and trappers. States will be required to develop strategies to define conservation in a manner that the general public can understand and will be willing to fund.

Specific challenges to the model facing agencies today are diverse. The lack of understanding of sound wildlife conservation principles and the North American model on which they are grounded has exposed agencies to attacks from various venues, including animal rights and anti-firearms activists. The commercial sale of both live and dead native wildlife, increasing restrictions on access to wildlife, and political pressure for regulatory changes that would result in private ownership of wildlife are a few of the other challenges impacting state agencies. This politicization of wildlife policy, in the face of sound science, may indeed be one of the biggest threats to the future of state fish and wildlife agencies, since each of these is also a direct threat to the North American model.

Increasing Hunter Recruitment, Retention, and Reactivation

Surveys indicate that hunter participation has declined steadily since the 1980s (US Department of the Interior et al. 2011). A number of potential causes, including lack of time, loss of hunting partners, lack of interest, aging demographics, lack of access, competition from other hobbies, and a lack of successful recruitment efforts, have likely led to the systemic decline in hunting license sales, which are currently the primary funding sources for most state conservation efforts. While the revenue generated from the sale of firearms and ammunition via the Pittman–Robertson Act has increased, the loss of license dollars proves to be an ongoing challenge for most state agencies and may jeopardize the future ability of these agencies to access their federal excise tax dollars. Ultimately, the potential for loss of traditional conservation revenues, in the form of hunting license dollars and the associated federal excise taxes, could result in the loss of critical funding for wildlife conservation that may prove difficult to replace with alternative sources.

Despite declines in hunting license sales, the importance of the economic impact and social support of hunters cannot be understated. Recent surveys indicate there are 13.8 million hunters and 40.8 million recreational shooters in the United States (Southwick Associates 2012b). These activities, when combined, result in almost 1.5 million jobs and $110 billion of economic output to the United States annually. Much of this impact occurs in rural areas, where many businesses are likely dependent on the annual influx of hunters for their long-term financial stability.

Many conservation NGOs and state fish and wildlife agencies are part of an ongoing effort to develop a national action plan to reverse the decline in participation among hunters and target shooters. Coordinated by the Wildlife Management Institute and the Council to Advance Hunting and the Shooting Sports, this plan is being designed to ensure that wildlife conservation remains fueled by hunters and shooting sports enthusiasts. As members of organizations keenly interested in promoting wildlife conservation, the partners have assembled a development work group of experts representing a cross section of agencies, conservation and shooting sports organizations, and representatives from industry to coordinate and optimize efforts to recruit, retain, and reactivate hunters and shooting sports enthusiasts.

The outcome of this national plan will be to inventory current efforts, coordinate resources, and develop customizable toolkits for agencies, conservation organizations, and industry partners. The plan will result in effective, proven strategies and tools to create more hunters and shooting sports participants, especially among nontraditional audiences. Stakeholders who adopt the plan will increase participation among likely hunters and shooting sports participants and see a groundswell of support from unlikely allies in the community.

Next Steps

State fish and wildlife agencies face numerous challenges and unprecedented needs in the future in order to remain a sustainable and effective force for managing wildlife resources and their habitats. While these challenges and needs are formidable, they are not insurmountable. Ultimately, state fish and wildlife agencies must continue to strive to remain relevant to their customers and to the public at large.

In light of the increasing disconnect between people and nature, state agencies will also face a growing problem of recruiting and staffing employees that have an interest in and a passion for fish and wildlife management and conservation. Identifying not only future employees but also future leaders and effective succession plans will be critical for agencies to succeed in overcoming the substantial challenges they face.

Agencies must also push to help modernize conservation laws and regulations, many of which were drafted more than 50 years ago. Updates to the PR, DJ, Endangered Species, Clean Water, Clean Air, and National Environmental Policy Acts and the Wallop–Breaux Amendment are necessary to ensure that these laws fit current needs of fish and wildlife conservation at regional, national, and continental scales. Likewise, passage of long-term sustainable funding, spearheaded by efforts of the Blue Ribbon Panel on Wildlife Conservation Funding for the benefit of more than 12,000 at-risk species, is of vital importance.

States have the wherewithal to meet all of these challenges, as evidenced by their monumental track record of state-level conservation. Ultimately, they will need to rely on the desires of the public in order to ensure that they remain relevant, that conservation strategies remain scientifically driven, that conservation funding grows and remains strong, and that the public not only helps to pay for but also receives the myriad of benefits derived from professional wildlife stewardship.

LITERATURE CITED

Brinkley, D. 2009. The wilderness warrior: Theodore Roosevelt and the crusade for America. HarperCollins, New York, New York, USA.

Day, J. C. 1996. Population projections of the United States by age, sex, race, and Hispanic origin: 1995 to 2050. US Bureau of the Census, Current Population Reports, US Government Printing Office, Washington, DC, USA.

Feldman, R. L., J. G. Titus, B. Poulter, J. DeBlieu, and A. S. Jones. 2009. State and local information on vulnerable species and coastal policies in the Mid-Atlantic: North Carolina. Pages 229–238 *in* Coastal sensitivity to sea-level rise: A focus on the Mid-Atlantic Region. US Environmental Protection Agency, Washington, DC, USA.

Geist, V. 1995. North American policies of wildlife conservation. Pages 75–129 *in* V. Geist and I. McTaggert-Cowan, editors. Wildlife conservation policy. Detselig Enterprises, Calgary, Alberta, Canada.

Geist, V., S. P. Mahoney, and J. F. Organ. 2001. Why hunting has defined the North American model of wildlife conservation. Transactions of the North American Wildlife and Natural Resources Conference 66:175–185.

Inkley, D. B., M. G. Anderson, A. R. Blaustein, V. R. Burkett, B. Felzer, B. Griffith, J. Price, and T. L. Root. 2004. Global climate change and wildlife in North America. Technical Review 04-2. The Wildlife Society, Bethesda, Maryland, USA.

Kellert, S. R., and E. O. Wilson, editors. 1995. The biophilia hypothesis. Island Press, Washington, DC, USA.

Kemp, A. C. 2009. High resolution studies of late Holocene relative sea-level change (North Carolina, USA). PhD diss., University of Pennsylvania, Philadelphia, Pennsylvania, USA.

Kemp, A. C., B. P. Horton, S. J. Culver, D. R. Corbett, O. van de Plassche, W. R. Gehrels, B. C. Douglas, and A. C. Parnell. 2009. Timing and magnitude of recent accelerated sea-level rise (North Carolina, USA). Geological Society of America 39:1035–1038.

Leopold, A. 1930. Report to the American game conference on an American game policy. Transactions of the American Game Conference 17:281–283.

Louv, R. 2008. Last child in the woods. Algonquin Books, New York, New York, USA.

NCCRC (North Carolina Coastal Resources Commission Science Panel on Coastal Hazards). 2010. North Carolina sea-level rise assessment report. Department of Environment and Natural Resources, Division of Coastal Management, Raleigh, North Carolina. USA.

Organ, J. F., V. Geist, S. P. Mahoney, S. Williams, P. R. Krausman, G. R. Batcheller, T. A. Decker, R. Carmichael, P. Nanjappa, R. Regan, R. A. Medellin, R. Cantu, R. E. McCabe, S. Craven, G. M. Vecellio, and D. J. Decker. 2012. The North American Model of Wildlife Conservation. Technical Review 12-04. The Wildlife Society, Bethesda, Maryland, USA.

Parry, M. L., O. F. Canziani, J. P. Palutikof, P. J. van der Linden, and C. E. Hanson, editors. 2007. Climate change 2007: Impacts, adaptation and vulnerability. Cambridge University Press, Cambridge, UK.

Poulter, B. 2005. Interactions between landscape disturbance and gradual environmental change: Plant community migration in response to fire and sea level rise. PhD diss., Duke University, Durham, North Carolina, USA.

Riess, S. A. 2013. Sport in industrial America, 1850–1920, 2nd edition. Wiley-Blackwell, Hoboken, New Jersey, USA.

Southwick Associates. 2012a. AFWA wildlife funding survey. Association of Fish and Wildlife Agencies, Washington, DC, USA.

———. 2012b. Hunting in America: An economic force for conservation. National Shooting Sports Foundation, Newtown, Connecticut, USA.

Titus, J. G., K. E. Anderson, D. R. Cahoon, D. B. Gesch, S. K. Gill, B. T. Gutierrez, E. R. Thieler, and S. J. Williams. 2009. Coastal sensitivity to sea-level rise: A focus on the Mid-Atlantic Region. US Environmental Protection Agency, Washington, DC, USA.

Union of Concerned Scientists. 2006. Climate change in the U.S. Northeast: A report of the northeast climate impacts assessment. Cambridge, Massachusetts, USA.

US Department of the Interior, US Fish and Wildlife Service, and US Department of Commerce, US Census Bureau. 2011. National survey of fishing, hunting, and wildlife-associated recreation.

USFWS (US Fish and Wildlife Service). 2015. Wildlife and sport fish restoration program. http://wsfrprograms.fws.gov/home.html.

Zervas, C. 2004. North Carolina bathymetry/topography sea level rise project: Determination of sea level trends (NOS CO-OPS 041). US Department of Commerce, Washington, DC, USA.

Index

Page numbers followed by "t" indicate tables.